Regular polygon

$$\text{Area} = \frac{1}{2}(\text{distance } ST)(\text{length one side})(\text{number of sides})$$

Equal sides and angles

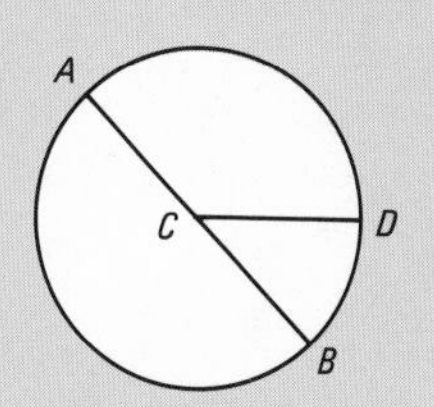

Circle

$$\text{Area} = \pi(CD)^2 = \frac{\pi(AB)^2}{4}$$

$$\pi = \frac{\text{circumference}}{\text{diameter}} = 3.14159...$$

$$\text{Circumference} = \pi(AB) = 2\pi(CD)$$

$$1 \text{ radian} = \frac{180^\circ}{\pi} = 57.2958^\circ$$

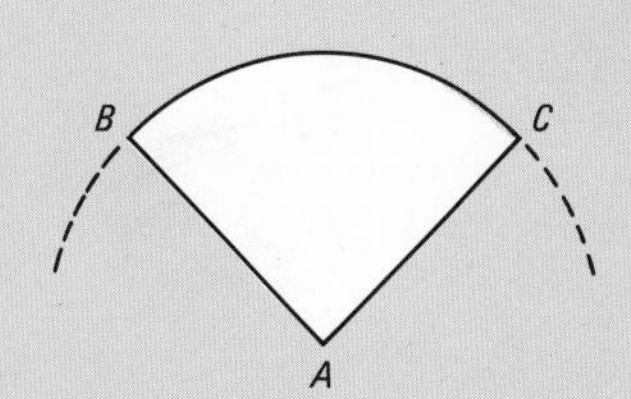

Sector of a circle

$$\text{Area} = \frac{(\text{arc length})(\text{radius})}{2} = \frac{\pi(AB)^2(\text{angle } BAC)}{360^\circ}$$

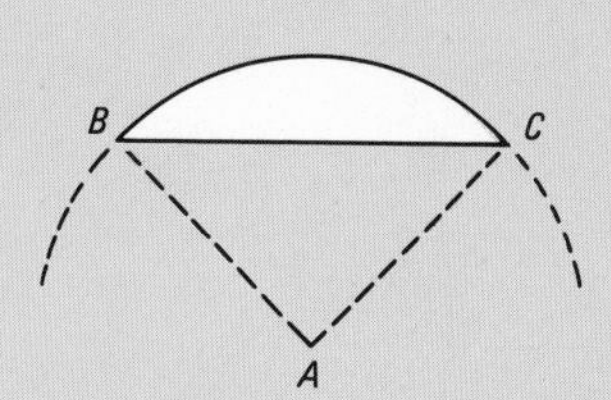

Segment of a circle

$$\text{Area} = \text{area of sector } (ABC) - \text{area of triangle } ABC = \frac{(\text{radius})^2}{2}\left[\frac{\pi(\text{angle } BAC^\circ)}{180^\circ} - \sin BAC^\circ\right]$$

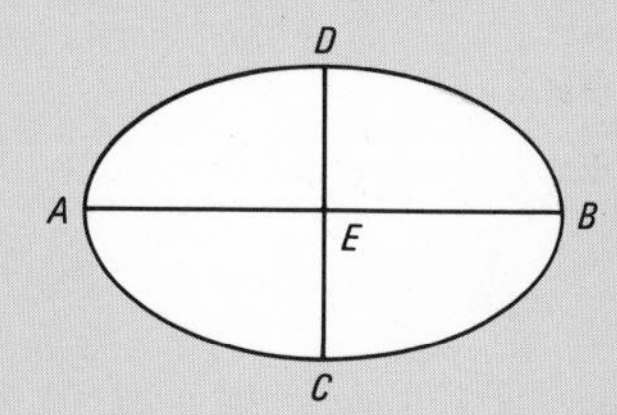

Ellipse

$$\text{Area} = \pi(AE)(ED) = \frac{\pi}{4}(AB)(CD)$$

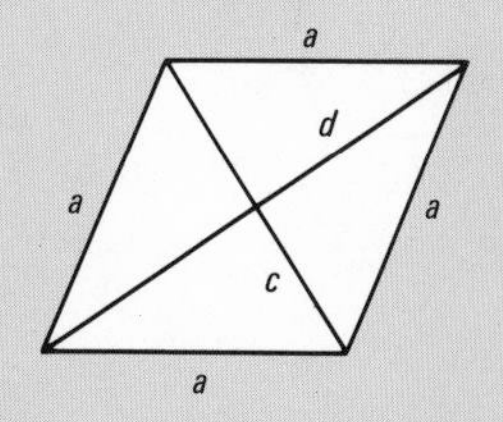

Rhombus

$$\text{Area} = \frac{1}{2}(c)(d)$$

$$4a^2 = c^2 + d^2$$

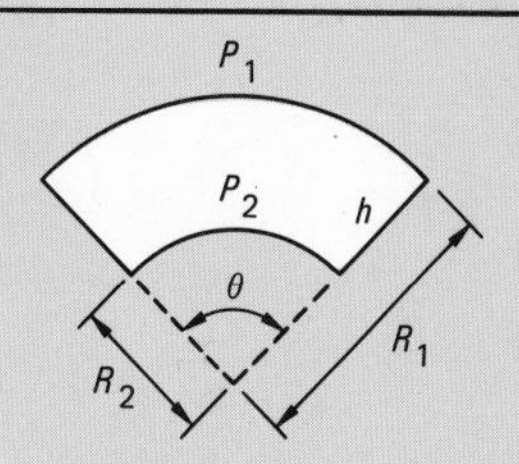

Sector of an annulus

$$\text{Area} = \frac{1}{2}\theta(R_1 + R_2)(R_1 - R_2) = \frac{1}{2}h(P_1 + P_2) = \frac{1}{2}h\theta(R_1 + R_2)$$

Engineering Fundamentals and Problem Solving

Second Edition

ENGINEERING FUNDAMENTALS AND PROBLEM SOLVING

Arvid R. Eide

Roland D. Jenison

Lane H. Mashaw

Larry L. Northup

Professors of Engineering
Iowa State University

McGraw-Hill Book Company

New York | St. Louis | San Francisco
Auckland | Bogotá | Hamburg | Johannesburg
London | Madrid | Mexico | Montreal
New Delhi | Panama | Paris | São Paulo
Singapore | Sydney | Tokyo | Toronto

ENGINEERING FUNDAMENTALS AND PROBLEM SOLVING

Printed in the United States of America.

1 2 3 4 5 6 7 8 9 0 VNHVNH 8 9 8 7 6 5

ISBN 0-07-019318-5

This book was set in Century Schoolbook by General Graphic Services.
The editors were Kiran Verma, Cydney C. Martin, and David A. Damstra;
the designer was Nicholas Krenitsky;
the production supervisor was Leroy A. Young.
New drawings were done by Fine Line Illustrations, Inc.
Von Hoffmann Press, Inc., was printer and binder.

Library of Congress Cataloging in Publication Data
Main entry under title:

Engineering fundamentals and problem solving.

Bibliography: p.
Includes index.
1. Engineering. I. Eide, Arvid R.
TA147.E52 1986 620 85-15191
ISBN 0-07-019318-5

Contents

PART TWO

MANIPULATING ENGINEERING DATA

PART THREE

COMPUTERS IN ENGINEERING

PART FOUR

APPLIED ENGINEERING CONCEPTS

Preface

TO THE STUDENT

As you begin the study of engineering you are no doubt filled with enthusiasm, curiosity, and a desire to succeed. Your first year will be spent primarily establishing a solid foundation in mathematics, physical and computer sciences, and communications. You may at times question what the benefits of this background material are and when the real engineering work and experience will begin. We believe that they begin now. We hope that the material in this book will motivate you in your educational pursuits as well as provide you with a basis for understanding how the engineer functions in today's technological world.

TO THE INSTRUCTOR

Engineering courses for freshman students are in a state of transition. The traditional engineering drawing and descriptive geometry courses have been pared considerably over the past decade and are now being supplemented with computer-graphics topics. Slide rules have disappeared, replaced by calculators and computers. As a result, courses in computations and computer programming have taken the place of formal slide-rule instruction. Of course, the emphasis placed on engineering graphics, computer graphics, computations, programming, orientation, etc., varies considerably among engineering schools.

Since 1974 students at Iowa State University have been taking a computations course that has as its major objective the improvement of problem-solving skills. Various computational aids have been used, from programmable calculators to mainframe computers with interactive terminals. The second edition of this text has evolved from 10 years of experience with teaching engineering-problem-solving skills to thousands of freshman students.

The second edition has the same four broad objectives as did the first edition: (1) to motivate engineering students in their first year, when exposure to engineering subject matter is limited; (2) to provide them with experience in solving problems (in both English and SI units) and in presenting solutions in a logical manner; (3) to introduce students to subject areas common to most engineering disciplines, areas which require the application of fundamental engineering concepts; and (4) to develop their basic skills for solving open-ended problems through a design process.

The material in this book is presented in a manner that allows you to emphasize certain aspects more than others without loss of continuity. The problems that follow most chapters vary in difficulty, so that students can experience success rather quickly and still be challenged as problems become more complex. Most problems in the second edition are new and there are more of them, in particular a greater number of complex problems. Some problems are suitable for solution using a personal or system computer should you wish to include that aspect.

There is sufficient material in the first 14 chapters for a three-credit semester course. By omitting selected chapters and/or varying coverage from term to term, you can present a sound computations course for a two- to four-credit quarter course or a two- to three-credit semester course. Expanded efforts in computer programming and coverage of Chap. 15 on design would provide sufficient material for a one-year sequence.

The book is conveniently divided into five parts. Part One, An Introduction to Engineering, begins with a description of the engineering profession. Material concerning the many disciplines in engineering has been added in the second edition. If a formal orientation course is given separately, Chap. 1 can be given simply as a reading assignment. Chapters 2 and 3 provide procedures for approaching an engineering problem, determining the necessary data and method of solution, and presenting the results. The authors have found from experience that emphasis in this area will reap benefits when the material and problems become more difficult later on.

Part Two, Manipulating Engineering Data, has been restructured in the second edition to combine engineering estimations, dimensions and units (including both English and SI units), and statistics. Throughout the book, discussions and example problems are presented primarily in SI metric, so that coverage of Chap. 5 is advisable. Other dimension systems do appear in the discussion, and many problems contain nonmetric units, so the students are exposed to conversion processes and to units that are still commonly used. Statistics is included in this section to strengthen the student's ability to deal with data manipulation and interpretation.

Part Three is an introduction to the use of computers in engineering. Unless you plan to provide some programming experience on available computers, this material can be omitted, although work with flow-charts, independent of a particular computing language, can improve the students' ability to logically think through problem solutions. Because of the variety of computational equipment available to students on engineering campuses, the authors felt that the treatment of the many computer languages would not be practical. This in no way implies that programming is not an important part of the engineering student's course of study. The student should be taught to obtain calculated results in the most economical, efficient manner. Chapters 7 and 8 bring the student to the point of computation by computer. You need only to introduce the language and available equipment for the student to complete the solution.

Part Four, Applied Engineering Concepts, allows you a great deal of flexibility. The time available and your personal interests and objectives will dictate to what depth any or all of these chapters are covered. Most of them have been strengthened by the addition of new topics and by the inclusion of many new and more difficult problems, some of which are suitable for computer solution. Material in Part Four may be covered in any order, since no chapter depends on another for background material.

The design process, covered in Part Five, is a logical extension of the fundamental problem-solving approach in engineering. A nine-step design process is explained, and the discussion is supplemented with an actual preliminary design performed by a freshman student team. The process as described allows you to enhance the text material with personal examples in order to bring design experience into the classroom.

Mathematical expertise beyond algebra, trigonometry, and analytical geometry is not required for any material in the book. The authors have found, however, that providing additional experience in precalculus mathematics is important at this stage of the student's education.

An instructor's manual is available that contains solutions to most chapter problems and provides suggestions for using the problems to their greatest advantage.

ACKNOWLEDGMENTS

The authors are indebted to many who assisted in the development of this new edition of the textbook. First we would like to thank the faculty of the Freshman Engineering Department at Iowa State University who have taught the engineering computations course over the past ten years. They, with support of engineering faculty from other departments, have made the course a success by their efforts. Several thousands of students have taken the course, and we want to thank them for their comments and ideas that have influenced this edition. The many suggestions of faculty and students alike have provided us with the information necessary to improve the first edition. We also express grateful appreciation to Vicky Eide who worked many lunch and evening hours to type the manuscript. Finally we thank our families for their constant support of our efforts.

Arvid R. Eide
Roland D. Jenison
Lane H. Mashaw
Larry L. Northup

Engineering Fundamentals and Problem Solving

An Introduction to Engineering

PART ONE

1

The Engineering Profession

1.1 Introduction

The rapidly expanding sphere of science and technology may seem overwhelming to the individual seeking a career in a technological field. A technical specialist today may be called either engineer, scientist, technologist, or technician, depending upon education, industrial affiliation, or specific work. For example, more than 250 colleges and universities offer engineering programs accredited by the Accreditation Board for Engineering and Technology (ABET). Included are such traditional specialties as aerospace, agricultural, ceramic, chemical, civil, electrical, industrial, and mechanical engineering—as well as the expanding areas of computer, energy, environmental, materials, and nuclear engineering. Programs in construction engineering, engineering science, mining engineering, and petroleum engineering add to a lengthy list of career options in engineering alone. Coupled with thousands of programs in science and technical training offered at hundreds of other schools, the task of choosing the right field no doubt seems formidable.

Since you are reading this book, we assume that you are interested in studying engineering or at least are trying to decide whether or not to do so. Up to this point in your academic life, you have probably had little experience with engineering and have gathered your impressions of engineering from advertising materials, counselors, educators, and perhaps a practicing engineer or two. Now you must investigate as many careers as you can as soon as possible to be sure of making the right choice.

The study of engineering requires a strong background in mathematics and the physical sciences. Section 1.5 discusses typical areas of study within an engineering program that lead to the bachelor's degree. You should also consult with your counselor about specific course requirements. If you are enrolled in an engineering college but have not chosen a specific discipline, consult with an adviser or someone on the engineering faculty about particular course requirements in your areas of interest.

Figure 1.1
Engineering accomplishments such as the space shuttle spawn technological advancements which benefit all of society. (*Control Data Corporation.*)

When considering a career in engineering or any closely related fields, you should explore the answers to several questions. What is engineering? What is an engineer? What are the functions of engineering? What are the engineering disciplines? Where does the engineer fit into the technical spectrum? How are engineers educated? What is meant by professionalism and engineering ethics? What have engineers done in the past? What are engineers doing now? What will engineers do in the future? Finding answers to such questions will assist you in assessing your educational goals and obtaining a clearer picture of the technological sphere.

Brief answers to some of these questions are given in the remainder of this chapter. By no means are they intended to be a complete discussion of engineering and related fields. You can find additional and more detailed technical career information in the reference materials listed in the bibliography at the end of the book.

1.2 The Technology Team

In 1876, 15 men led by Thomas Alva Edison gathered in Menlo Park, New Jersey, to work on "inventions." By 1887, the group had secured over 400 patents, including ones for the electric light bulb and the phonograph. Edison's approach typified that used for early engineering developments. Usually one person possessed nearly all the knowledge in one field and directed the research, development, design, and manufacture of new products in this field.

Today, however, technology has become so advanced and sophisticated that one person cannot possibly be aware of all the intricacies of a single device or process. The concept of systems engineering has thus evolved; that is, technological problems are studied and solved by a technology team.

Scientists, engineers, technologists, technicians, and artisans form the *technology team*. The abilities of the team range across what is often called the *technical spectrum*. At one end of the spectrum are individuals with an understanding of scientific and engineering principles. They possess the ability to apply these principles for the benefit of mankind. At the other end of this technical spectrum are persons skilled in the trades. They are the individuals who bring the actual ideas into reality.

Each of the technology team members has a specific function in the technical spectrum, and it is of utmost importance that each specialist understand the role of all team members. It is not difficult to find instances where the education and tasks of team members overlap. For any engineering accomplishment, successful team performance requires cooperation that can be realized only through an understanding of the functions of the technology team. We will now investigate each of the team specialists in more detail.

1.2.1 Scientist

Scientists have as their prime objective increased knowledge of nature (see Fig. 1.2). In the quest for new knowledge, the scientist conducts research in a systematic manner. The research steps referred to as the *scientific method* are often summarized as follows:

1. Formulate a hypothesis to explain a natural phenomenon.
2. Conceive and execute experiments to test the hypothesis.
3. Analyze test results and state conclusions.
4. Generalize the hypothesis into the form of a law or theory if experimental results are in harmony with the hypothesis.
5. Publish the new knowledge.

Figure 1.2
Many scientists perform their work in a laboratory. Here a chemist investigates the effect of a ruthenium catalyst on the reaction of hydrogen and carbon dioxide. (*Ames Laboratory, U.S. Department of Energy.*)

An open and inquisitive mind is an obvious characteristic of a scientist. Although the scientist's primary objective is that of obtaining an increased knowledge of nature, many scientists are also engaged in the development of their ideas into new and useful creations. But to differentiate quite simply between the scientist and engineer, we might say that the true scientist seeks to understand more about natural phenomena, whereas the engineer primarily engages in applying new knowledge.

1.2.2 Engineer

In the 1982 *Annual Report* of ABET, the following definition of engineering appears.

Engineering *is the profession in which a knowledge of the mathematical and natural sciences gained by study, experience, and practice is applied with judgment to develop ways to utilize, economically, the materials and forces of nature for the benefit of mankind.*

In the National Council of Engineering Examiners' *Model Law*, the following statement is found.

Engineer *shall mean a person who, by reason of his special knowledge and use of mathematical, physical, and engineering sciences and the principles and methods of engineering analysis and design, acquired by education and experience, is qualified to practice engineering.*

Both the engineer and scientist are thoroughly educated in the mathematical and physical sciences, but the scientist primarily uses this knowledge to acquire new knowledge, whereas the engineer applies the knowledge to design and develop usable devices, structures, and processes. In other words, the scientist seeks to know, the engineer aims to do.

You might conclude that the engineer is totally dependent on the scientist for the knowledge to develop ideas for human benefit. Such is not always the case. Scientists learn a great deal from the work of engineers. For example, the science of thermodynamics was developed by a physicist from studies of practical steam engines built by engineers who had no science to guide them. On the other hand, engineers have applied the principles of nuclear fission discovered by scientists to develop nuclear power plants and numerous other devices and systems requiring nuclear reactions for their operation. The scientist's and engineer's functions frequently overlap, leading at times to a somewhat blurred image of the engineer. What distinguishes the engineer from the scientist in broad terms, however, is that the engineer often conducts research, but does so for the purpose of solving a problem.

The end result of an engineering effort—generally referred to as *design*—is a device, structure, or process which satisfies a need. A successful design is achieved when a logical procedure is followed to meet a specific need. The procedure, called the *design process*, is similar to the scientific method with respect to a step-by-step routine, but it differs in objectives and end results. The design process encompasses the following activities, all of which must be completed.

1. Identification
2. Definition
3. Search
4. Establishment of criteria and constraints
5. Consideration of alternatives
6. Analysis
7. Decision
8. Specification
9. Communication

In the majority of cases, designs are not accomplished by an engineer's simply completing the nine steps shown in the order given. As the designer proceeds through each step, new information may be discovered and new objectives may be specified for the design. If so, the designer must backtrack and repeat steps. For example, if none of the alternatives appear to be economically feasible when the final solution is to be selected, the designer must redefine the problem or possibly relax some of the criteria to admit less expensive alternatives. Thus, because decisions must frequently be made at each step as a result of new developments or unexpected outcomes, the design process becomes iterative.

It is very important that you begin your engineering studies with an appreciation of the thinking process used to arrive at a solution to a problem and ultimately to produce a competitive design. During your studies you will learn many techniques of analysis. *Analysis* is the taking of a given system, the operating environment of the system, and determining the response of the system. An example will illustrate.

Example problem 1.1 A protective liner exactly 12 m wide is available to line a channel for conveying water from a reservoir to downstream areas. If a trapezoidal-shaped channel (see Fig. 1.3) is constructed so that the liner will cover the surface completely, what is the flow area for $x = 2$ m and $\theta = 45°$? Flow area multiplied by average flow velocity will yield volume rate of flow, an important parameter in the study of open-channel flows.

Figure 1.3

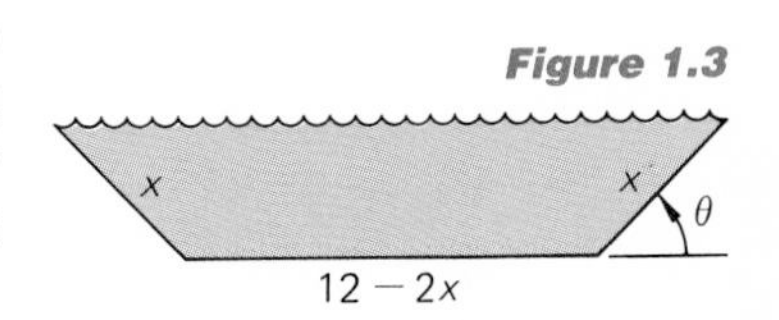

Solution The geometry is defined in Fig. 1.3. The flow area is given by the expression for the area of a trapezoid:

$$A = \tfrac{1}{2}(b_1 + b_2)h$$

where $b_1 = 12 - 2x$

$$b_2 = 12 - 2x + 2x \cos \theta$$

$$h = x \sin \theta$$

Therefore,

$$A = 12x \sin \theta - 2x^2 \sin \theta + x^2 \sin \theta \cos \theta$$

For the situation when $x = 2$ and $\theta = 45°$, the flow area is

$$A = (12)(2)(\sin 45°) - 2(2)^2(\sin 45°) + (2)^2(\sin 45°)(\cos 45°)$$

$$= 13.3 \text{ m}^2$$

You have solved many problems of this nature by analysis; that is, a system is given (the channel as shown in Fig. 1.3), the operating environment is specified (the channel is flowing full), and you must find the system performance (determine the flow area). Analysis usually yields a unique solution. It would be very easy to set up a computer program to calculate areas of flow for many combinations of x and θ. What would be the purpose of doing this? Would we choose random values for x and θ? Are there values for x and θ for which our analysis routine would not yield a physically possible answer? The next two examples will provide some guidance.

Example problem 1.2 A protective liner exactly 12 m wide is available to line a channel conveying water from a reservoir to downstream areas. For the trapezoidal cross section shown in Fig. 1.3, what are the values of x and θ for a flow area of 16 m^2?

Solution Based on our work in Example prob. 1.1, we would have

$$16 = 12x \sin \theta - 2x^2 \sin \theta + x^2 \sin \theta \cos \theta$$

The solution procedure is not direct, and the solution is not unique, as it was in Example prob. 1.1. A graphical approach is used to illustrate some of the possible solutions. Figure 1.4 shows flow area A plotted against the side-slope angle θ for several values of x. Note that this process does not involve solving the above equation for x and θ when $A = 16$. However, from Fig. 1.4 we can find combinations for x and θ which yield $A = 16$. Approximate numerical values of some of the possible combinations are $x = 5$, $\theta = 31°$; $x = 5$, $\theta = 74°$; $x = 4$, $\theta = 33°$; $x = 4$, $\theta = 90°$. Note that for $x = 1$ it is not possible to achieve a flow area of 16 m^2.

You probably have not solved many problems of this nature.

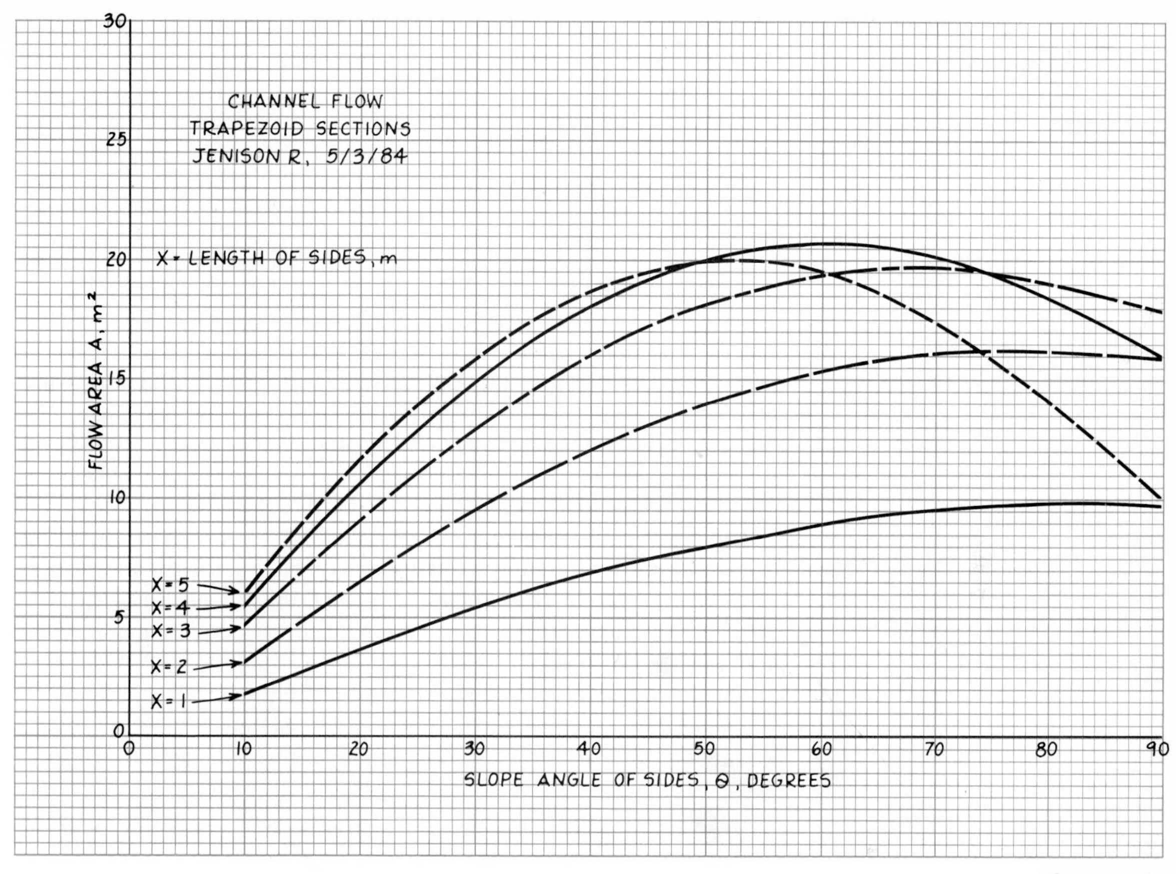

Figure 1.4

Example prob. 1.2 is a synthesis problem; that is, the operating environment is specified (channel flowing full), the performance is known (flow area is 16 m^2), and you must determine the system. Example prob. 1.2 is the inverse problem to Example prob. 1.1. In general, synthesis problems do not have a unique solution, as can be seen from Example prob. 1.2.

Most of us have difficulty synthesizing. We cannot "see" a direct method to find an x and θ that yield a flow area of 16 m^2. Our solution to Example prob. 1.2 involved repeated analysis to "synthesize" the solution. We constructed a family of curves (x is constant) of A versus θ, which enabled us to pick off combinations of x and θ that yield a specified flow area.

Example problem 1.3 For the situation described in Example prob. 1.2, find values of x and θ that yield a maximum flow area.

Solution This is a design problem in which a "best" solution is sought, in this case a maximum flow area for the trapezoidal cross section shown in Fig. 1.3. We can determine the solution from the repeated analysis we did for Example prob. 1.2. With reference to

Fig. 1.4, we can see that for the curves representing $x = 1, 2, 3, 4$, the flow area increases with increasing θ to a maximum and then decreases. The maximum for each x occurs at a different θ. If we look at only the maximum value of flow area for the different values of x, the maximum area increases for $x = 1, 2, 3, 4$ but drops off for $x = 5$. For those values of x considered, the maximum flow area is 20.8 m^2 for a configuration of $x = 4$ and $\theta = 60°$.

Analysis and synthesis are very important to the engineering design effort, and a majority of your engineering education will involve techniques of analysis and synthesis in problem solving. We must not, however, forget the engineer's role in the entire design process. In an industrial setting, the objective is to correctly assess a need, determine the best solution to the need, and market the solution more quickly and less expensively than the competition. This demands careful adherence to the design process.

The successful engineer in a technology team will take advantage of computers and computer graphics. Today, with the aid of computers and computer graphics, it is possible to perform analysis, decide among alternatives, and communicate results far more quickly and with more accuracy than ever before. This translates to better engineering and an improved quality of living.

Terms like "computer-aided design" (CAD) and "computer-aided manufacturing" (CAM) label the modern engineering activities that continue to make engineering a challenging profession. Your work with analysis and synthesis techniques will require the use of a computer to a large extent in your education.

1.2.3
Technologist and Technician

Much of the actual work of converting the ideas of scientists and engineers into tangible results is performed by technologists and technicians (see Fig. 1.5). A technologist generally possesses a baccalaureate degree and a technician an associated degree. Technologists are involved in the direct application of their education and experience to make appropriate modifications in designs as the need arises. Technicians primarily perform routine computations and experiments and prepare design drawings as requested by engineers and scientists. Thus, technicians (typically) are educated in mathematics and science but not to the depth required of scientists and engineers. Technologists and technicians obtain a basic knowledge of engineering and scientific principles in a specific field and develop certain manual skills that enable them to communicate technically with all members of the technology team. Some tasks commonly performed by technologists and technicians include drafting, estimating, model building, data recording and reduction, troubleshooting, servicing, and specification. Often they are the vital link between the idea on paper and the idea in practice.

Figure 1.5
Two members of a technology team discuss the procedures for a test of a manufacturing process. (*Digital Equipment Corporation.*)

1.2.4 Artisan

The artisan possesses manual skills necessary to produce parts specified by scientists, engineers, technologists, and technicians. Artisans need not be particularly concerned with the principles of science and engineering incorporated in a design (see Fig. 1.6). They are usually trained on the job, serving an apprenticeship during which the skills and abilities to build and operate specialized equipment are developed. Some of the specialized jobs of artisans include those of welder, machinist, electrician, carpenter, plumber, and mason.

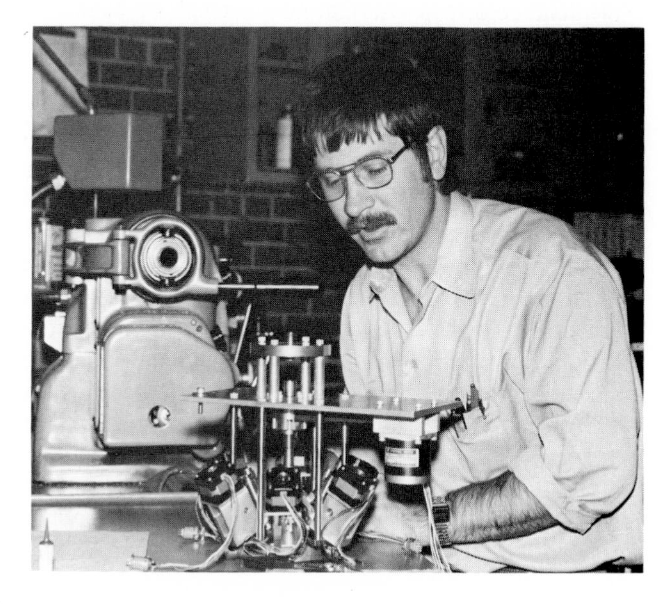

Figure 1.6
A machinist puts the finishing touches to a multiviewing transducer which will assist in locating the characterizing flaws in metals. (*Ames Laboratory, U.S. Department of Energy.*)

1.3 The Functions of the Engineer

As we alluded to in the previous section, engineering feats accomplished from earliest recorded history up to the industrial revolution could best be described as individual accomplishments. The various pyramids of Egypt were usually designed by one individual, who directed tens of thousands of laborers during construction. The person in charge called every move, made every decision, and took the credit if the project was successful or the consequences if the project failed.

With the industrial revolution, there was a rapid increase in scientific findings and technological advances. One-person engineering teams were no longer practical or desirable. We know that today no single aerospace engineer is responsible for the jumbo jets and no one civil engineer completely designs a bridge. Automobile manufacturers assign several thousand engineers to the design of a new model. So we not only have the technology team as described earlier, but we have engineers from many disciplines working together on single projects.

One approach to an explanation of an engineer's role in the technology spectrum is to describe the different types of work that engineers do. For example, civil, electrical, mechanical, and other engineers become involved in design, which is an engineering function. The *engineering functions*, which are discussed briefly in this section, are research, development, design, production, testing, construction, operations, sales, management, consulting, and teaching. Many of the *engineering disciplines* will be discussed in Sec. 1.4.

To avoid confusion between the meaning of engineering disciplines and engineering functions, let us consider the following. Normally a student selects a curriculum (aerospace, chemical, mechanical, etc.) either before or soon after admission to an engineering college. When and how the choice is made varies with each school. The point is, the student does not choose a function, but a discipline. To illustrate further, consider a student who has chosen mechanical engineering. This student will, during an undergraduate education, learn how mechanical engineers are involved in the engineering functions of research, development, design, etc. Some program options allow a student to pursue an interest in a specific subdivision within the curriculum, such as energy conversion in a mechanical engineering program. Most other curricula have similar options.

Upon graduation, when you accept a job with a company, you will be assigned to a functional team performing in a specific area such as research, or design, or sales. Within some companies, particularly smaller ones, you may become involved in more than one function—design *and* testing, for example. It is important to realize that regardless of your choice of discipline, you may become involved in one or more of the functions to be discussed in the following paragraphs.

1.3.1 Research

Successful research is one catalyst for starting the activities of a technology team or, in many cases, the activities of an entire industry. The research engineer seeks new findings, as does the scientist; but it must be kept in mind that the research engineer also seeks a way to use the discovery.

Some qualities of a successful research engineer are intelligence, perceptiveness, cleverness, patience, and self-confidence. Most students interested in research will pursue the master's and doctor's degrees in order to develop their intellectual abilities and the necessary research skills. An alert and perceptive mind is needed to recognize nature's truths when they are encountered. When attempting to reproduce natural phenomena in the laboratory, cleverness and patience are prime attributes. Research often involves tests, failures, retests, etc., for long periods of time. Research engineers are therefore often discouraged and frustrated and must strain their abilities and rely on their self-confidence in order to sustain their efforts to a successful conclusion.

Billions of dollars are spent each year on research at colleges and universities, industrial research laboratories, government installations, and independent research institutes. The team approach to research is predominant today primarily because of the need to incorporate a vast amount of technical information into the research effort. Individual research is also carried out but not to the extent it was several years ago. A large share of research monies are channeled into the areas of energy, environment, health, defense, and space exploration. Research funding from federal agencies is very sensitive to national and international priorities. During a career as a research engineer, you might expect to work in many diverse, seemingly unrelated areas, but your qualifications will allow you to adapt to many different research efforts.

Figure 1.7
Surface qualities of metals are studied by a research engineer using an electron-probe microanalyzer. (*Pratt and Whitney Aircraft.*)

1.3.2 Development

Using existing knowledge and new discoveries from research, the development engineer attempts to produce a device, structure, or process that is functional. Building and testing scale or pilot models is the primary means by which the development engineer evaluates ideas. A major portion of the development work requires use of well-known devices and processes in conjunction with established theories. Thus reading of available literature and a solid background in the sciences and in engineering principles are necessary for the development engineer's success.

Many people who suffer from heart irregularities are able to function normally today because of the pacemaker, an electronic device that maintains a regular heartbeat. The pacemaker is an excellent example of the work of development engineers.

The first model, conceived by medical personnel and developed by engineers at the Electrodyne Company, was an external device that sent pulses of energy through electrodes to the heart. However, the power requirement for stimulus was so great that patients got severe burns on their chests. As improvements were being studied, research in surgery and electronics enabled development engineers to devise an external pacemaker with electrodes through the chest attached directly to the heart. Although more efficient from the standpoint of power requirements, the devices were uncomfortable, and patients frequently suffered infection where the wires entered the chest. Finally two independent teams developed the first internal pacemaker, 8 years after the original pacemaker had been tested. Their experience and research with tiny pulse generators for spacecraft led to this achievement. But the very fine wire used in these early models proved to be inadequate and quite often failed, forcing patients to have the entire pacemaker replaced. A team of engineers at General Electric developed a pacemaker that incorporated a new wire, called a *helicable*. The helicable consisted of 49 strands of wire coiled together and then wound into a spring. The spring diameter was about 46 μm, half the diameter of a human hair. Thus, with doctors and development engineers working together, an effective, comfortable device was perfected that has enabled many heart patients to enjoy a more active life. Today pacemakers have been developed that operate at more than one speed, enabling the patient to speed up or slow down heart rate depending on physical activity.

We have discussed the pacemaker in detail to point out that changes in technology can be in part owing to development engineers. Only 13 years to develop an efficient, dependable pacemaker; 5 years to develop the transistor; 25 years to develop the digital computer—it only indicates that modern engineering methods generate and improve products nearly as fast as research generates new knowledge.

Successful development engineers are ingenious and creative. As-

Figure 1.8
Development engineers often use full-scale models to predict the performance of a new product. (*International Business Machines Corporation.*)

tute judgment is often required in devising models that can be used to determine whether a project will be successful in performance and economical in production. Obtaining an advanced degree is helpful, but it is not as important as it is for an engineer who will be working in research. Practical experience more than anything else produces the qualities necessary for a career as a development engineer.

Development engineers are often asked to demonstrate that an idea will work. Within certain limits, they do not work out the exact specifications that a final product should possess. Such matters are usually left to the design engineer if the idea is deemed feasible.

1.3.3 Design

The development engineer produces a concept or model that is passed on to the design engineer for converting into a device, process, or structure (see Fig. 1.9). The designer relies on education and experience to evaluate many possible solutions, keeping in mind cost of manufacture, ease of production, availability of materials, and performance requirements. Usually several designs and redesigns will be undertaken before the product is brought before the general public.

To illustrate the role the design engineer plays, we will discuss the development of the over-the-shoulder seat belts for added safety in automobiles, which created something of a design problem. De-

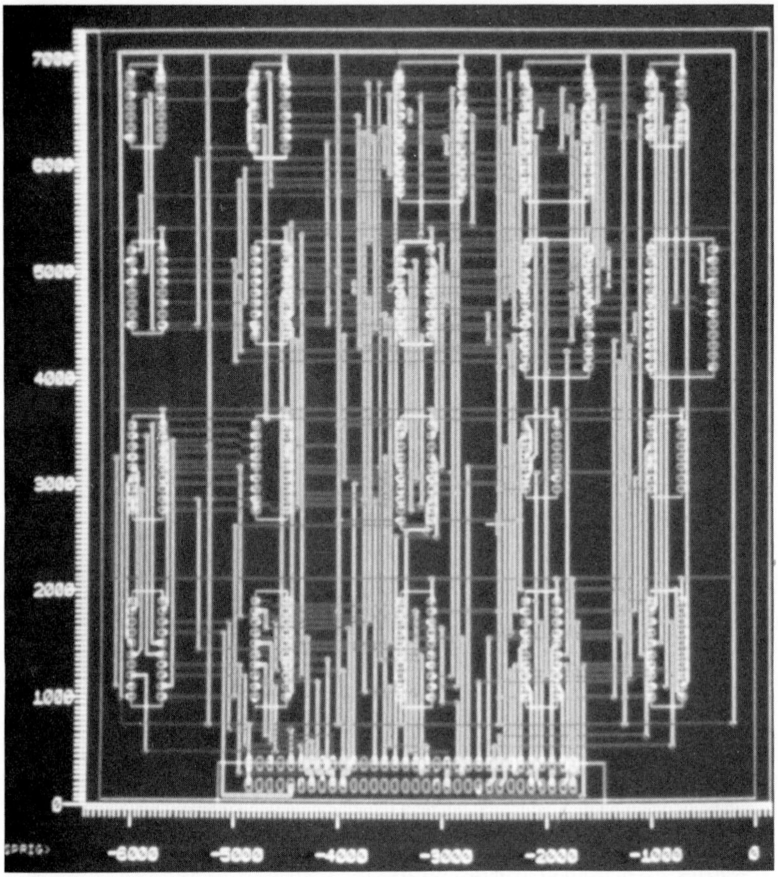

Figure 1.9
This computer graphics display of the connections on a printed circuit board represents a database which will be passed to design from development engineering. (*International Business Machines Corporation.*)

signers had to decide where and how the anchors for the belts would be fastened to the car body. They had to determine what standard parts could be used and what parts had to be designed from scratch. Consideration was given to passenger comfort, inasmuch as awkward positioning could deter usage. Materials to be used for the anchors and the belt had to be selected. A retraction device had to be designed that would give flawless performance.

From one such part of a car, one can extrapolate the numerous considerations that must be given to the approximately 12,000 other parts that form the modern automobile: optimum placement of engine accessories, comfortable design of seats, maximization of trunk space, and aesthetically pleasing body design all require thousands of engineering hours to be successful in a highly competitive industry.

Like the development engineer, the designer is creative. However, unlike the development engineer, who is usually concerned only with a prototype or model, the designer is restricted by the state of the art in engineering materials, production facilities, and, perhaps most important, economic considerations. An excellent design from the standpoint of performance may be completely impractical when viewed from a monetary point of view. To make the necessary decisions, the designer must have a fundamental knowledge of many engineering specialty subjects as well as an understanding of economics and people.

1.3.4 Production and Testing

When research, development, and design have created a device for use by the public, the production and testing facilities are geared for mass production (see Figs. 1.10 and 1.11). The first step in

Figure 1.10
Test engineers use a laboratory model to analyze the performance of a proposed design. (*Proctor and Gamble.*)

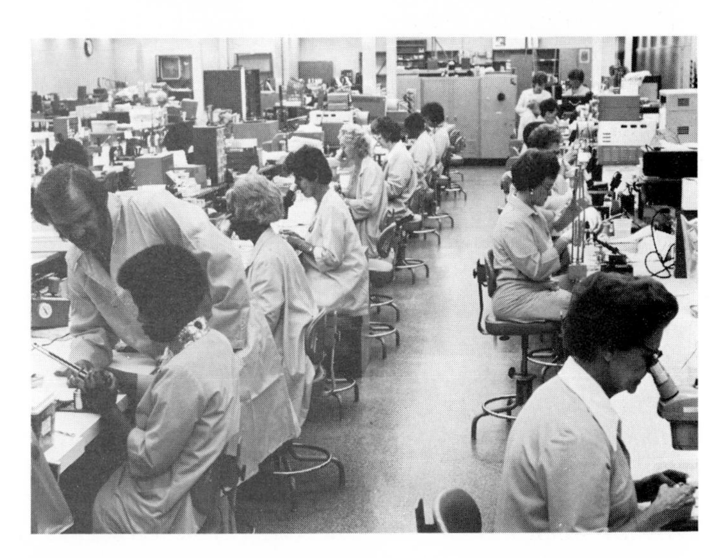

Figure 1.11
A production engineer supervises employees assembling miniature electronic components. (*Bourns, Inc.*)

production is to devise a schedule that will efficiently coordinate materials and personnel. The production engineer is responsible for such tasks as ordering raw materials at the optimum times, setting up the assembly line, and handling and shipping the finished product. The individual who chooses this field must possess the ability to visualize the overall operation of a particular project as well as know each step of the production effort. Knowledge of design, economics, and psychology is of particular importance for production engineers.

Test engineers work with a product from the time it is conceived by the development engineer until such time as it may no longer be manufactured. In the automobile industry, for example, test engineers evaluate new devices and materials that may not appear in automobiles until several years from now. At the same time, they test component parts and completed cars currently coming off the assembly line. They are usually responsible for quality control of the manufacturing process. In addition to the education requirements of the design and production engineers, a fundamental knowledge of statistics is beneficial to the test engineer.

1.3.5
Construction

The counterpart of the production engineer in manufacturing is the construction engineer in the building industry (see Fig. 1.12). When an organization bids on a competitive construction project, the construction engineer begins the process by estimating material, labor, and overhead costs. If the bid is successful, a construction engineer assumes the responsibility of coordinating the project. On large projects, a team of construction engineers may supervise the individual segments of construction such as mechanical (plumbing), electrical (lighting), and civil (building). In addition to a strong background in engineering fundamentals, the construction engineer needs on-the-job experience and an understanding of labor relations.

Figure 1.12
A construction engineer studies the plans as construction proceeds on a nuclear-energy facility. Construction engineers become involved with cost estimation, site planning, facility design, and inspection. (*Union Carbide.*)

Figure 1.13
An operations engineer is responsible for putting together appropriate facilities such as this computer system used to direct the manufacturing operations. (*Digital Equipment Corporation.*)

1.3.6 Operations

Up to this point, discussion has centered around the results of engineering efforts to discover, develop, design, and produce products that are of benefit to human beings. For such work, engineers obviously must have offices, laboratories, and production facilities in which to accomplish it. The major responsibility for supplying such facilities falls on the operations engineer (see Fig. 1.13). Sometimes called a plant engineer, this individual selects sites for facilities, specifies the layout for all facets of the operation, and selects the fixed equipment for climate control, lighting, and communication. Once the facility is in operation, the plant engineer is responsible for maintenance and modifications as requirements demand. Because this phase of engineering comes under the economic category of overhead, the operations engineer must be very conscious of cost and keep up with new developments in equipment so that overhead is maintained at the lowest possible level. A knowledge of basic engineering, industrial engineering principles, economics, and law are prime educational requirements of the operations engineer.

1.3.7 Sales

In many respects, all engineers are involved in selling. To the research, development, design, production, construction, and operations engineers, selling means convincing management that money should be allocated for development of particular concepts or expansion of facilities. This is, in essence, selling one's own ideas. Sales engineering, however, means finding or creating a market for a product. The complexity of today's products requires an individual thoroughly familiar with materials in and operational procedures for consumer products to demonstrate to the consumer in layperson's terms how the products can be of benefit. The sales engineer is thus the liaison between the company and the consumer, a very important means of influencing a company's reputation. An engineering background plus a sincere interest in people and a desire to be helpful are the primary attributes of a sales engineer. The sales engineer usually spends a great deal of time in the plant learning about the product to be sold. After a customer purchases a product, the sales engineer is responsible for coordinating service and maintaining customer satisfaction. As important as sales engineering is to a company, it still has not received the interest from new engineering graduates that other engineering functions have. See Fig. 1.14.

1.3.8 Management

Traditionally, management has consisted of individuals trained in business and groomed to assume positions leading to the top of the

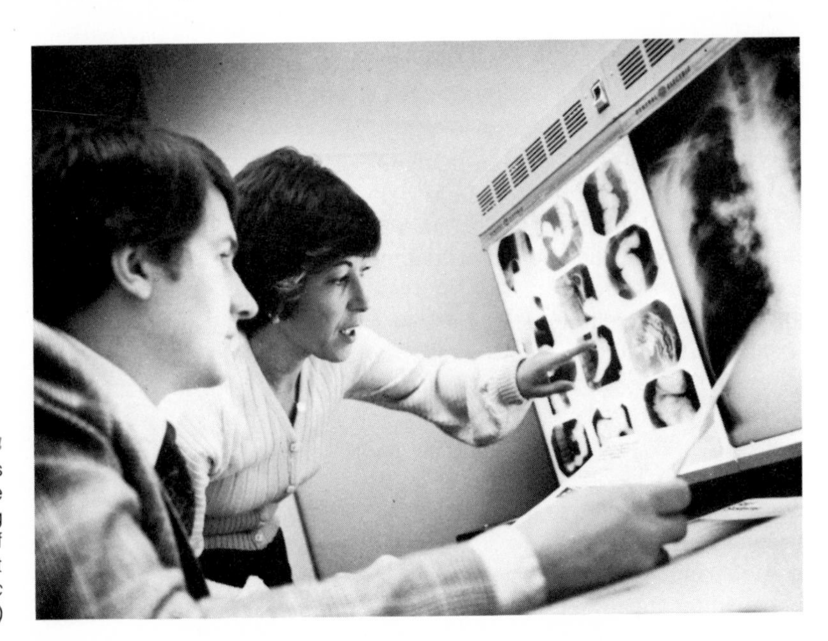

Figure 1.14
A sales engineer describes features of a new product to a prospective customer. Sales engineering requires thorough knowledge of company operations and product performance. (*General Electric Company.*)

Figure 1.15
A management engineer uses computer technology to assist in the planning operation. (*International Business Machines Corporation.*)

corporate ladder. However, with the influx of scientific and technological data being used in business plans and decisions, there has been a need for people in management with knowledge and experience in engineering and science. Recent trends indicate that a growing percentage of management positions are being assumed by engineers and scientists. Inasmuch as one of the principal functions of management is to use company facilities to produce an economically feasible product, and decisions must often be made that may affect thousands of people and involve millions of dollars over periods of several years, a balanced education of engineering or science and business seems to produce the best managerial potential.

At some time during your career as a engineer, a decision must be made about whether to remain with the technical functions of research, development, design, etc., or to obtain business education or experience and pursue the managerial route. This will require an honest self-evaluation and a commitment to abide totally by the decision. Your future success will depend on this.

1.3.9 Consulting

For someone interested in self-employment, a consulting position may be an attractive one (see Fig. 1.16). Consulting engineers op-

Figure 1.16
Consulting engineers discuss a plan for inner-city development. (*Digital Equipment Corporation.*)

erate alone or in partnership, furnishing specialized help to clients who request it. Of course, as in any business, risks must be taken. Moreover, a sense of integrity and a knack for correct engineering judgment are primary necessities in such work.

A consulting engineer must possess a professional engineer's license before beginning practice. Consultants usually spend many years in a specific area before going on their own. A successful consulting engineer maintains a business primarily by being able to solve unique problems for which other companies have neither the time nor capability. In many cases, large consulting firms maintain a staff of engineers of diverse background so that a wide range of engineering problems can be contracted.

1.3.10 Teaching

Individuals interested in a career that involves helping others to become engineers will find teaching very rewarding (see Fig. 1.17). The engineering teacher must possess an ability to communicate abstract principles and engineering experiences in a manner that young people can understand and appreciate. By merely following

Figure 1.17
A modern engineering graphics classroom where television and computers assist the teacher in the education process.

general guidelines, the teacher is usually free to develop his or her own method of teaching and means of evaluating its effectiveness. In addition to teaching, the engineering educator can also become involved in student advising and research.

Engineering teachers today must have a mastery of fundamental engineering and science principles and a knowledge of applications. Customarily, they must obtain an advanced degree in order to improve their understanding of basic principles, to perform research in a specialized area, and perhaps to gain teaching experience on a part-time basis.

1.4 The Engineering Disciplines

There are 22 engineering disciplines listed in the summary of engineering enrollments presented in the October 1983 issue of *Engineering Education*. Engineering colleges offer a combination of these disciplines as 4-year programs or, in a few instances, 5-year programs leading to the baccalaureate. In some schools two or more disciplines are combined within one department which may offer separate degrees or include a discipline as an area of specialty within another discipline. For the latter case a degree in the area of specialty is not offered. Examples of combinations of engineering disciplines include civil/construction, electrical/computer, and industrial/general.

Including the combined disciplines, the top (largest) six curricula are shown in Fig. 1.18. The percentages represent senior students in each curriculum for the fall of 1982. It becomes more difficult to determine the percentage of engineers by discipline in industry because engineers are more likely to be identified by function rather than discipline. For example, many mechanical, electrical, and civil

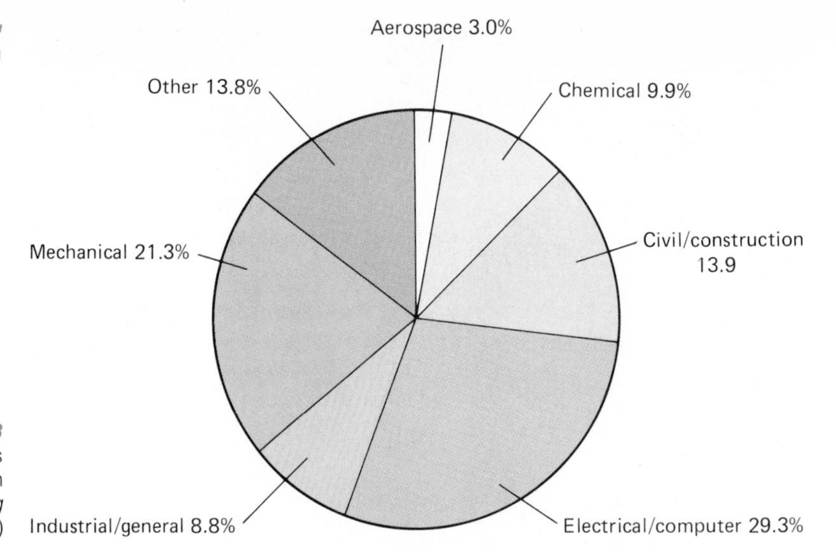

Figure 1.18
The top six engineering disciplines based on enrollment of seniors in the fall of 1983. (*Engineering Education, October 1984.*)

engineers work in the aerospace industry and would be referred to as design engineers, research engineers, etc. In Fig. 1.18, the "other" category includes such engineering disciplines as agricultural, bioengineering, ceramic, engineering science, materials, metallurgical, mining, nuclear, and petroleum.

Each of the top six disciplines will be discussed in the remainder of this section. Those areas which pique your interest should be investigated in more detail by contacting the appropriate department at your university.

1.4.1 Aerospace Engineering

Aerospace engineers study flight of all types of vehicles in all environments. They design, develop, and test aircraft, missiles, space vehicles, helicopters, hydrofoils, ships, and submerging ocean vehicles. The particular areas of specialty include aerodynamics, propulsion, orbital mechanics, stability and control, structures, design, and testing.

Aerodynamics is the study of the effects of moving a vehicle through the earth's atmosphere. The air produces forces that have both a positive effect on a properly designed vehicle (lift) and a negative effect (drag). In addition, at very high speeds the air generates heat on the vehicle which must be dissipated to protect crews, passengers, and cargo. Aerospace engineering students learn to determine such things as optimum wing and body shapes, vehicle performance, and environmental impact.

The operation and construction of turboprops, turbo and fan jets, rockets, ram and pulse jets, nuclear and ion propulsion are part of the aerospace engineering student's study of propulsion. Such con-

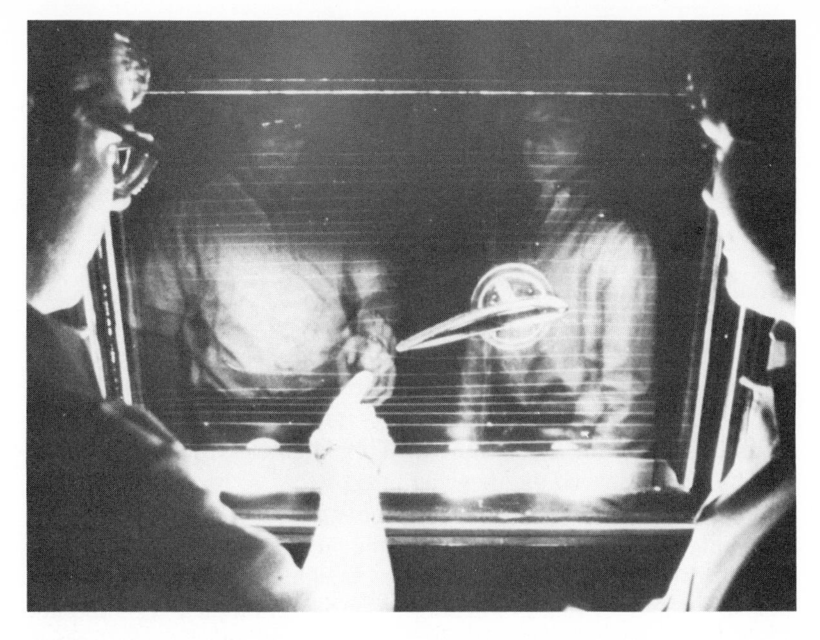

Figure 1.19
Aerospace engineering students observe air-flow phenomena in a smoke tunnel.

straints as efficiency, noise levels, and flight distance enter into the selection of a propulsion system for a flight vehicle.

The aerospace engineer develops plans for interplanetary missions based on a knowledge of orbital mechanics. The problems encountered include determination of trajectories, stabilization, rendezvous with other vehicles, changes in orbit, and interception.

Stability and control involves the study of techniques for maintaining stability and establishing control of vehicles operating in the atmosphere or in space. Automatic control systems for autopilots and unmanned vehicles are part of the study of stability and control.

The study of structures is primarily involved with thin-shelled, flexible structures that can withstand high stresses and extreme temperature ranges. The structural engineer works closely with the aerodynamics engineer to determine the geometry of wings, fuselages, and control surfaces. The study of structures also involves thick-shelled structures that must withstand extreme pressures at ocean depths.

The aerospace design engineer combines all the aspects of aerodynamics, propulsion, orbital mechanics, stability and control, and structures into the optimum vehicle. Design engineers work in a team and must learn to compromise in order to determine the best design satisfying all criteria and constraints.

The final proofing of a design involves the physical testing of a prototype. Aerospace test engineers learn to use testing devices such as wind tunnels, lasers, strain gauges, and microcomputers. The testing takes place in structural laboratories, propulsion facilities, and in the flight medium with the actual vehicle.

1.4.2 Chemical Engineering

Chemical engineers deal with the chemical and physical principles that allow us to maintain a suitable environment. They create, design, and operate processes that produce useful materials including fuels, plastics, structural materials, food products, health products, fibers, and fertilizers. As our natural resources grow short, chemical engineers are creating substitutes or finding ways to extend our remaining resources.

The chemical engineer, in the development of new products, in designing processes, and in operating plants, may work in a laboratory, pilot plant, or full-scale plant. In the laboratory, the chemical engineer searches for new products and materials that benefit humankind and the environment. This laboratory work would be classified as research engineering.

In a pilot plant, the chemical engineer is trying to determine the feasibility of carrying on a process on a large scale. There is a great deal of difference between a process working in a test tube in the laboratory and a process working in a production facility. The pilot plant is constructed to develop the necessary unit operations to carry out the process. Unit operations are fundamental chemical and physical processes that are uniquely combined by the chemical engineer to produce the desired product. A unit operation may involve separation of components by mechanical means such as filtering, settling, and floating. Separation may also take place by changing the

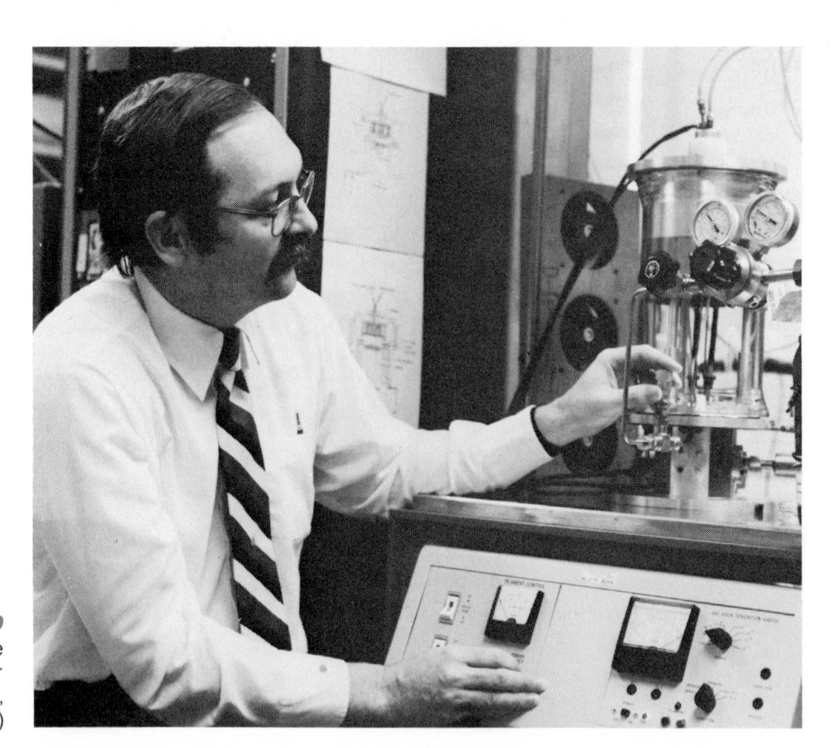

Figure 1.20
Chemical properties of materials are studied in this research facility for microelectronics. (*Ames Laboratory, U.S. Department of Energy.*)

form of a component, for example, through evaporation, absorption, or crystallization. Unit operations also involve chemical reactions such as oxidation and reduction. Certain chemical processes require addition or removal of heat or the transfer of mass. The chemical engineer thus works with heat exchangers, furnaces, evaporators, condensers, and refrigeration units in developing large-scale processes.

In a full-scale plant, the chemical engineer will continue to "fine tune" the unit operations to produce the optimum process based on the lowest cost. The day-by-day operations problems in a chemical plant such as piping, storage, and material handling are the responsibility of chemical engineers.

1.4.3 Civil/Construction Engineering

Civil engineering is the oldest branch of the engineering profession. It involves application of the laws, forces, and materials of nature to the design, construction, operation, and maintenance of facilities that serve our needs in an efficient, economical manner. Civil engineers work for consulting firms engaged in private practice, for manufacturing firms, and for federal, state, and local governments. Because of the nature of their work, civil engineers assume a great deal of responsibility, which means that professional registration is an important goal for the civil engineer beginning practice.

The specialties within civil engineering include structures, transportation, sanitary and water resources, geotechnical, and surveying.

Structural engineers design bridges, buildings, dams, tunnels, and supporting structures. The designs include consideration of mass, winds, temperature extremes, and other natural phenomena such as earthquakes. Civil engineers with a strong structural background are often found in aerospace and manufacturing firms, playing an integral role in the design of vehicular structures.

Civil engineers in transportation plan, design, construct, operate, and maintain facilities that move people and goods throughout the world. For example, they make the decisions on where a freeway system should be located and describe the economic impact of the system on the affected public. They plan for growth of residential and industrial sectors of the nation. The modern rapid-transit systems are another example of the solution to a public need satisfied by transportation engineers.

Sanitary engineers are concerned with maintaining a healthful environment by proper treatment and distribution of water, treatment of wastewater, and control of all forms of pollution. The water resources engineer specializes in the evaluation of potential sources of new water for increasing or shifting populations, irrigation, and industrial needs.

Before any structure can be erected, a careful study of the soil,

Figure 1.21
This skeletal steel structure of a large storage facility was designed by civil engineers and erected under the supervision of a construction engineer.

rock, and groundwater conditions must be undertaken to ensure stability. In addition to these studies, the geotechnical engineer analyzes building materials such as sand, gravel, and cement to determine proper constituency for concrete and other products.

Surveying engineers develop maps for any type of engineering project. For example, if a road is to be built through a mountain range, the surveyors will determine the exact route and develop the topographical survey which is then used by the transportation engineer to lay out the roadway.

Construction engineering is a significant portion of civil engineering and many engineering colleges offer a separate degree in this area. Generally, construction engineers will work outside at the actual construction site. They become involved with the initial estimating of construction costs for surveying, excavation, and construction. They will supervise the construction, start-up, and initial operation of the facility until the client is ready to assume operational responsibility. Construction engineers work around the world on many construction projects such as highways, skyscrapers, and power plants.

1.4.4 Electrical/Computer Engineering

Electrical/computer engineering is the largest branch of engineering, representing about 30 percent of the graduates entering the engineering profession. Because of the rapid advances in technology associated with electronics and computers, this branch of engineering is also the fastest growing.

The areas of specialty include communications, power, electronics, measurement, and computers.

We depend almost every minute of our lives on communication equipment developed by electrical engineers. Telephones, television, radio, and radar are common communications devices that we often take for granted. Our national defense system depends heavily on the communications engineer and on the hardware used for our early warning and detection systems.

The power engineer is responsible for producing and distributing the electricity demanded by residential, business, and industrial users throughout the world. The production of electricity requires a generating source such as fossil fuels, nuclear reactions, or hydroelectric dams. The power engineer may be involved with research and development of alternative generation sources such as sun, wind, and fuel cells. Transmission of electricity involves conductors and insulating materials. On the receiving end, appliances are designed by power engineers to be highly efficient in order to reduce both electrical demand and costs.

The area of electronics is the fastest-growing specialty in electrical engineering. The development of solid-state circuits (functional electronic circuits manufactured as one part rather than wired together) has produced high reliability in electronic devices. Microelectronics has revolutionized the computer industry and electronic controls. Circuit components on the order of 1 micrometer (μm) wide enable reduced costs and higher electronic speeds to be attained in circuitry. The microprocessor, the principal component of a digital computer, is a major result of solid-state circuitry and microelectronics technology. The home computer, automobile control systems, and a multitude of electrical-applications devices conceived, designed, and produced by electronics engineers have greatly improved our standard of living.

Figure 1.22 An electronics engineer completes the design of an integrated circuit using a design software package on a computer graphics system. (*Digital Equipment Corporation.*)

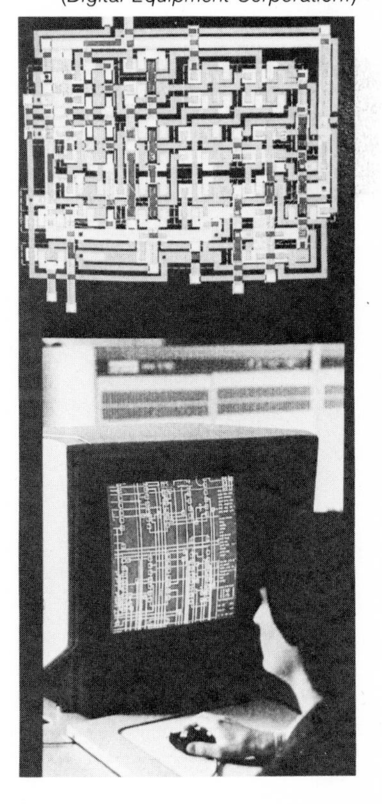

Great studies have been made in the control and measurement of phenomena that occur in all types of processes. Physical quantities such as temperature, flow rate, stress, voltage, and acceleration are detected and displayed rapidly and accurately for optimal control of processes. In some cases, the data must be sensed at a remote location and accurately transmitted long distances to receiving stations. The determination of radiation levels is an example of the electrical process called *telemetry*.

The impact of microelectronics on the computer industry has created a multibillion dollar annual business that in turn has enhanced all other industries. The design, construction, and operation of computer systems is the task of computer engineers. This specialty within electrical engineering has in many schools become a separate degree program. Computer engineers deal with both hardware and software problems in the design and application of computer systems. The areas of application include research, education, design engineering, scheduling, accounting, control of manufacturing operations, process control, and home computing needs. No single development in history has had as great an impact on our lives in as short a time span as has the computer.

1.4.5 Industrial/General Engineering

Industrial engineering covers a broad spectrum of activities in organizations of all sizes. The principal efforts of industrial engineers are directed to the design of production systems for all the goods and services. As an example, consider the procedures and processes necessary to produce and market a power lawn mower. When the design of the lawnmower is complete, industrial engineers establish the manufacturing sequence from the point of bringing the materials to the manufacturing center to the final step of shipping the assembled lawnmowers to the marketing agencies. Industrial engineers develop a production schedule, oversee the ordering of standard parts (engines, wheels, bolts), develop a plant layout (assembly line) for production of nonstandard parts (frame, height adjustment mechanism), and perform a cost analysis for all phases of production.

As production is ongoing, industrial engineers will perform various studies, called *time and methods studies*, which assist in optimizing the handling of material, the shop processes, and the overall assembly line. In a large organization, industrial engineers will likely specialize in one of the many areas involved in the operation of a plant. In a smaller organization, industrial engineers are likely to be involved in all the plant activities. Because of their general study in many areas of engineering and their knowledge of the overall plant operations, industrial engineers are frequent choices for promotion into management-level positions.

The study of human factors is an important area of industrial engineering. In product design, for example, industrial engineers involved in fashioning automobile interiors study the comfort and fatigue factors of seats and instrumentation. And in the factory they develop training programs for operators and supervisors of new machinery or for new assembly-line operators.

With the rapid development of computer-aided manufacturing (CAM) techniques and of computer-integrated manufacturing (CIM), the industrial engineer will play a large role in the factories of the future. The industrial engineer of the future will also be involved in retraining the labor force to work in a high-technology environment.

1.4.6 Mechanical Engineering

Mechanical engineering originated when people first began to use levers, ropes, and pulleys to multiply their own strength and to use wind and falling water as a source of energy. Today mechanical

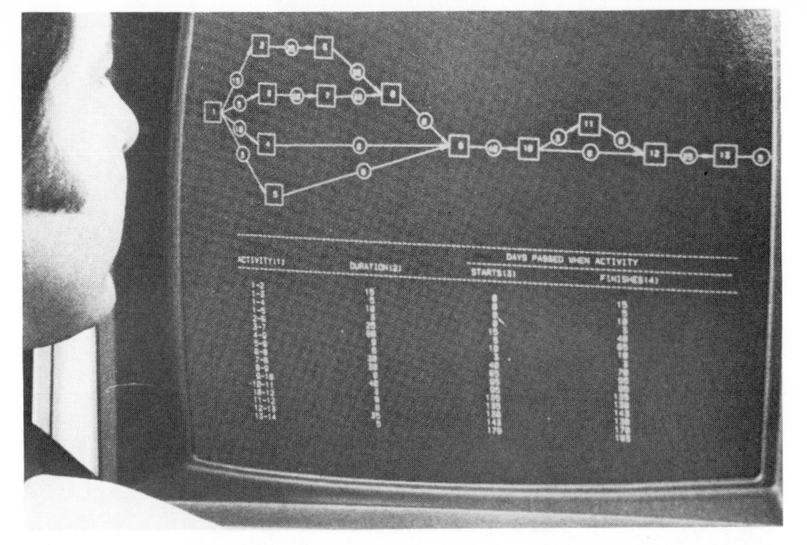

Figure 1.23
An industrial engineer studies a critical path chart to determine areas of a project which may need additional manpower or other assistance in order to remain on schedule. (*Tektronix, Inc.*)

engineers are involved with all forms of energy utilization and conversion, machines, manufacturing materials and processes, and engines.

Mechanical engineers utilize energy in many ways for our benefit. Refrigeration systems keep perishable goods for long periods of time, air condition our homes and offices, and aid in various forms of chemical processing. Heating and ventilating systems keep us comfortable when the environment around us changes with the seasons. Ventilating systems help keep the air around us breathable by removing undesirable fumes. Mechanical engineers analyze heat transfer from one object to another and design heat exchangers to effect a desirable heat transfer.

The energy crisis of the 1970s brought to focus a need for new sources of energy as well as new and improved methods of energy conversion. Mechanical engineers are involved in research in solar, geothermal, and wind energy sources, along with research to increase the efficiency of producing electricity from fossil fuel, hydroelectric, and nuclear sources.

Machines and mechanisms used in all forms of manufacturing and transportation are designed and developed by mechanical engineers. Automobiles, airplanes, and trains combine a source of power and a machine to provide transportation. Tractors, combines, and other implements aid the agricultural community. Lathes, milling machines, grinders, and drills assist in the manufacture of goods. Sorting devices, typewriters, staplers, and mechanical pencils are part of the office environment. Machine design requires a strong mechanical engineering background and a vivid imagination.

In order to drive the machines, a source of power is needed. The mechanical engineer is involved with the generation of electricity by converting chemical energy in fuels to thermal energy in the

Figure 1.24
Mechanical engineers using computer graphics to assist their design effort. (*Tektronix, Inc.*)

form of steam, then to mechanical energy through a turbine to drive the electric generator. Internal-combustion devices such as gasoline, turbine, and diesel engines are designed for use in all areas of transportation. The mechanical engineer studies engine cycles, fuel requirements, ignition performance, power output, cooling systems, engine geometry, and lubrication in order to develop high-performance–low-energy-consuming engines.

The engines and machines designed by mechanical engineers require many types of materials for construction. The tools that are needed to process the raw material for other machines must be designed. For example, a very strong material is needed for a drill bit that must cut a hole in a steel plate. If the tool is made from steel it must be a higher-quality steel than that found in the plate. Methods of heat treating, tempering, and other metallurgical processes are applied by the mechanical engineer.

Manufacturing processes such as electric-discharge machining, laser cutting, and modern welding methods are used by mechanical engineers in the development of improved products. Mechanical engineers are also involved in the testing of new materials and products in the search for better techniques.

1.5 Education of the Engineer

The amount of information coming from the academic and business world is increasing exponentially, and at the current rate, it will double in less than 20 years. More than any other group, engineers are using this knowledge to shape civilization. To keep pace with a changing world, engineers must be educated to solve problems that are as yet unheard of. A large share of the responsibility for this

Accreditation Board for Engineering and Technology*

CODE OF ETHICS OF ENGINEERS

THE FUNDAMENTAL PRINCIPLES

Engineers uphold and advance the integrity, honor and dignity of the engineering profession by:

I. using their knowledge and skill for the enhancement of human welfare;

II. being honest and impartial, and serving with fidelity the public, their employers and clients;

III. striving to increase the competence and prestige of the engineering profession; and

IV. supporting the professional and technical societies of their disciplines.

THE FUNDAMENTAL CANONS

1. Engineers shall hold paramount the safety, health and welfare of the public in the performance of their professional duties.

2. Engineers shall perform services only in the areas of their competence.

3. Engineers shall issue public statements only in an objective and truthful manner.

4. Engineers shall act in professional matters for each employer or client as faithful agents or trustees, and shall avoid conflicts of interest.

5. Engineers shall build their professional reputation on the merit of their services and shall not compete unfairly with others.

6. Engineers shall act in such a manner as to uphold and enhance the honor, integrity and dignity of the profession.

7. Engineers shall continue their professional development throughout their careers and shall provide opportunities for the professional development of those engineers under their supervision.

345 East 47th Street New York, NY 10017

*Formerly Engineers' Council for Professional Development. (Approved by the ECPD Board of Directors, October 5, 1977)

AB-54 2/85

Figure 1.28
Code of Ethics for Engineers. (*Accreditation Board for Engineering and Technology.*)

Engineers' Creed

As a Professional Engineer, I dedicate my professional knowledge and skill to the advancement and betterment of human welfare.

I pledge—:

To give the utmost of performance;

To participate in none but honest enterprise;

To live and work according to the laws of man and the highest standards of professional conduct;

To place service before profit, the honor and standing of the profession before personal advantage, and the public welfare above all other considerations.

In humility and with need for Divine Guidance, I make this pledge.

NATIONAL SOCIETY OF PROFESSIONAL ENGINEERS
FOUNDED 1934

Figure 1.29
Engineer's Creed. (*National Society of Professional Engineers.*)

We do not mean to imply that our fossil fuel resources will be gone in a short time. However, as demand increases and supplies become scarcer, the cost of obtaining the energy increases and places additional burdens on already financially strapped regions and individuals. Engineers with great vision are needed to develop alternative sources of energy from the sun, wind, and ocean and to improve the efficiency of existing energy consumption devices.

Along with the production and consumption of energy come the secondary problems of pollution. Such pollutants as smog and acid rain, carbon monoxide, and radiation must receive attention in order to maintain the balance of nature.

1.7.2 Water

The basic water cycle—from evaporation to cloud formation, then to rain, runoff, and evaporation again—is taken for granted by most people. However, if the rain or the runoff is polluted, then the cycle is interrupted and our water supply becomes a crucial problem. In addition, some highly populated areas have a limited water supply and must rely on water distribution systems from other areas of the country. Many formerly undeveloped agricultural regions are now productive because of irrigation systems. However, the irri-

Figure 1.30
Solar panels collect energy from the sun. More effort is needed to improve conversion efficiency and to make large-scale collectors economically feasible.

gation systems deplete the underground streams of water that are needed downstream.

These problems must be solved in order for life to continue to exist as we know it. Because of the regional water distribution patterns, the federal government must be a part of the decision-making process for water distribution. One of the concerns that must be eased is the amount of time required to bring a water distribution plan into effect. Government agencies and the private sector are strapped by regulations that cause delays in planning and construction of several years. Greater cooperation and a better-informed public are goals that public works engineers must strive to achieve.

1.7.3 A Competitive Edge in the World Marketplace

We have all purchased or used products that were manufactured outside the country. Many of these products incorporate technology that was developed in the United States. In order to maintain our strong industrial base, we must develop practices and processes that enable us to compete, not just with other U.S. industries, but with international industries.

The goal of any industry is to generate a profit. In today's marketplace this means creating the best product in the shortest time at a lesser price than the competition. A modern design process incorporating sophisticated analysis procedures and supported by high-speed computers with graphical displays increases the capability for developing the "best" product. The concept of integrating the design and manufacturing functions with CAD/CAM and CIM promises to shorten the design-to-market time for new products and for upgraded versions of existing products. The development of the automated factory is an exciting concept that is receiving a great deal of attention from manufacturing engineers today. Remaining competitive by producing at a lesser price requires a national effort involving labor, government, and distribution factors. In any case, engineers are going to have a significant role in the future of our industrial sector.

1.7.4 Public Works

In the winter 1982 issue of *Professional Engineer*, H. C. Heldenfels, then president of the Associated General Contractors, cited evidence of serious deterioration in America's infrastructure, the facilities that enable us to receive our goods and services, drink pure water, dispose of waste properly, and move about as needed for our jobs and recreation. Some of the particular problems cited were:

1. Nearly 50 percent of our sewage treatment plants cannot handle the demand.
2. One of every two miles of paved roads needs major repair or complete reconstruction.
3. Nearly 50 percent of our bridges are potentially dangerous to today's traffic load.
4. Over half of our rail network faces replacing by the late 1980s.
5. Many deep-water ports are in need of major repairs and expansion in order to continue to handle the water traffic involved in importing and exporting.
6. Municipal water systems will require $100 billion of repairs and expansion in the next several years.

Figure 1.31
A major engineering effort is needed to effect repair of existing roads and construct new ones to handle the massive transportation requirements. New and stronger materials and improved design are needed to effectively maintain our road and bridge system.

It is estimated that the total value of the public works facilities is $1.3 trillion. To protect this investment, innovative thinking and creative funding must be fostered. Some of this is already occurring in road design and repair. Within the past few years, a method has been applied successfully that recycles asphalt pavement and actually produces a stronger product. Engineering research is producing extended-life pavement with new additives and structural designs. New, relatively inexpensive methods of strengthening old bridges have been used successfully.

1.7.5 The Potential of Space

The exploration of the space around earth and a portion of the solar system has had a profound effect on our lifestyles. Much of the electronic technology we have available in our homes and workplaces today was a spin-off from the technology developed to put humans into orbit and on the moon and to send vehicles to explore nearby planets. For example, satellite technology enables live television coverage of international happenings.

The space shuttle has demonstrated that humans can perform work in a zero-gravity situation—for example, use of the arm to recover objects in orbit. With this technology, a space station could be constructed which would enable experiments of all kinds to be conducted. Exotic methods of forming materials that function best in a zero-gravity situation can be developed. Biological experiments under zero gravity or controlled gravity can be carried out. Improved methods of communication are possible.

A vast potential for engineers exists in the application of newly developed technology from space exploration to the improvement of our everyday standard of living.

1.8 Conclusion

We have touched only briefly on the possibilities for exciting and rewarding work in all engineering areas. The first step is to obtain the knowledge during your college education that is necessary for your first technical position. After that, you must continue your education, either formally by seeking an advanced degree or degrees, or informally through continuing education courses or appropriate reading to maintain pace with the technology, an absolute necessity for a professional.

Many challenges await you. Prepare to meet them well.

Problems

1.1 Compare the definitions of an engineer and a scientist from at least three different sources that discuss engineering career opportunities.

1.2 Compare the definitions of an engineer and a technologist from at least two sources.

1.3 Find three textbooks that introduce the design process. Copy the steps in the process from each textbook. Note similarities and differences and write a paragraph describing your conclusions.

1.4 Find the name of a pioneer engineer in the field of your choice and write a brief paper on the accomplishments of this individual.

1.5 Select a specific branch of engineering and list at least 20 different industrial organizations that utilize engineers from this field.

1.6 Select a branch of engineering, e.g., mechanical engineering, and an engineering function, e.g., design. Write a brief paper on some typical activities that are undertaken by the engineer performing the specified function. Sources of information can include books on engineering career opportunities and practicing engineers in the particular branch.

1.7 For a particular branch of engineering, e.g., electrical engineering, find the program of study for the first 2 years and compare it with the program offered at your school approximately 20 years ago. Comment on the major differences.

1.8 Do Prob. 1.7 for the last 2 years of study in a particular branch of engineering.

1.9 List five of your own personal characteristics and compare that list with the list in Sec. 1.5.1.

1.10 Prepare a brief paper on the requirements for professional registration in your state. Include the type and content of the required examinations.

1.11 Prepare a 5-min talk to present to your class describing one of the technical societies and how it can benefit you as a student.

1.12 Choose one of the following topics (or one suggested by your instructor) and write a paper that discusses technological changes that have occurred in this area in the past 15 years. Include commentary on the societal impact of the changes and on new problems that may have arisen because of the changes.

(*a*) Passenger automobiles
(*b*) Electric power generating plants
(*c*) Computer graphics
(*d*) Heart surgery
(*e*) Heating systems (furnaces)
(*f*) Microprocessors
(*g*) Water treatment
(*h*) Road paving (both concrete and asphalt)
(*i*) Computer-controlled metal fabrication processes
(*j*) Robotics

2

Engineering Solutions

2.1 Introduction

This chapter provides a basic guide to problem analysis, organization, and presentation. Early in your education, you must develop an ability to solve and present simple or complex problems in an orderly, logical, and systematic way. Material presented here will be clarified by use in examples throughout the text. Problems at the end of each chapter will provide an opportunity for each of you to build essential problem-solving skills.

2.2 Problem Analysis

A distinguishing characteristic of a qualified engineer is the ability to solve problems. Mastery involves a combination of art and science. By *science* we mean a knowledge of the principles of mathematics, chemistry, physics, mechanics, and other technical subjects that must be learned so that they can be applied correctly when appropriate. By *art* we mean the proper judgment, common sense, and know-how that must be used to reduce a real-life problem to such a form that science can be applied to its solution. To know when and how rigorously science should be applied and whether the resulting answer reasonably satisfies the original problem is an art.

Much of the science of successful problem solving comes from formal training in school or from continuing education after graduation. But most of the art of problem solving cannot be learned in a formal course; it is rather a result of experience and common sense. Its application can be more effective, however, if problem solving is approached in a logical and organized method.

To clarify the distinction, let us suppose that a manufacturing engineer working for an electronics company is given the task of recommending whether a new personal computer can be profitably produced. At the time the engineering task is assigned, the competitive selling price has already been established by the marketing division. Also, the design group has developed working models of the personal computer with specifications of all components, which means that the cost of these components is known. The question of profit thus rests on the cost of assembly. The theory of engineering economy (the science portion of problem solving) is well known by the engineer and is applicable to the cost factors and time frame involved. Once the production methods have been established, the cost of assembly can be computed using standard techniques, such as methods time measurement analysis. Selection of production methods (the art portion of problem solving) depends largely on the experience of the engineer. Knowing what will or will not work in each part of the manufacturing process is a must in the cost estimate, but that data cannot be found in a handbook. It is in the head of the engineer. It is an art originating from experience, common sense, and good judgment.

Before the solution to any problem is undertaken, whether by a student or by a practicing professional engineer, a number of important ideas must be considered. Consider the following questions: How important is the answer to a given problem? Would a rough preliminary estimate be satisfactory, or is a high degree of accuracy demanded? How much time do you have and what resources are at your disposal? In a real-world situation the answers you arrive at may depend on the amount of data collected, the sophistication of equipment used, the accuracy of the data collected, the number and training of people available to assist, and several other factors.

Most complex problems require some level of electronic digital computational support. What about the theory you intend to use? Is it state of the art? Is it valid for this particular application? Do you currently understand the theory or must time be allocated for review and learning? Can you make assumptions that simplify without sacrificing needed accuracy? Are other assumptions valid and applicable?

The art of problem solving is a skill developed with practice. It is the ability to arrive at a proper balance between the time and resources expended on a problem and the accuracy and validity obtained in the solution. When you can optimize time and resources versus reliability, then problem-solving skills will serve you well.

2.3 The Engineering Method

The design steps introduced in Chap. 1 and expanded upon in Chap. 15 are simply the overall thought process that an engineer goes through when solving a problem. One significant portion of this design procedure is called *the analysis phase.*

Analysis is the use of mathematical and scientific principles to verify the performance of alternative solutions. Analysis conducted by engineers in many design projects may involve three areas: application of the laws of nature, application of the laws of economics, and application of common sense.

This analysis procedure will hereinafter be called *the engineering method.* It consists of six basic steps:

1. **Recognize and understand the problem.** Perhaps the most difficult part of problem solving is developing the ability to recognize and define the problem precisely. Many academic problems that you will be asked to solve

Figure 2.1
Problem solving in a modern engineering office. (*Digital Equipment Corporation.*)

have this step completed by the instructor. For example, if your instructor asks you to solve a quadratic algebraic equation but provides you with all the coefficients, the problem has been completely defined before it is given to you and little doubt remains about what the problem is.

If the problem is not totally defined, considerable effort must be expanded in studying the problem, eliminating the things that are unimportant, and zeroing in on the root problem. Effort at this step pays great dividends by eliminating or reducing false trials and thereby shortening the time taken to complete later steps.

2. **Accumulate facts.** All pertinent physical facts such as sizes, temperatures, voltages, currents, costs, concentrations, weights, times, etc., must be ascertained. Some problems require that steps 1 and 2 be done simultaneously. In others, step 1 might automatically produce some of the physical facts. Do not mix or confuse these details with data that are suspect or only assumed to be accurate. Deal only with items that can be verified. Sometimes it will pay to actually verify data that you believe to be factual but that may actually be in error.

3. **Select appropriate theory or principle.** Select appropriate theories or scientific principles that apply to the solution of the problem. Understand and identify limitations or constraints that apply to the selected theory.

4. **Make necessary assumptions.** Perfect solutions do not exist to real problems. Simplifications need to be made if they are to be solved. Certain assumptions can be made that do not significantly affect the accuracy of the solution, yet other assumptions may result in a large reduction in accuracy.

Although the selection of a theory or principle is stated in the engineering method as preceding the introduction of simplifying assumptions, there are cases where the order of these two steps should be reversed. For example, you will see in Chap. 11 (Material Balance) that you will often need to assume that a process is steady, uniform, and without chemical reactions, so that the applicable theory can be written simply as input = output. The point is that steps 3 and 4 should be considered for every problem, although the logical order may differ from problem to problem.

5. **Solve the problem.** If steps 3 and 4 have resulted in a mathematical equation (model), it is normally solved by application of mathematical theory, a trial-and-error solution, or some form of graphical solution. The results will normally be in numerical form with appropriate units.

6. **Verify and check results.** In engineering practice, the work is not finished merely because a solution has been obtained. It must be checked to ensure that it is mathematically correct and that units have been properly specified. Correctness can be checked by reworking the problem, by using a different technique, or by performing the calculations in a different order to be certain that the numbers agree in both trials. The units can be examined to see that all equations are dimensionally correct.

The answer must then be examined to see if it makes sense. An experienced engineer will generally have a good idea of the order of magnitude to expect. If the answer doesn't seem reasonable, there is probably an error in the mathematics or in the assumptions and theory used. For example, suppose that you are asked to compute the monthly payment required to repay a car loan of $5 000 over a 3-year period at an annual interest rate of 12 percent. After solving this problem, you have an answer of $11 000 per month. Even if you are inexperienced in engineering economy, you know the answer is not reasonable, so you should reexamine your theory and computations. Examination and evaluation of the reasonableness of an answer is a habit that you should strive to acquire. Your instructor and employer alike will find it unacceptable to be given results which you have indicated to be correct but which are obviously incorrect by a significant percentage. Otherwise the instructor or employer might conclude that you have not developed good judgment or, worse yet, have not taken the time to do the necessary checking.

2.4 Organization of Problem Presentation

The engineering method of problem solving as presented in the previous section is an adaptation of the well-known *scientific problem-solving method.* It is an overall approach to problem solving that should become an everyday part of the engineer's thought process. All engineers should follow this global approach to the solution of any problem and at the same time should learn to translate the information accumulated into a well-documented problem solution.

The steps listed below parallel the engineering method and provide reasonable documentation of the solution. If the steps listed are properly executed during the solution of problems in this text, it is our belief that you will enhance your ability to solve a wide range of engineering problems.

1. **Problem statement.** State the problem to be solved. The statement can often be simply a summary of the problem, but it should contain all the essential information, including what is to be determined.

2. **Diagram.** Prepare a diagram (sketch) with all pertinent dimensions, flow rates, currents, voltages, weights, etc. A diagram is a very efficient method

of showing given and required information. It also is a simple way of showing the physical setup, which may be difficult to describe adequately in words. Data that cannot properly be placed in a diagram should be separately listed.

3. **Theory.** If a theory has to be derived, developed, or modified, present it next. In some cases, a properly referenced equation is sufficient. At other times, an extensive theoretical derivation may be necessary.

4. **Assumptions.** Explicitly list in sufficient detail any and all pertinent assumptions that must be made to obtain a solution or that you have arbitrarily placed on the problem. This step is vitally important for the reader's understanding of the solution and its limitations. Recall that steps 3 and 4 might be reversed in some problems.

5. **Solution steps.** Show completely all steps taken in obtaining the solution. This is particularly important in an academic situation because your reader, the instructor, must have the means of judging your understanding of the solution technique. Steps completed but not shown make it difficult for instructors to evaluate this aspect of your work and, therefore, difficult for them to grade or critique the work.

6. **Verify and identify results.** Check solution accuracy and, if possible, verify the results. Clearly indicate the final answer by underlining it with a double line. Assign proper units. An answer without units (when it should have units) is meaningless.

Remember, the final step of the engineering method requires that the answer be examined to determine if it is realistic. This step should not be overlooked.

2.5 Standards of Problem Presentation

Once the problem has been solved and checked, it is necessary to present the solution according to some standard. The standard will vary from school to school and industry to industry.

On many occasions your solution will be presented to other individuals who are technically trained but who do not have an intimate knowledge of the problem itself. Presenting technical information to persons with nontechnical backgrounds may require methods different from those used to communicate with other engineers, so it is all-important that the information be clearly presented.

The objective is usually to furnish a technical presentation designed for an instructor or supervisor who does understand such data. A mark of an engineer is the ability to present information with great clarity in a neat, careful manner. In short, the information must be communicated wholly to the reader. (Discussion of drawings or simple sketches will not be included in this chapter, although they are important in many presentations. See App. C.)

Employers insist on carefully done presentations that completely document all work involved in solving the problems. Thorough documentation may be important in the event of a lawsuit, for which the details of the work might be introduced into the court proceedings as evidence. Lack of such documentation may result in the loss of a case that might otherwise have been won. Moreover, internal company use of the work is easier and more efficient if all aspects of it have been carefully supported and substantiated by data and theory.

Each industrial company, consulting firm, governmental agency, and university has established standards for presenting technical information. These standards vary slightly, but all fall into a basic pattern, which will be discussed below. Each organization expects its employees to follow its standards. Details can be easily learned in a particular situation once you are familiar with the general pattern that exists in all of these standards.

It is not possible to specify a single problem layout or format that will accommodate all types of engineering-problem solutions. Such a wide variety of solutions exists that the technique used must be adapted to fit the information to be communicated. In all cases, however, one must lay out a given problem in such a fashion that it can be easily grasped by the reader. No matter what technique is used, it must be logical and understandable.

Guidelines for problem presentation are suggested below. Acceptable layouts for problems in engineering are also illustrated. The guidelines are not intended as a precise format that must be followed but rather as a suggestion that should be considered and incorporated whenever applicable.

1. The most common type of paper used is that which is ruled horizontally and vertically on the reverse side, with only heading and margin rulings on the front. It is often called *engineering-problems paper.* The rulings on the reverse side, which are faintly visible through the paper, help one maintain horizontal lines of lettering and provide guides for sketching and simple graph construction. Moreover, the lines on the back of the paper will not be lost as a result of erasures.

2. The completed top heading of the problems paper should include such information as name, date, course number, and sheet number. The upper right-hand block should normally contain a notation such as *a/b,* where *a* is the page number of the sheet and *b* is the total number of sheets in the set.

3. Work should ordinarily be done in pencil with a lead that is hard enough (approximately H or 2H) that the linework is crisp and unsmudged. Erasures should always be complete, with all eraser particles removed.

4. Either vertical or slant letters may be selected as long as they are not mixed. Care should be taken to produce good, legible lettering but without such care that little work is accomplished. (See App. C for more information about lettering.)

5. Spelling should be checked for correctness. There is no reasonable excuse for incorrect spelling in a properly done problem solution.

6. Work must be easy to follow and uncrowded. Making an effort to keep it so contributes greatly to readability and ease of interpretation.

7. If several problems are included in a set, they must be distinctly sep-

arated, usually by a horizontal line drawn completely across the page between problems. Never begin a second problem on the same page if it cannot be completed there. It is usually better to begin each problem on a fresh sheet, except in cases where two or more problems can be completed on one sheet. It is not necessary to use a horizontal separation line if the next problem in a series begins at the top of a new page.

8. Diagrams that are an essential part of a problem presentation should be clear and understandable. Students should strive for neatness as a mark of a professional. (Refer to App. C for details of graphical techniques.) Often a good rough sketch is adequate, but using a straightedge can greatly improve the appearance and accuracy of the sketch. A little effort in preparing a sketch to approximate scale can pay great dividends when it is necessary to judge the reasonableness of an answer, particularly if the answer is a physical dimension that can be seen on the sketch.

9. The proper use of symbols is always important, particularly when the International System (SI) of units is used. It involves a strict set of rules that must be followed so that absolutely no confusion of meaning can result. (Details concerning the use of units can be found in Chap. 5.) There are also symbols in common and accepted use for engineering quantities that can be found in most engineering handbooks. These symbols should be used whenever possible. It is important that symbols be consistent throughout a solution and that all be defined for the benefit of the reader and also for your own reference.

The physical layout of a problem solution logically follows steps similar to those of the engineering method. You should attempt to present the process by which the problem was solved in addition to the solution, so that any reader can readily understand all aspects of the solution. Figure 2.2 illustrates the placement of the information.

Figures 2.3 and 2.4 are examples of typical engineering-problem solutions. You may find that they are helpful guides as you prepare your problem presentations.

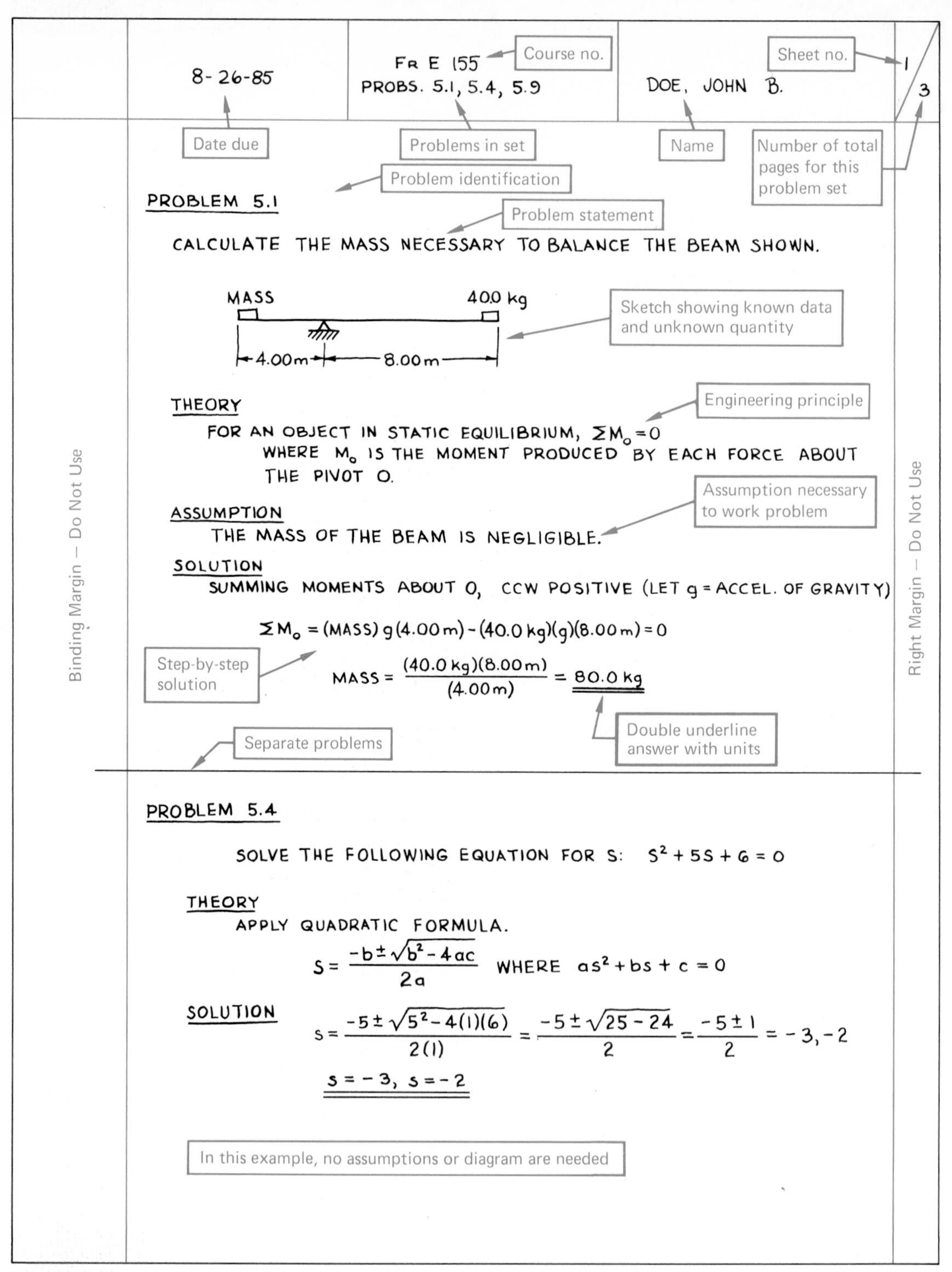

8-26-85
FR E 155
PROBS. 5.1, 5.4, 5.9
DOE, JOHN B.
1
3
Course no.
Sheet no.
Date due
Problems in set
Name
Number of total pages for this problem set
Problem identification
Problem statement
Binding Margin — Do Not Use
Right Margin — Do Not Use
PROBLEM 5.1
CALCULATE THE MASS NECESSARY TO BALANCE THE BEAM SHOWN.
MASS
40.0 kg
4.00 m
8.00 m
Sketch showing known data and unknown quantity
THEORY
Engineering principle
FOR AN OBJECT IN STATIC EQUILIBRIUM, ΣM_O = 0
WHERE M_O IS THE MOMENT PRODUCED BY EACH FORCE ABOUT THE PIVOT O.
ASSUMPTION
Assumption necessary to work problem
THE MASS OF THE BEAM IS NEGLIGIBLE.
SOLUTION
SUMMING MOMENTS ABOUT O, CCW POSITIVE (LET g = ACCEL. OF GRAVITY)
ΣM_O = (MASS) g (4.00 m) − (40.0 kg)(g)(8.00 m) = 0
Step-by-step solution
MASS = (40.0 kg)(8.00 m) / (4.00 m) = 80.0 kg
Double underline answer with units
Separate problems
PROBLEM 5.4
SOLVE THE FOLLOWING EQUATION FOR S: S² + 5S + 6 = 0
THEORY
APPLY QUADRATIC FORMULA.
S = (−b ± √(b² − 4ac)) / 2a WHERE as² + bs + c = 0
SOLUTION
s = (−5 ± √(5² − 4(1)(6))) / 2(1) = (−5 ± √(25 − 24)) / 2 = (−5 ± 1) / 2 = −3, −2
s = −3, s = −2
In this example, no assumptions or diagram are needed

Figure 2.2
Elements of a problem layout.

	8-22-85	FR E 160 PROBLEM 13.1	DOE, JANE A.	1/2

PROBLEM 13.1 SOLVE FOR THE VALUE OF RESISTANCE R IN THE CIRCUIT SHOWN BELOW.

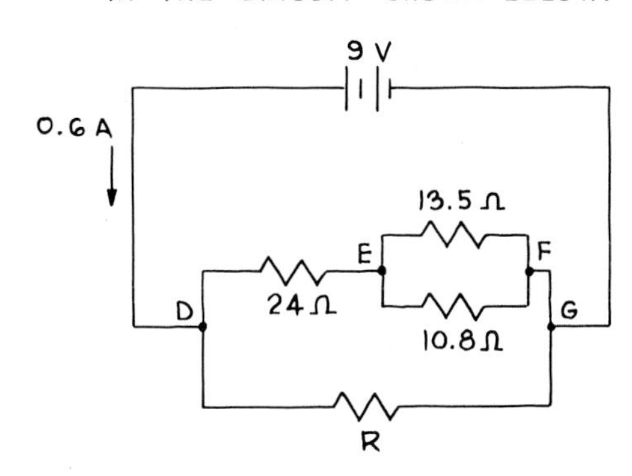

THEORY

- FOR RESISTANCES IN PARALLEL: $\frac{1}{R_{TOTAL}} = \frac{1}{R_1} + \frac{1}{R_2} + \frac{1}{R_3} + \ldots$

  THUS FOR 2 RESISTANCES IN PARALLEL

  $$R_{TOTAL} = \frac{R_1 R_2}{R_1 + R_2}$$

- FOR RESISTANCES IN SERIES: $R_{TOTAL} = R_1 + R_2 + R_3 + \ldots$
- OHM'S LAW: $E = RI$ WHERE E = ELECT. POTENTIAL IN VOLTS
  I = CURRENT IN AMPERES
  R = RESISTANCE IN OHMS

SOLUTION

- CALCULATE EQUIVALENT RESISTANCE BETWEEN POINTS E AND F. RESISTORS ARE IN PARALLEL.

  $$\therefore R_{EF} = \frac{R_1 R_2}{R_1 + R_2} = \frac{(13.5)(10.8)}{13.5 + 10.8} = \frac{145.8}{24.3} = 6.00\ \Omega$$

- CALCULATE EQUIVALENT RESISTANCE OF UPPER LEG BETWEEN D AND G.

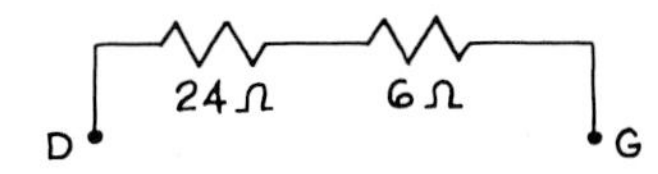

SERIES CIRCUIT

$$\therefore R'_{DG} = R_{24} + R_6 = 24 + 6 = 30\ \Omega$$

In this example, no assumptions were necessary

Figure 2.3
Sample problem presentation.

- CALCULATE EQUIVALENT RESISTANCE BETWEEN D AND G.

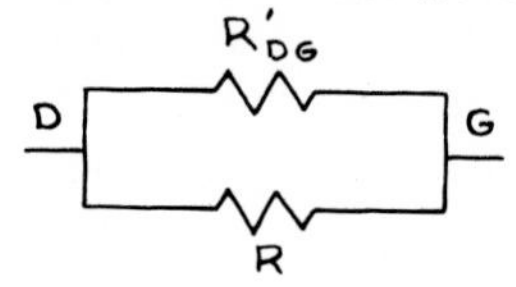

PARALLEL RESISTORS, SO

$$R_{DG} = \frac{(R'_{DG})(R)}{R'_{DG} + R}$$

- CALCULATE TOTAL RESISTANCE OF CIRCUIT USING OHM'S LAW.

$$R_{DG} = \frac{E}{I} = \frac{9\ V}{0.6\ A} = 15\ \Omega$$

- CALCULATE VALUE OF R FROM PREVIOUS EQUATIONS.

$$R_{DG} = 15\ \Omega = \frac{(R'_{DG})(R)}{R'_{DG} + R} = \frac{(30)(R)}{30 + R}$$

SOLVING FOR R:

$$(30 + R)(15) = 30R$$

$$30 + R = 2R$$

$$\underline{\underline{R = 30\ \Omega}}$$

Figure 2.3 (cont.)

2.5 A small aircraft has a glide ratio of 15:1. (This glide ratio means the plane moves 15 units horizontally for each 1 unit in elevation.) You are exactly in the middle of a 3-mi-diameter lake at 500 ft when the fuel supply is exhausted. You see a gravel road by the dock (Fig. 2.9) and must decide whether to try a ground or water landing. Show appropriate assumptions and calculations to support your decision.

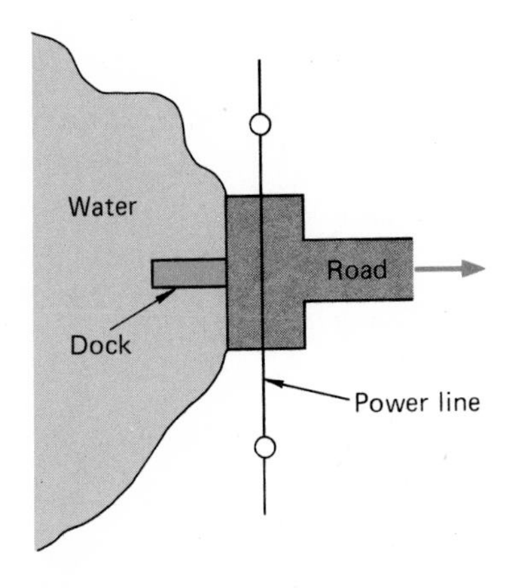

Figure 2.9

2.6 A young engineering student in a stationary hot-air balloon is momentarily fixed at 1 325-ft elevation above a level piece of land. He looks down (60° from horizontal) and turns laterally 360°. How many acres are contained within the cone generated by his line of sight? How high would he be if, when performing the same procedure, an area four times greater is encompassed?

2.7 A pilot in an ultralight knows that her aircraft in landing configuration will glide 2.0×10^1 km from a height of 2.0×10^3 m. A TV transmitting tower is located in a direct line with the local runway. If the pilot glides over the tower with 3.0×10^1 m to spare and touches down on the runway at a point 6.5 km from the base of the tower, how high is the tower?

2.8 A simple roof truss design is shown in Fig. 2.10. The lower section *VWXY* is made from three equal-length segments. *UW* and *XZ* are perpendicular to *VT* and *TY*, respectively. If *VWXY* is 2.0×10^1 m and the height of the truss is 2.5 m, determine the lengths of *XT* and *XZ*.

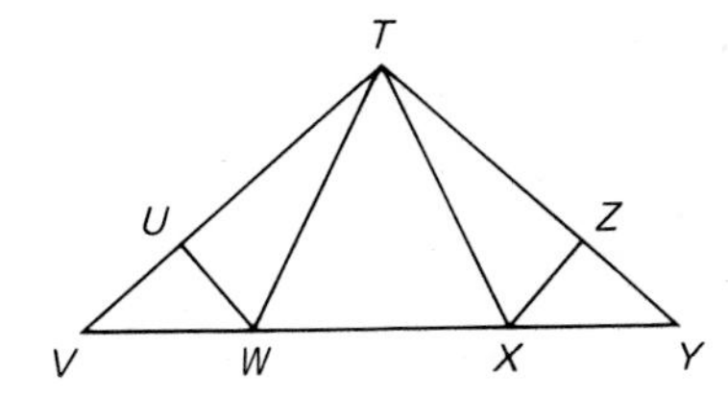

Figure 2.10

2.9 The wheel on an automobile turns at the rate of 145 r/min. Express this angular speed in (*a*) revolutions per second and (*b*) radians per second. If the wheel has a 30.0-in diameter, what is the velocity of the auto in miles per hour?

2.10 A bicycle wheel has a 28.0-in diameter wheel and is rotating at 225 r/min. Express this angular speed in radians per second. How far will the bicycle travel (in meters) in 15.0 min and what will be the velocity in miles per hour?

2.11 Assume the earth's orbit to be circular at 93.0×10^6 mi about the sun. Determine the speed of the earth (in miles per second) around the sun if there are exactly 365 days per year.

2.12 A child swinging on a tree rope 8.00 m long reaches a point 2.00 m above the lowest point. Through what total arc in degrees has the child passed? What distance in feet has the child swung?

2.13 Two engineering students were assigned the job of measuring the height of an inaccessible cliff (Fig. 2.11). The angles and distances shown were measured on a level beach in a vertical plane due south of the cliff. Determine height *AB*.

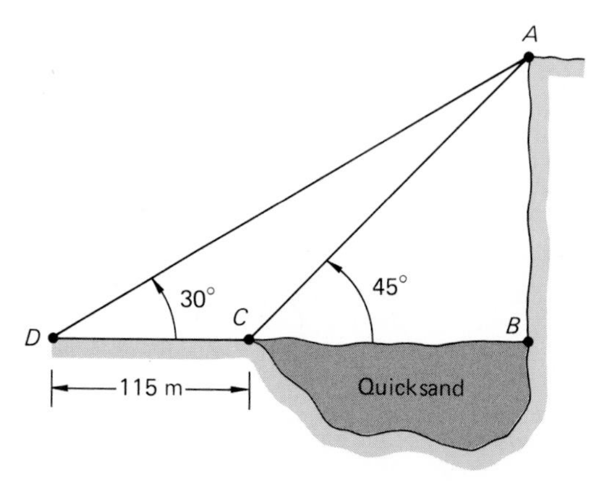

Figure 2.11

2.14 A survey team with appropriate equipment has been asked to measure a nonrectangular plot of land *ABCD* (Fig. 2.12). The following data were recorded: *CD* = 150.0 m, *DA* = 145.0 m, and *AB* = 110.0 m. Angle *DAB* = 115° and angle *DCB* = 100°. Calculate the length of side *BC* and the area of the plot.

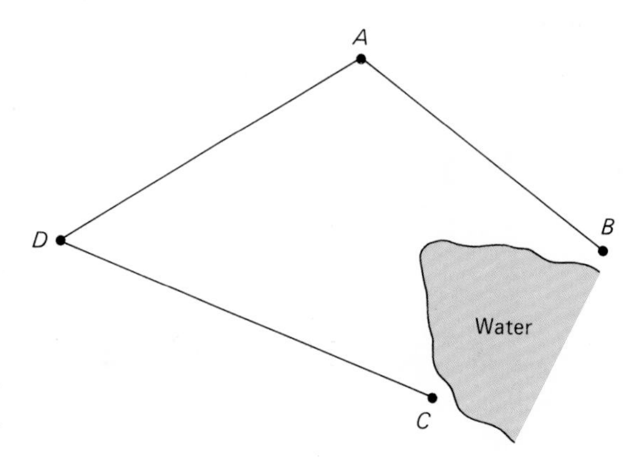

Figure 2.12

2.15 In Fig. 2.13 locations *D, E, F,* and *G* are surveyed points in a land development on level terrain. Leg *DG* is measured as 500.0 m. Angles at other stations are recorded as $GDE = 55°$, $DEF = 92°$, $FGD = 134°$, and $DGE = 87°$. Compute the lengths *DE, EF, FG,* and *EG*.

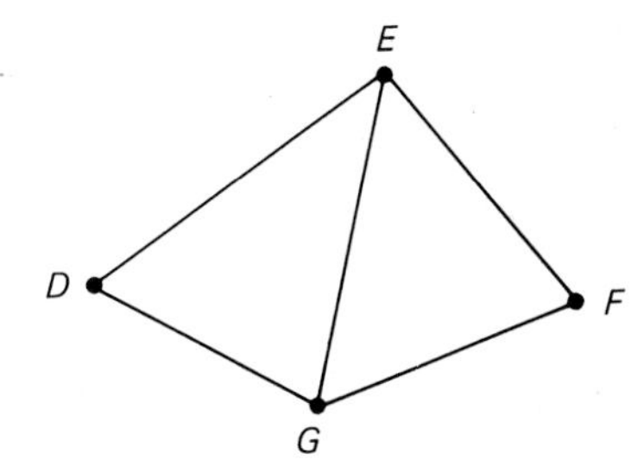

Figure 2.13

2.16 An aircraft moves through the air with a relative velocity of 2.75×10^2 km/h at a heading of N20°E. If there is a 45-km/h wind from the west calculate:

(*a*) The true ground speed and heading of the aircraft
(*b*) The heading the pilot should fly so that the true heading is N20°E

2.17 What heading must a pilot fly in compensating for a 125-km/h west wind to have a ground track of due south? The aircraft cruise speed is 6.00×10^2 km/h. What is the actual ground speed?

2.18 A circular piece of sheet metal 80.0 cm in diameter is to be used to manufacture (stamped, perhaps) several new highway signs. Each shape is to be inscribed with as little waste as possible. Calculate the percent of waste material when the inscribed shape is (*a*) a square; (*b*) an octagon; (*c*) a polygon with 32 sides.

2.19 Three circles are tangent to each other as in Fig. 2.14. The respective radii are 1.00×10^3, 8.00×10^2, and 5.00×10^2 mm. Find the area of the triangle (in square millimeters) formed by joining the three centers. Then determine the area within the triangle that is outside of the circles.

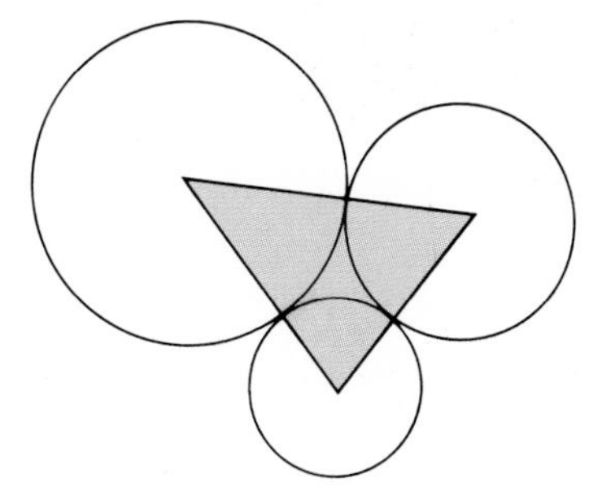

Figure 2.14

2.20 A waterwheel turns a belt on a drive wheel for a flour mill. The pulley on the waterwheel is 2.00 m in diameter, and the drive pulley is 0.500 m in diameter. If the centers of the pulleys are 4.00 m apart, calculate the length of belt needed.

2.21 A narrow, flat belt is used to drive a 50.00-cm (diameter) pulley from a 10.00-cm (diameter) pulley. The centers of the two pulleys are 50.00 cm apart. How long must the belt be if the pulleys rotate in the same direction? In opposite directions?

2.22 A homeowner decides to install a family swimming pool. It is 24.0 ft in diameter with a 4.00-ft water level. It has a conical-shaped bottom from edge to center at such an angle as to make the water at the apex of the cone 10.0 ft deep. Consider the following problems:

(*a*) If the bottom of the cylindrical portion of the pool is at ground level (Fig. 2.15), how many cubic feet of soil must be excavated for the cone?

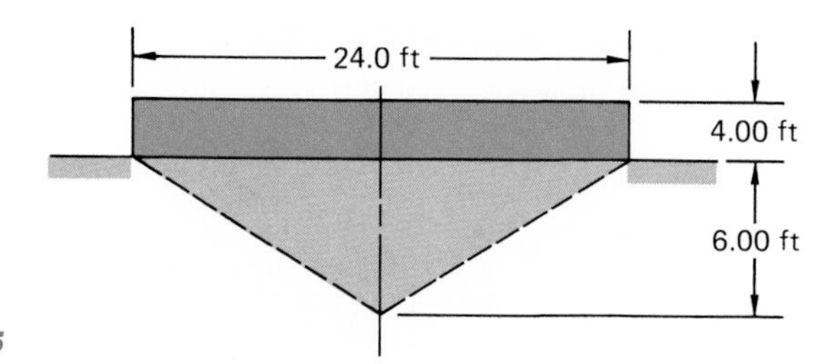

Figure 2.15

(*b*) How many gallons of water will be required to fill the pool to 4 ft above ground?

(*c*) If the owner moves the water in pails from a nearby lake, how many tons will be carried? (Density of water is 62.4 lbm/ft^3.)

(*d*) If he carries two 5-gal pails per trip and makes 20 trips per evening, how many days will it take to fill the pool?

2.23 A block of metal has a 90° notch cut from its lower surface. The notched part rests on a circular cylinder 2.0 cm in diameter as shown in Fig. 2.16. If the lower surface is 1.3 cm above the base plate, how deep is the notch?

Figure 2.16

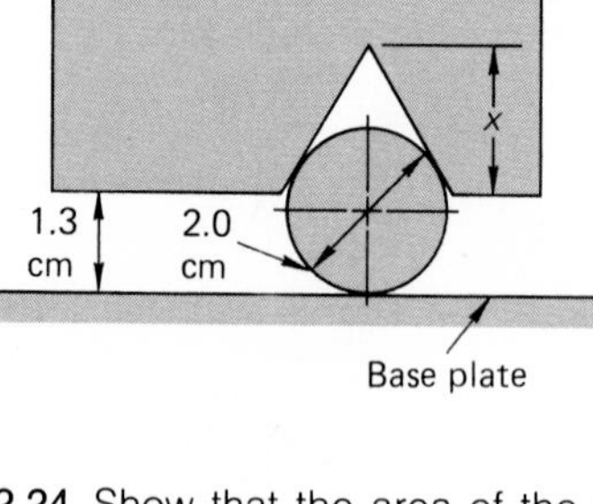

Figure 2.17

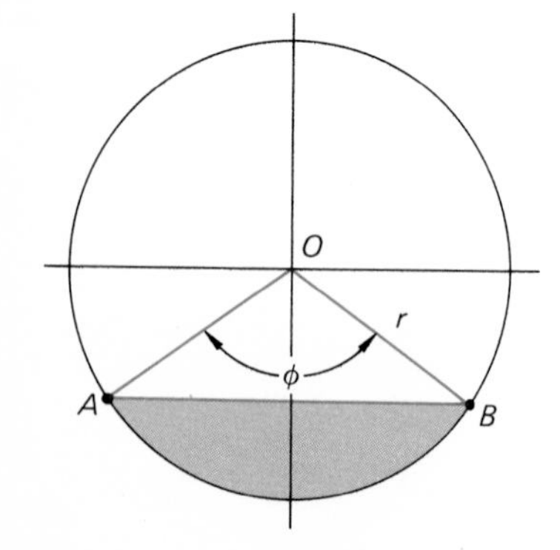

2.24 Show that the area of the shaded segment in Fig. 2.17 is given by the expression

$$A_s = \frac{r^2}{2}(\phi - \sin\phi)$$

2.25 A diesel fuel tank is 4.00 ft in diameter and 7.00 ft long. If the tank is positioned on its circular base (Fig. 2.18), how many gallons of fuel are in the tank when it is filled to a depth of 4.00 ft? Develop an expression for volume (v) in gallons for any depth (h) in feet for this configuration. Density of diesel fuel is 815 kg/m^3.

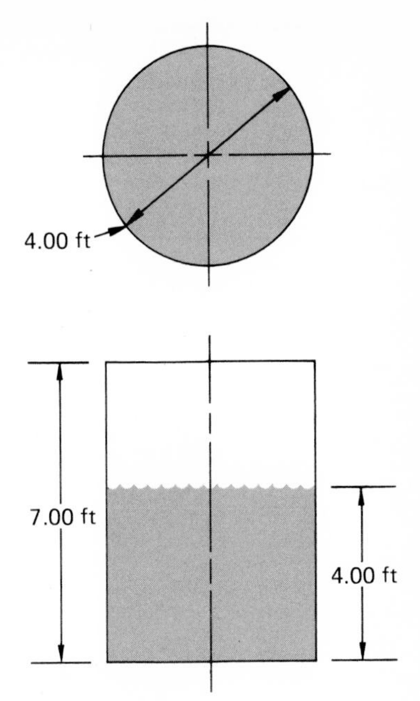

Figure 2.18

2.26 Consider the diesel tank in Prob. 2.25 as it would be if it were positioned on its side instead of base:

(*a*) Develop an expression for volume (v) in gallons as a function of height (h) in feet.

(*b*) If the empty tank has a mass of 3.00×10^2 lbm, develop an expression for the mass of tank and fuel (in pounds-mass) as a function of height (h) in feet.

(*c*) How many gallons of fuel are in the tank at a height (h) of 1.50 ft? At that depth, what is the total mass of tank and fuel?

2.27 Can tops are punched from a triangular piece of sheet metal. The pattern used requires 60.0-mm-diameter circular tops (see Fig. 2.19).

(*a*) What is the area of one can top in square millimeters?

(*b*) What is the area of triangle *ABC* in square millimeters?

(*c*) What is the area of trim *TPS* (shaded) in square millimeters?

(*d*) If angle *TPS* is fixed by the position of *A*, *B*, and *C* as shown and four can tops are cut along the base, how many circular tops can be cut from the entire sheet?

(*e*) What is the percent waste?

(*f*) What is the area of waste contained in triangle *ABC* (in square millimeters)?

(*g*) What is the area of waste in the rectangle *BQTC* (in square millimeters)?

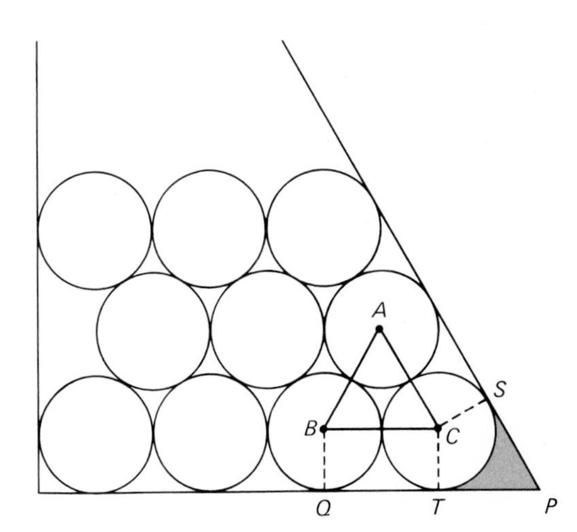

Figure 2.19

3

Representation of Technical Data

3.1 Introduction

This chapter provides information and guidelines that will be helpful when collecting, plotting, and interpreting technical and scientific data. Two areas in particular will be discussed in some detail: (1) graphical presentation of data and (2) graphical analysis.

Proper graphical presentation of data is necessary because calculated or experimental results are frequently collected in tabular form. Presentation in such form is generally not an optimal method of demonstrating the relationships between numerical values, since columns of numbers can sometimes be difficult to interpret. A system of graphing is thus needed. A visual impression, that is, something carefully and correctly presented in graph or chart form, is a much easier way to compare the rate of change or relative magnitude of variables.

In contrast, graphical analysis involves calculation and interpretation of data after it has been plotted. At times visual impressions are not adequate, so determination of an equation is required.

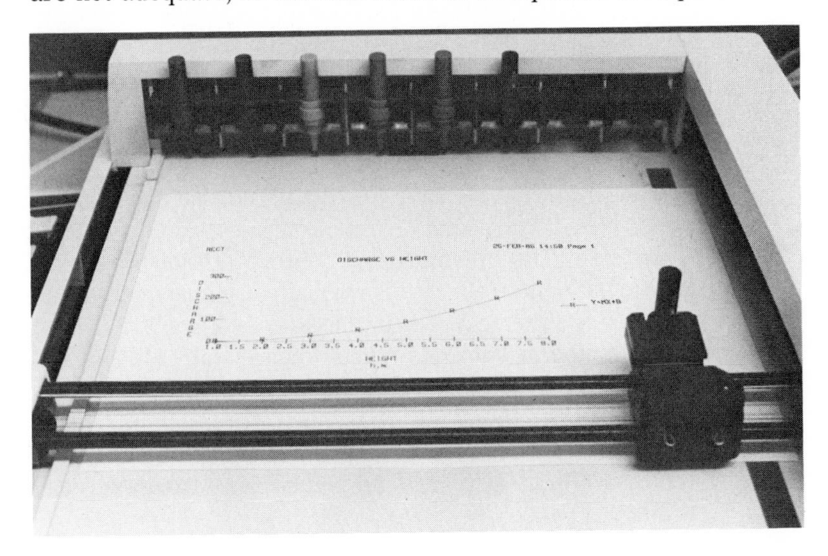

Figure 3.1
Hard-copy display of technical data.

Even though it is important for engineers to be able to interpret, evaluate, and communicate different forms of data, it is virtually impossible to include in one chapter all types of graphs and charts that they may encounter. Popular-appeal, or advertising, charts such as bar charts, pie diagrams, and distribution charts, although useful to the engineer, will not be discussed here. Such types are summarized in App. C.

Numerous examples will be used throughout the chapter to illustrate methods of representation because the effectiveness of graphs depends to a large extent on the details of their construction and appearance.

3.2 Collecting and Recording Data

Modern science as we know it today was founded on scientific measurement. Meticulously designed experiments, carefully analyzed, have produced volumes of scientific data that have been collected, recorded, and documented. For such data to be meaningful, however, certain laboratory procedures must be followed. Formal data sheets, such as those shown in Fig. 3.2, or notebooks should be used to record all observations. Information about all instruments and experimental apparatus used should also be recorded. Sketches illustrating the physical arrangement of equipment can be very helpful. Under no circumstances should observations be recorded elsewhere or data points erased. The data sheet is the "notebook of original entry." If there is reason for doubting the value of any entry, it may be canceled (that is, not considered) by drawing a line through it. The cancellation should be done in such a manner that the original entry is not obscured, in case you want to refer to it later.

Sometimes a measurement requires minimal accuracy, so time can be saved by making rough estimates. As a general rule, however, it is advantageous to make all measurements as accurately and reasonably as economics will allow. Unfortunately, as different observations are made throughout any experiment, some degree of inconsistency will develop. Errors will enter into all experimental work regardless of the amount of care exercised. A more complete discussion of error is covered in Chap. 4.

It can be seen from what we have just discussed that the analysis of experimental data involves not only measurements and collection of data but also careful documentation and interpretation of results.

Experimental data once collected is normally organized into some tabular form, which is the next step in the process of analysis. Data, such as that tabulated in Table 3.1, should be carefully labeled and neatly lettered so that results are not misinterpreted. This particular collection of data represents atmospheric pressure and temperature measurements recorded at various altitudes by students during a flight in a light aircraft. The points, once tabulated, can

Electrical Engineering Laboratory, Iowa State University
of Science and Technology

Title of Test ________ No. ________

Test made by ________
FOREMAN

Apparatus tested: ________ No. ________ Set ________

Rating ________

Date ________ Wiring Checked by ________ Data Checked by ________

	1	2	3	4	5	6	7	8	9	10	11	12	13	14
INSTRUMENT NO.														
SHUNT OR C.T. NO														
MULT. OR P.T. NO.														
RANGE USED														
SCALE READ														
FACTOR														
1														
2														
3														
4														
5														
6														
7														
8														
9														
10														
11														
12														
13														
14														
15														
16														
17														
18														
19														
20														
21														
22														
23														
24														
25														
26														
27														
28														
29														

(*a*)

DEPARTMENT OF MECHANICAL ENGINEERING
IOWA STATE UNIVERSITY
OF SCIENCE AND TECHNOLOGY
AMES, IOWA

OBSERVERS ________

TEST OF ________ DATE ________

(*b*)

Figure 3.2
Data sheets used by engineering departments.

Table 3.1

Height H, m	Temperature T, °C	Pressure P, kPa
0	15.0	101.3
300	12.8	97.7
600	11.1	94.2
900	8.9	90.8
1 200	6.7	87.5
1 500	5.0	84.3
1 800	2.8	81.2
2 100	1.1	78.2
2 400	−1.1	75.3
2 700	−2.8	72.4
3 000	−5.0	68.7
3 300	−7.2	66.9
3 600	−8.9	64.4
3 900	−11.1	61.9

be plotted and compared with standard atmospheric conditions, thereby making possible numerous aerodynamic calculations.

Although the tabulation of data is a necessary step, you may sometimes find it difficult to visualize a relation between variables when simply viewing a column of numbers. A most important step in the sequence from collection to analysis is, therefore, the construction of appropriate graphs or charts.

In recent years, modern laboratory equipment has been developed which will automatically sample and plot data for analysis. In the future, we expect to see expansion of these techniques along with continuous visual displays that will allow us to interactively control the experiments.

3.3 General Graphing Procedures

The proper construction of a graph from tabulated data can be generalized into a series of steps. Each of these steps will be discussed and illustrated in considerable detail in the following subsections.

1. Select the correct type of graph paper and grid spacing.
2. Choose the proper location of the horizontal and vertical axes.
3. Determine the scale units for each axis so that the data can be appropriately displayed.
4. Graduate and calibrate the axes.
5. Identify each axis completely.
6. Plot points and use permissible symbols (that is, ones commonly used and easily understood).
7. Draw the curve or curves.
8. Identify each curve and add titles and other necessary notes.
9. Darken lines for good reproduction.

3.3.1 Graph Paper

Printed coordinate graph paper is commercially available in various sizes with a variety of grid spacing. Rectilinear ruling can be purchased in a range of lines per inch or lines per centimeter, with an overall paper size of 8.5 × 11 in considered most typical. Figure 3.3 *a* is an illustration of graph paper having 10 lines per centimeter.

Closely spaced coordinate ruling is generally avoided for results that are to be printed or photoreduced. However, for accurate engineering analyses requiring some amount of interpolation, data are normally plotted on closely spaced, printed coordinate paper. Graph paper is available in a variety of colors, weights, and grades. Translucent paper can be used when the reproduction system requires a material that is not opaque.

If the data requires the use of log-log or semilog paper, such paper can also be purchased in different formats, styles, weights, and grades. Both log-log and semilog grids are available in from one to five cycles per axis. (A later section will discuss different applications of log-log and semilog paper.) Polar-coordinate paper is available in various sizes and graduations. A typical sheet is shown in Fig. 3.3 *b*. Examples of commercially available logarithmic paper are given in Fig. 3.3 *c* and *d*.

3.3.2 Axes Location and Breaks

The axes of a graph consist of two intersecting straight lines. The horizontal axis, normally called the *x axis*, is the *abscissa*. The vertical axis, denoted the *y axis*, is the *ordinate*. Common practice is to place the independent values along the abscissa and the dependent values along the ordinate, as illustrated in Fig. 3.4.

Many times, mathematical graphs contain both positive and negative values of the variables. This necessitates the division of the coordinate field into four quadrants, as shown in Fig. 3.5. Positive values increase toward the right and upward from the origin.

On any graph, a full range of values is desirable, normally beginning at zero and extending slightly beyond the largest value. To avoid crowding, the entire coordinate area should be used as completely as possible. However, certain circumstances require special consideration to avoid wasted space. For example, if values to be plotted along the axis do not range near zero, a "break" in the grid or the axis may be used, as shown in Fig. 3.6 *a* and *b*.

When judgments concerning relative amounts of change in a variable are required, the axis or grid should not be broken or the zero line omitted, with the exception of time in years, such as 1970, 1971, etc., since that designation normally has little relation to zero.

Since most commercially prepared grids do not include sufficient

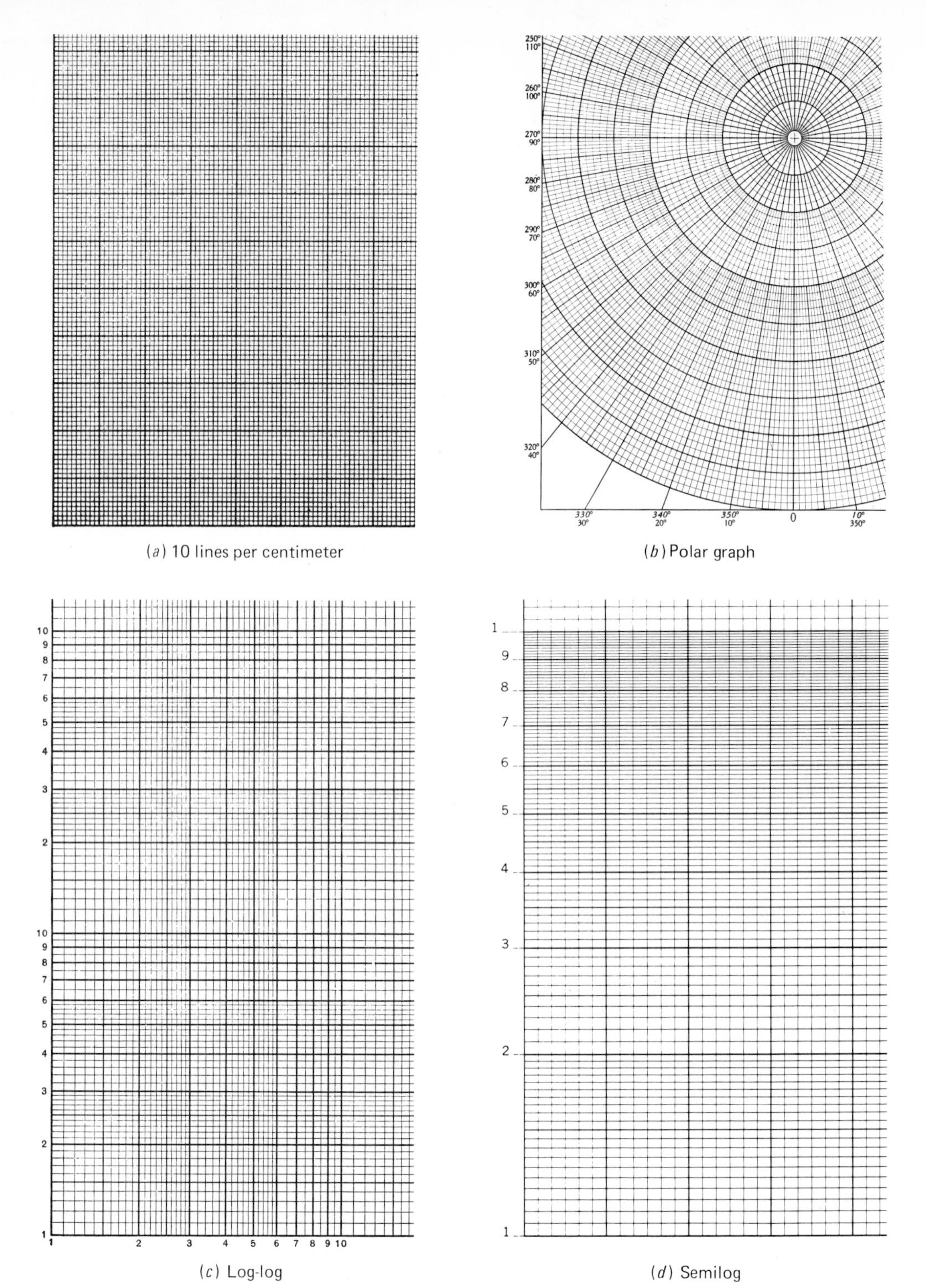

(*a*) 10 lines per centimeter

(*b*) Polar graph

(*c*) Log-log

(*d*) Semilog

Figure 3.3
Commercial graph paper.

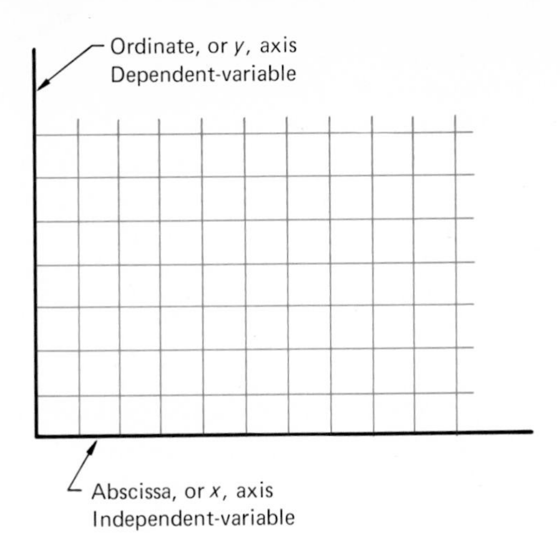

Figure 3.4
Abscissa (x) and ordinate (y) axes.

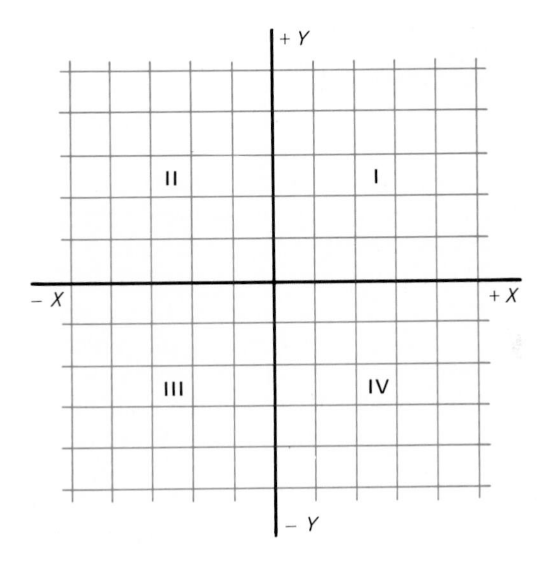

Figure 3.5
Coordinate axes.

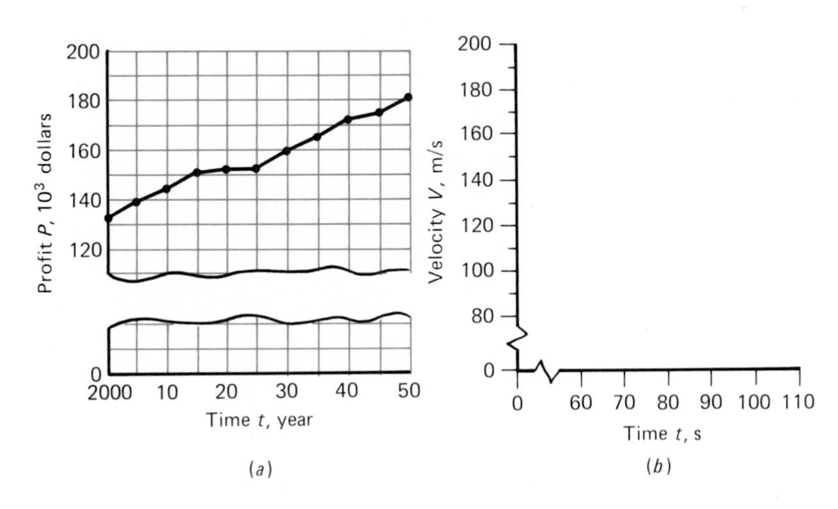

Figure 3.6
Typical axes breaks.

border space for proper labeling, the axes should preferably be placed 20 to 25 mm (approximately 1 in) inside the edge of the printed grid in order to allow ample room for graduations, calibrations, axes labels, reproduction, and binding. The edge of the grid may have to be used on log-log paper, since it is not always feasible to move the axis in. However, with careful planning, the vertical and horizontal axes can be repositioned in most cases, depending on the range of the variables.

3.3.3 Scale Graduations, Calibrations, and Designations

The scale is a series of marks, called *graduations*, laid down at predetermined distances along the axis. Numerical values assigned to significant graduations are called *calibrations*.

A scale can be uniform, with equal spacing along the stem, as found on the metric, or engineer's, scales. If the scale represents a variable whose exponent is not equal to 1 or a variable that contains trigonometric or logarithmic functions, the scale is called a *nonuniform, or functional, scale*. Examples of both these scales are shown in Fig. 3.7.

Many computer software packages have been written to produce graphs. The quality and accuracy of these computer-made graphs vary widely, depending on the sophistication of the software as well as on the plotter or printer employed. Typically, the software will produce an axis scale to accommodate the range of data values. This may or may not produce an easily interpreted scale. Alternatively, you may be able to specify the range the axis scale should have, allowing greater control of the scale drawn. An example of a computer-generated graph is shown in Fig. 3.8.

When hand-plotting data, one of the most important considerations is the proper selection of scale graduations. A basic guide to follow is the *1, 2, 5 rule*, which can be stated as follows:

Scale graduations are to be selected so that the smallest division of the axis is a positive or negative integer power of 10 times 1, 2, or 5.

The justification and logic for this rule are clear. Graduation of an axis by this procedure makes possible interpolation of data be-

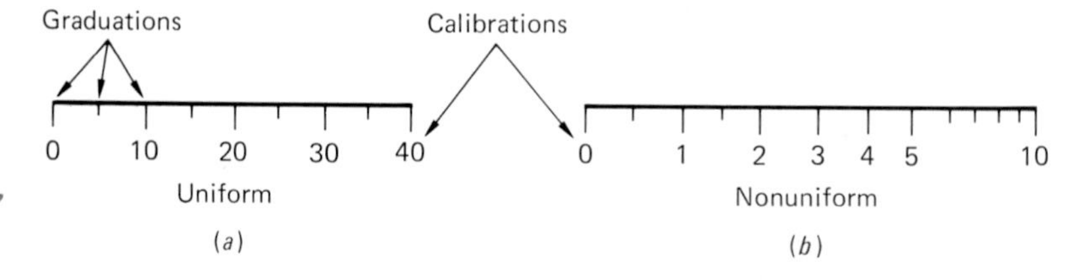

Figure 3.7
Scale graduations and calibrations.

Figure 3.8
Computer-plotted graph.

tween graduations when plotting or reading a graph. Figure 3.9 illustrates both acceptable and nonacceptable examples of scale graduations.

Violations of the 1, 2, 5 rule that are acceptable involve certain units of time as a variable. Days, months, and years can be graduated and calibrated as illustrated in Fig. 3.10.

Scale graduations normally follow a definite rule, but the number of calibrations to be included is primarily a matter of good judgment. Each application requires consideration based on the scale length

Figure 3.9
Scale graduations.

Figure 3.10
Time as a variable.

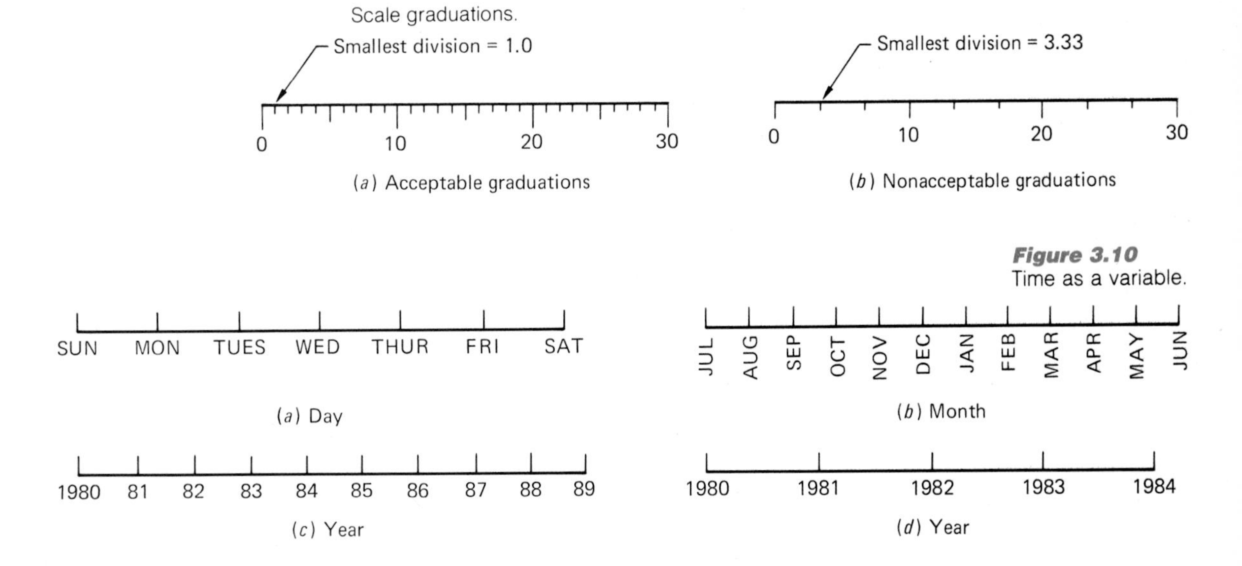

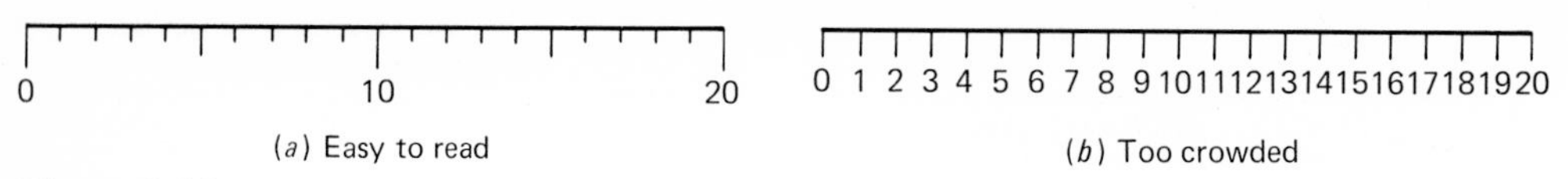

Figure 3.11
Scale calibrations.

and range as well as the eventual use. Figure 3.11 demonstrates how calibrations can differ on a scale with the same length and range. Both examples obey the 1, 2, 5 rule, but as you can see, too many closely spaced calibrations make the axis difficult to read.

Computer-produced graphics with uniform scales may not follow the 1, 2, 5 rule, particularly if the software calculates a scale based on the range of values of the data. If the software has the option of separately specifying the range, you may be able to obtain a scale that does follow the 1, 2, 5 rule, making it easier to read values from the graph.

The selection of a scale deserves attention from another point of view. If the rate of change is to be depicted accurately, then the slope of the curve should represent a true picture of the data. By contracting or expanding the axis or axes, an incorrect impression of the data could be implied. Such a procedure is to be avoided. Figure 3.12 demonstrates how the equation $Y = X$ can be misleading if not properly plotted. Occasionally distortion is desirable, but it should always be carefully labeled and explained to avoid misleading conclusions.

Here again, a computer plot may not produce scales appropriate for interpreting the slope of a curve unless you can specifically control the range plotted on each axis.

If plotted data consist of very large or small numbers, the SI prefix names (milli-, kilo-, mega-, etc.) may be used to simplify calibrations. As a guide, if the numbers to be plotted and calibrated consist of more than three digits, it is customary to use the appropriate prefix; an example is illustrated in Fig. 3.13.

Figure 3.12
Proper representation of data.

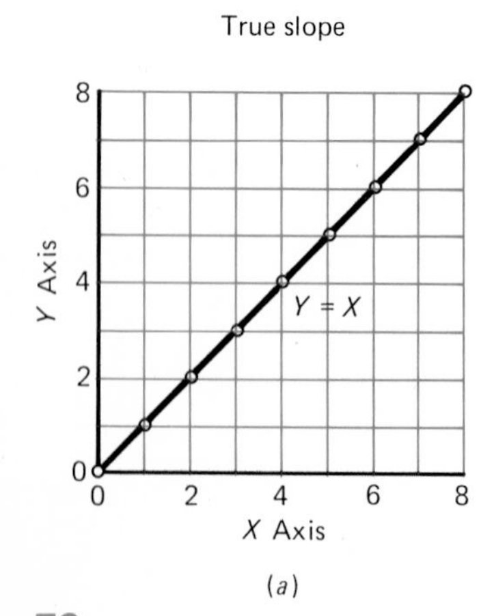

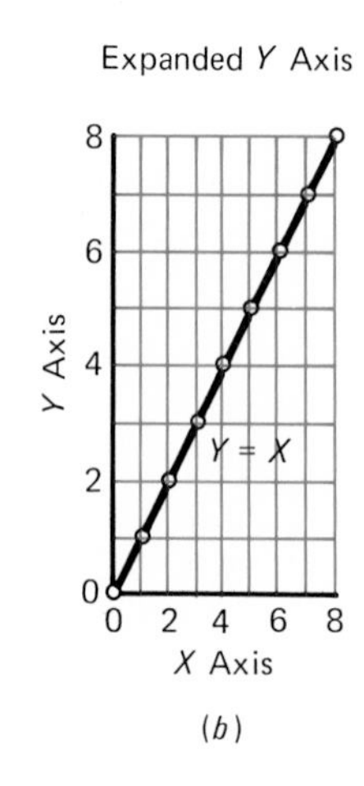

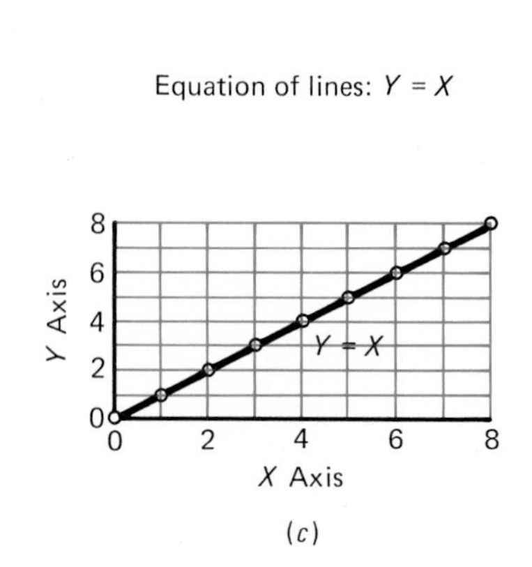

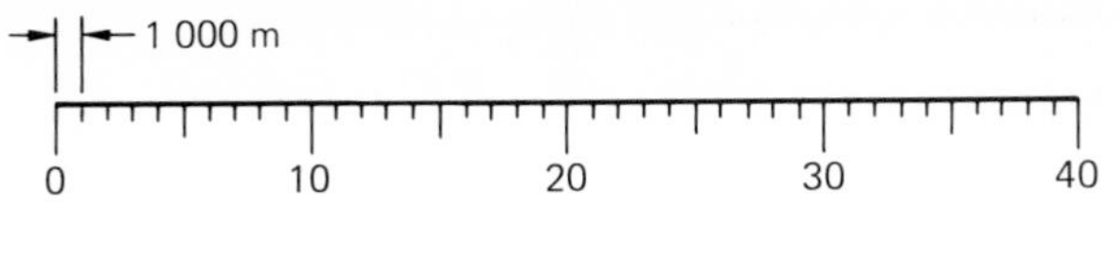

Figure 3.13
Reading the scale.

The length scale calibrations in Fig. 3.13 contain only two digits, but the scale can be read by understanding that the distance between the first and second graduation (0 to 1) is a kilometer; therefore, the calibration at 10 represents 10 km.

Certain quantities, such as temperature in degrees Celsius and altitude in meters, have traditionally been tabulated without the use of prefix multipliers. Figure 3.14 depicts a procedure by which these quantities can be conveniently calibrated. Note in particular that the distance between 0 and 1 on the scale represents 1 000°C.

This is another example of how the SI notation is convenient, since the prefix multipliers (micro-, milli-, kilo-, mega-, etc.) allow the calibrations to stay within the three-digit guideline.

The calibration of logarithmic scales is illustrated in Fig. 3.15. Since log-cycle designations start and end with powers of 10 (i.e., 10^{-1}, 10^0, 10^1, 10^2, etc.) and since commercially purchased paper is normally available with each cycle printed 1 through 10, Fig. 3.15*a* and *b* demonstrates two preferred methods of calibration.

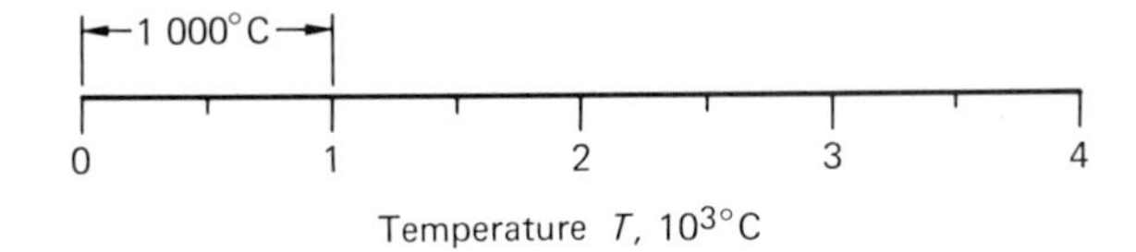

Figure 3.14
Reading the scale.

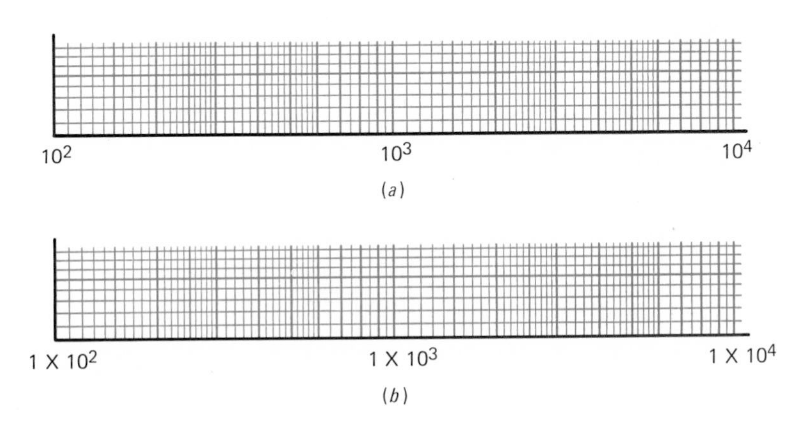

Figure 3.15
Calibration of log scales.

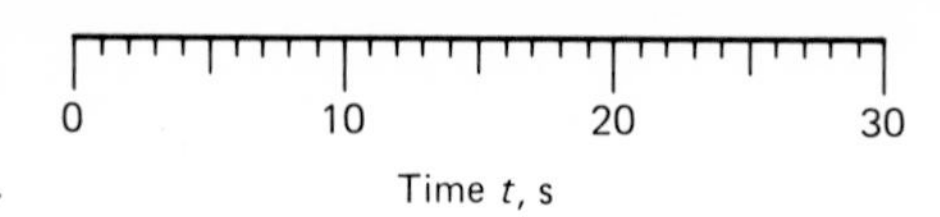

Figure 3.16
Axis identification.

3.3.4
Axis Labeling

Each axis should be clearly identified. At a minimum, the axis label should contain the name of the variable, its symbol, and its units. Since time is frequently the independent variable and is plotted on the x axis, it has been selected as an illustration in Fig. 3.16.

Scale designations should preferably be placed outside the axes, where they can be shown clearly. Labels should be lettered parallel to the axis and positioned so that they can be read from the bottom or right side of the page.

3.3.5
Point-Plotting Procedure

Data can normally be categorized in one of three general ways: as

Figure 3.17
Plotting data points.

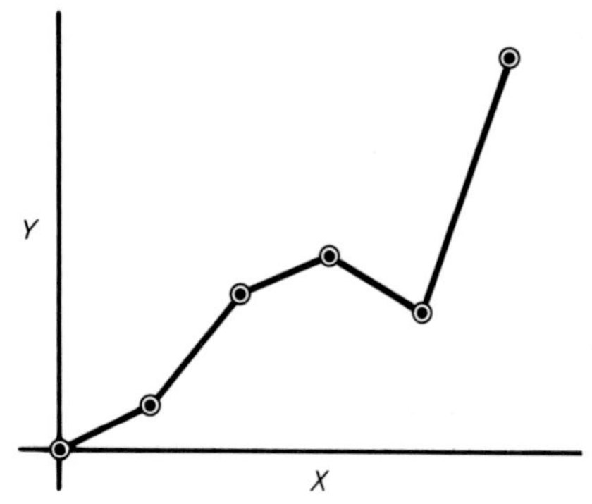

(*a*) Observed: Usually plotted with observed data points connected by straight, irregular line segments. Line does not penetrate the circles.

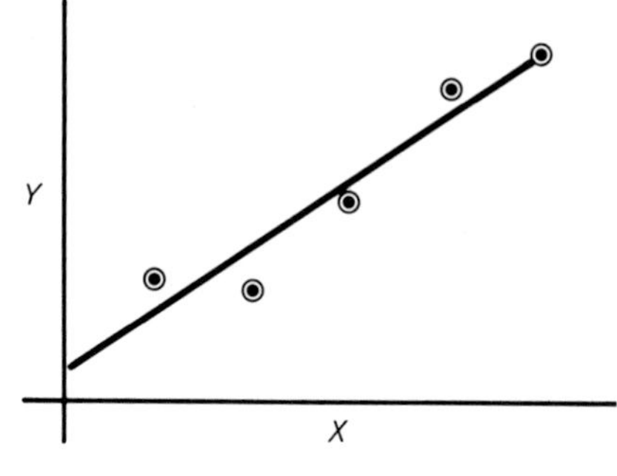

(*b*) Empirical: Reflects the author's interpretation of what occurs between known data points. Normally represented as a smooth curve or straight line fitted to data. Data points may or may not fall on curve.

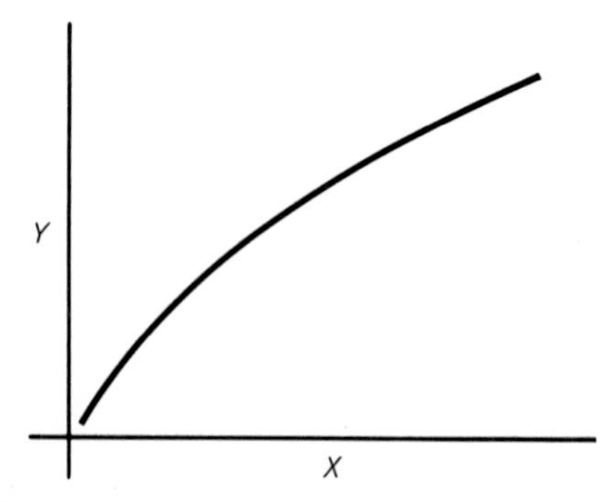

(*c*) Theoretical: Graph of an equation. Curves or lines are smooth and without symbols. Every point on the curve is a data point.

observed, empirical, or theoretical. Observed and empirical data points are usually located by various symbols, such as a small circle or square around each data point, whereas graphs of theoretical relations (equations) are normally constructed smooth, without use of symbol designation. Figure 3.17 illustrates each type.

3.3.6 Curves and Symbols

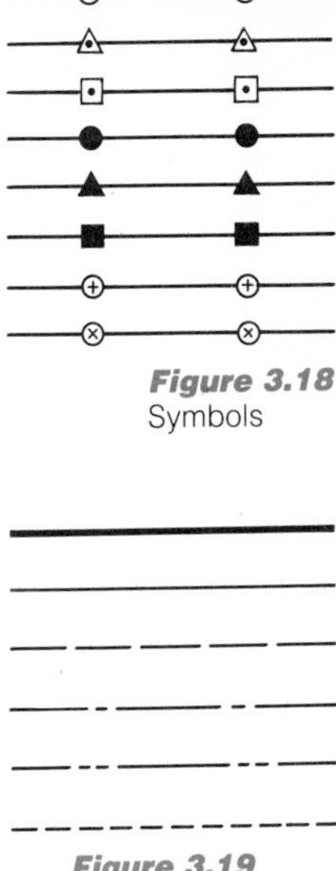

Figure 3.18 Symbols

Figure 3.19 Line representation.

On graphs prepared from observed data resulting from laboratory experiments, points are usually designated by various symbols (see Fig. 3.18).

If more than one curve is plotted on the same grid, several of these symbols may be used (one type for each curve). To avoid confusion, however, it is good practice to label each curve.

When several curves are plotted on the same grid, another way in which they can be distinguished from one another is by the use of different types of lines, as illustrated in Fig. 3.19. Solid lines are normally reserved for single curves, and dashed lines are commonly used for extensions. The curves should be heavier than the grid ruling.

A key, or legend, should be placed in an isolated portion of the grid, preferably enclosed in a border, to define point symbols or line types that are used for curves. Remember that the lines representing each curve should never be drawn through the symbols, so that the precise point is always identifiable. Figure 3.20 demonstrates the use of a key and the practice of breaking the line at each symbol.

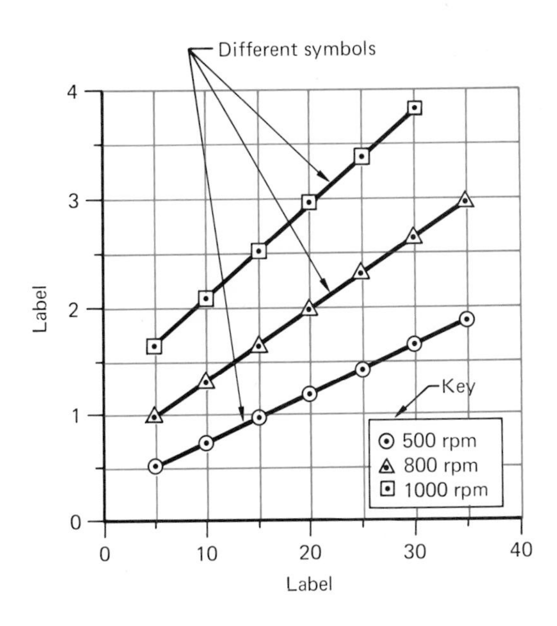

Figure 3.20 Key.

3.3.7 Titles

Each graph must be identified with a complete title. The title should include a clear, concise statement of the data being represented, along with items such as the name of the author, the date of the experiment, and any and all information concerning the plot, including the name of the institution or company. Titles are normally enclosed in a border.

All lettering, the axes, and the curves should be sufficiently bold to stand out on the graph paper. Letters should be neat and of standard size. Figure 3.21 is an illustration of plotted experimental data incorporating many of the items discussed in the chapter.

3.4 Empirical Functions

An empirical function is generally described as one based on values obtained by experimentation. It is often identified as any function for which no analytic equation has previously been defined. Needless to say, a great many mathematical equations might fit such a function. Fortunately, it is possible to classify many empirical results into one of four general categories: (1) linear, (2) exponential, (3) power, or (4) periodic.

Figure 3.21
Sample plot.

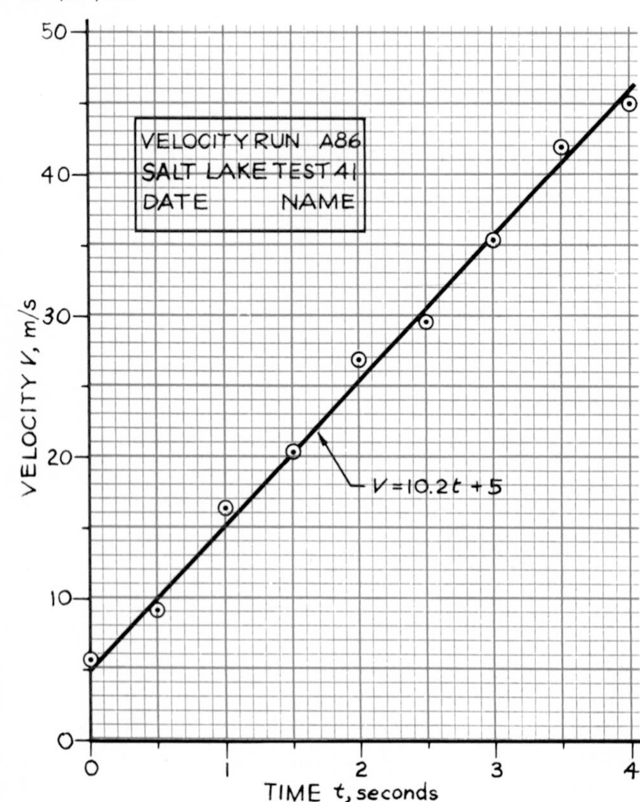

Necessary steps to follow when plotting a graph.

1 Select the type of graph paper (rectilinear, semilog, log-log, etc.) and grid spacing for best representation of the given data.
2 Choose the proper location of the horizontal and vertical axes.
3 Determine the scale units (range) for each axis to display the data appropriately.
4 Graduate and calibrate the axes (1, 2, 5 rule).
5 Identify each axis completely.
6 Plot points and use permissible symbols.
7 Draw the curve or curves.
8 Identify each curve and add title and necessary notes.
9 Darken lines for good reproduction.

As the name suggests, a linear function will plot as a straight line on uniform rectangular coordinate paper. Likewise, when a curve representing experimental data is a straight line or a close approximation to a straight line, the relationship of the variables can be expressed by a linear equation.

Correspondingly, exponential equations, when plotted on semilog paper, will be linear. The basic form of the equation is $y = be^{mx}$. This can be written in log form and becomes $\log y = mx \log e + \log b$. Alternatively, using natural logarithms, the equation becomes $\ln y = mx + \ln b$ because $\ln e = 1$. The independent variable x is plotted against $\ln y$.

The power equation has the form of $y = bx^m$. Written in log form it becomes $\log y = m \log x + \log b$. This data will plot straight on log-log paper, since the log of the independent variable x is plotted against the log of y.

When the data represents experimental results and a series of points are plotted to represent the relationship between the variables, it is improbable that a straight line can be constructed through every point, since some error is inevitable. If all points do not lie on the same line, an approximation scheme or averaging method may be used to arrive at the best possible fit.

3.5 Curve Fitting

Different methods or techniques are available to arrive at the best "straight-line" fit. The time and expense involved increases, however, with the degree of accuracy and reliability of the method selected.

Three methods commonly employed for finding the best fit are listed below:

1. Method of selected points
2. Method of averages
3. Method of least squares

Figure 3.22
An orientation session where engineers are discussing methods of productivity improvement with a department manager. (*Allen-Bradley.*)

Each of these techniques is progressively more accurate. The first method will be briefly described in Sec. 3.6. The most accurate method, method 3, is discussed in Chap. 6.

Method 2, the method of averages, is based on the idea that the line location is positioned to make the algebraic sum of the differences between observed and calculated values of the ordinate equal to zero.

In both methods 2 and 3, the procedure involves minimizing what are called *residuals*, or the difference between an observed ordinate and the corresponding computed ordinate. The method of averages will not be applied in this book, but there are any number of reference texts available that adequately cover the concept.

3.6 Method of Selected Points

The method of selected points is a valid method of determining the equation that best fits data that exhibit a linear relationship. Once the data have been plotted and determined to be linear, a line is selected that appears to fit the data best. This is most often accomplished by visually selecting a line that goes through as many data points as possible and has approximately the same number of data points on either side of the line.

Once the line has been constructed, two points, such as A and B, are selected *on the line* and at a reasonable distance apart. The coordinates of both points A (x_1, y_1) and B (x_2, y_2) must satisfy the equation of the line, since both are points on the line.

3.7 Empirical Equations—Linear

When experimental data plot as a straight line on rectangular grid paper, the equation of the line belongs to a family of curves whose basic equation is given as

$$y = mx + b \qquad 3.1$$

where m is the slope of the line, a constant, and b is a constant referred to as the *y intercept* (the value of y when $x = 0$).

To demonstrate how the method of selected points works, consider the following example.

Example problem 3.1 A racing automobile is clocked at various times t and velocities V. Determine the equation of a straight line constructed through the points recorded in Table 3.2. Once an analytic expression has been determined, velocities at intermediate values can be computed.

Table 3.2

Time t, s	0	5	10	15	20	25	30	35	40
Velocity V, m/s	24	33	62	77	105	123	151	170	188

Procedure

1. Plot the data on rectangular paper. If the results form a straight line (see Fig. 3.23), the function is linear and the general equation is of the form

$$V = mt + b$$

where m and b are constants.

2. Select two points on the line, $A\ (t_1, V_1)$ and $B\ (t_2, V_2)$, and record the value of these points. Points A and B should be widely separated to reduce the effect on m and b of errors in reading values from the graph. Points A and B are identified on Fig. 3.23 for instructional reasons. They should not be shown on a completed graph that is to be displayed.

A (10, 60)

B (35, 164)

3. Substitute the points A and B into $V = mt + b$.

$$60 = m(10) + b$$

$$164 = m(35) + b$$

4. The equations are solved simultaneously for the two unknowns.

$$m = 4.16$$

$$b = 18.4$$

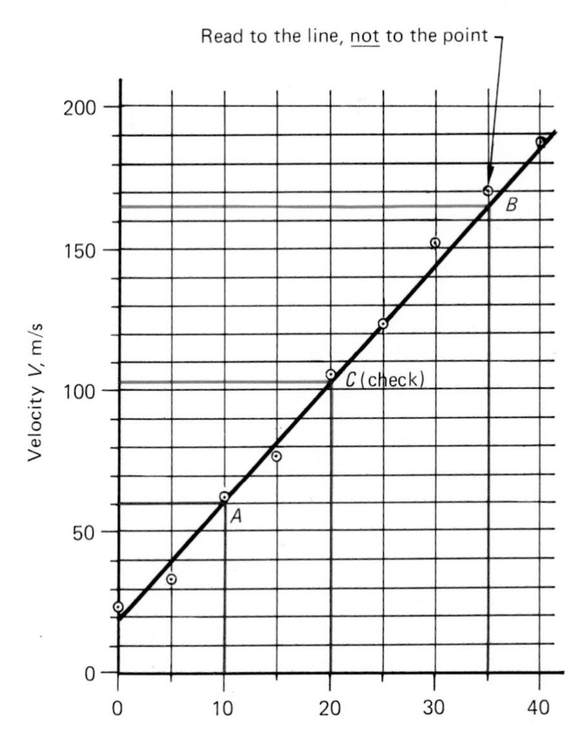

Figure 3.23
Data plot.

5. The general equation of the line can then be written as

$V = 4.16t + 18.4$

6. Using another point, $C\ (t_3, V_3)$, check for verification:

$C\ (20, 102)$

$102 \approx 4.16(20) + 18.4$

$102 \approx 83.2 + 18.4$

It is also possible by discreet selection of points to simplify the solution. For example, if point A is selected at (0, 18.4) and this coordinate is substituted into the general equation

$18.4 = m(0) + b$

the constant b is immediately known.

3.8 Empirical Equations—Power Curves

When experimentally collected data are plotted on rectangular coordinate graph paper and the points do not form a straight line, then you must determine which family of curves the line may most closely approximate. (Recognizing the correct form of the equation comes primarily from experience.) Consider the following familiar example.

Example problem 3.2 Suppose that a solid object is dropped from a tall building, and the values are as recorded in Table 3.3.

Table 3.3

Time, t, s	Distance, s, m
0	0
1	4.9
2	19.6
3	44.1
4	78.4
5	122.5
6	176.4

Solution To anyone who has studied fundamental physics, it is apparent that these values should correspond to the general equation for a free-falling body (neglecting air friction):

$$s = \tfrac{1}{2}gt^2$$

But assume for a moment that all we have is the table of values.

First it is helpful to make a freehand plot to observe the data visually. See Fig. 3.24.

From this quick plot, the data points are more easily recognized as belonging to a family of curves whose general equation can be written

$$y = bx^m \qquad 3.2$$

Remember that before the method of selected points can be applied to determine the equation of the line in this example problem, the plotted line must be straight, because two points on a curved line do not uniquely identify the line.

Mathematically, this general equation can be put in linear form by taking the logarithm of both sides,

$$\log y = m \log x + \log b$$

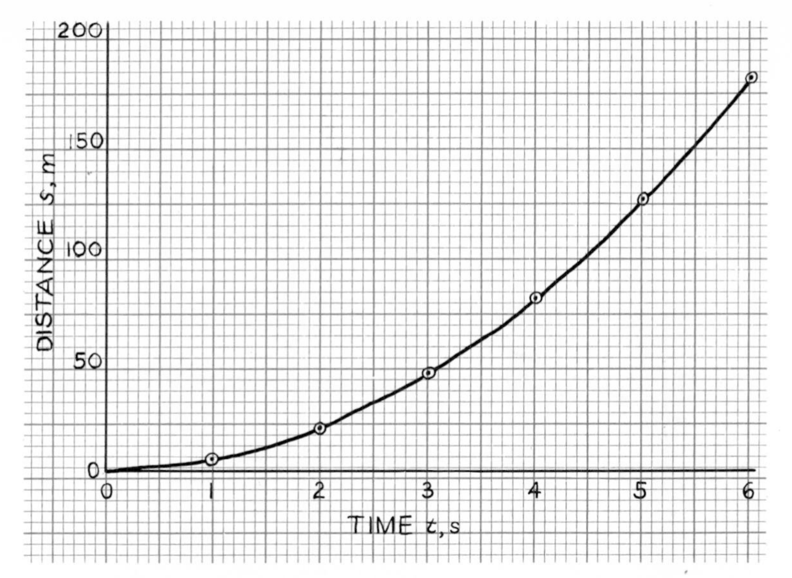

Figure 3.24
Rectilinear plot (freehand).

This relationship indicates that if the logs of both y and x were recorded and the results plotted on rectangular paper, the line would be straight.

Realizing that the log of zero is undefined and plotting the remaining points (see Table 3.4), we get the graph shown in Fig. 3.25.

Since the graph of log s versus log t plots as a straight line, it is possible to use the general form of the equation

$$\log y = m \log x + \log b$$

and apply the method of selected points.

When reading values for points A and B from the graph, we must remember that the logarithm of each variable has already been determined and plotted.

A (0.200 0, 1.090 0)

B (0.600 0, 1.890 0)

Points A and B can now be substituted into the general equation $\log s = m \log t + \log b$ and solved simultaneously.

Table 3.4

Time, t	Distance, s	Log t	Log s
0	0		
1	4.9	0.0000	0.6902
2	19.6	0.3010	1.2923
3	44.1	0.4771	1.6444
4	78.4	0.6021	1.8943
5	122.5	0.6990	2.0881
6	176.4	0.7782	2.2465

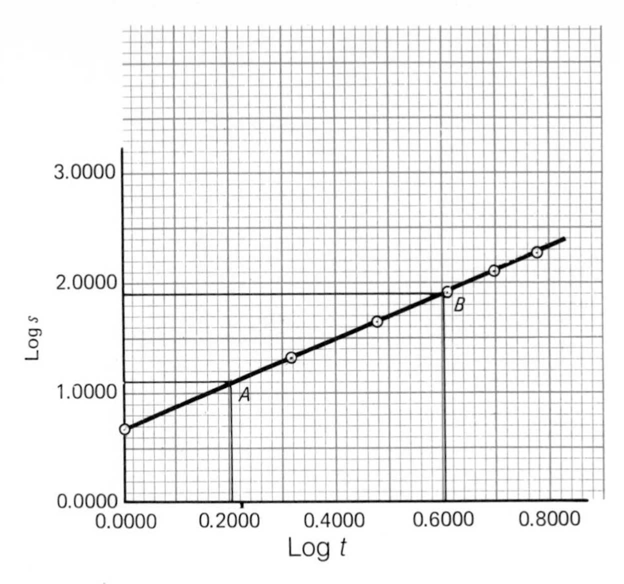

Figure 3.25
Log-log plot on rectilinear grid.

$$1.890\,0 = m(0.600\,0) + \log b$$

$$1.090\,0 = m(0.200\,0) + \log b$$

$$m = 2.0$$

$$\log b = 0.69$$

$$b = 4.9$$

An examination of Fig. 3.25 shows that the value of log b (0.69) can be read from the graph where log $t = 0.0$. This, of course, is where $t = 1$ and is the y intercept for log-log plots.

The general equation can then be written as

$$s = 4.9t^2$$

Or, $s = \frac{1}{2}gt^2$, where $g = 9.8\ \text{m/s}^2$.

The only inconvenience that results from determining an equation from the originally collected data derives from the necessity of taking logarithms of each variable and then plotting the logs of these variables.

Obviously, this step is not necessary, since functional paper is commercially available with log x and log y scales already constructed. Log-log paper allows the variables themselves to be plotted directly without the need of computing the log of each value.

In the preceding example, once the general form of the equation is selected [that is, Eq. (3.2)], the data can be plotted directly on log-log paper. Since the resulting curve will be a straight line, the method of selected points can be used directly. (See Fig. 3.26.)

The log form of the equation is again used:

$$\log s = m \log t + \log b$$

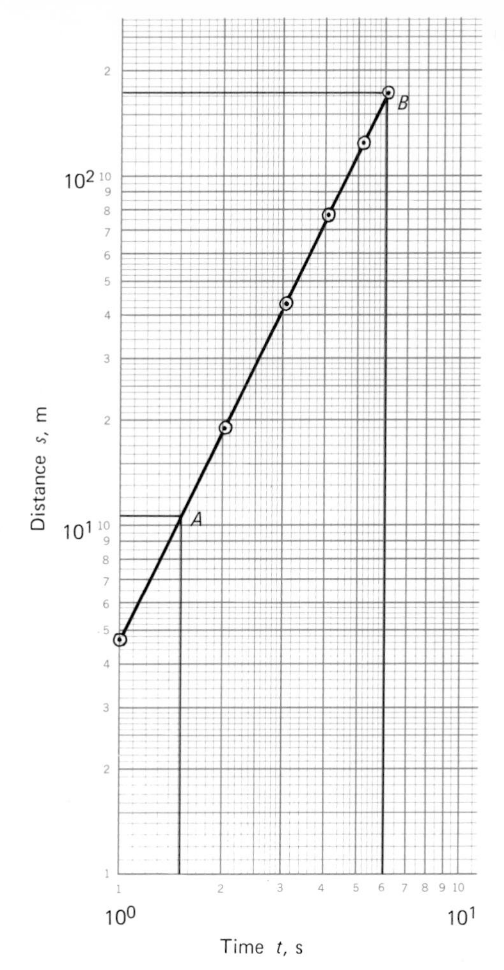

Figure 3.26
Log-log plot.

Select points A and B on the line:

A (1.5, 10.9)

B (6, 175)

Substitute the values into the general equation $\log s = m \log t + \log b$, taking careful note that the numbers are the variables and *not* the logs of the variables.

$$\log 175 = m \log 6 + \log b$$

$$\log 10.9 = m \log 1.5 + \log b$$

Again, solving these two equations simultaneously results in the following approximate values for the constants b and m:

$$b = 4.9$$

$$m = 2.0$$

Identical conclusions can be reached:

$$s = \frac{1}{2}gt^2$$

This time, however, one can use functional scales rather than calculate the log of each number.

3.9 Empirical Equations—Exponential Curves

Suppose your data do not produce a straight line (or nearly straight) on rectangular coordinate paper nor is the line approximately straight on log-log paper. Without experience in analyzing experimental data, you may feel lost as to how to proceed. Normally, when experiments are conducted you have an idea as to how the parameters are related and you are merely trying to quantify that relationship. If you plot your data on semilog graph paper and it produces a reasonably straight line, then it has the general form $y = be^{mx}$.

Assume that an experiment produces the data shown in Table 3.5. The data when plotted produce the graph shown as Fig. 3.27. To determine the constants in the equation $y = be^{mx}$, write it in linear form, either as

$$\log y = mx \log e + \log b$$

$$\text{or } \ln y = mx + \ln b$$

The method of selected points can now be employed for $\ln Q = mV + \ln b$ (choosing the natural log form). Points A (5, 20) and B (70, 600) are carefully selected on the line, so they must satisfy the equation. Substituting the values of V and Q at points A and B, we get

$$\ln 600 = 70m + \ln b$$

$$\text{and } \ln 20 = 5m + \ln b$$

Solving simultaneously for m and b, we have

$$m = 0.052\,3$$

$$\text{and } \ln b = 2.735\,93$$

$$\text{or } b = 15.42$$

The desired equation then is determined to be $Q = 15.4e^{0.052V}$. This determination can be checked by choosing a third point, substituting the value for V, and solving for Q. It can also be seen from the graph that $Q \cong 15.5$ when $V = 0$, which is a reasonable check on the value of b.

Table 3.5

Fuel consumption, mm^3/s	Velocity, m/s
25.2	10.0
44.6	20.0
71.7	30.0
115	40.0
202	50.0
367	60.0
608	70.0

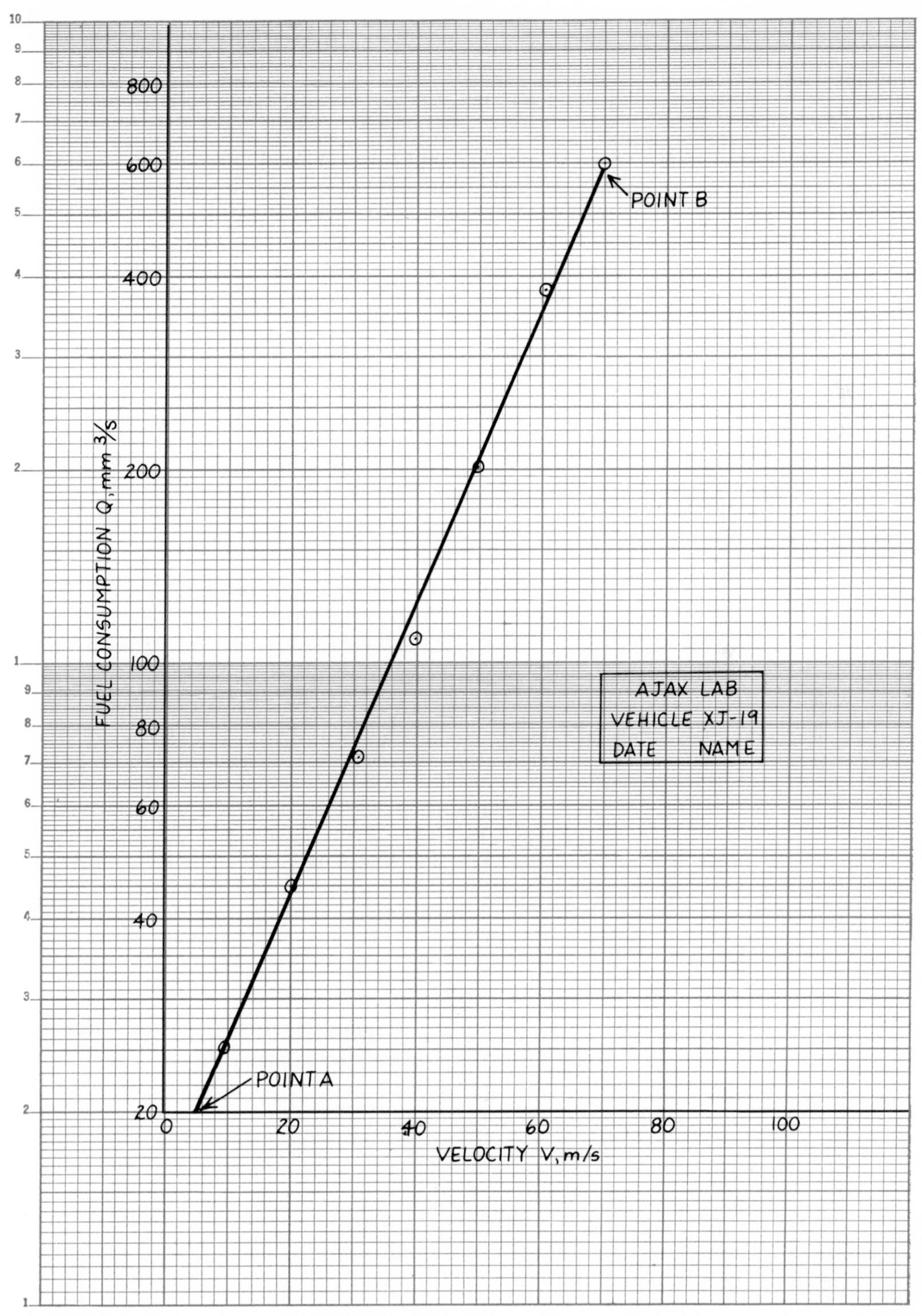

Figure 3.27
Semilog plot

3.10 Polar Charts

The relationship of certain parameters can best be displayed by plotting the data on polar-coordinate paper. An example of a polar plot is shown on Fig. 3.28.

You will note that polar graph paper is composed of equally spaced concentric circles cut by radii with equal angular intervals. The center, or pole, is always zero for the dependent variable, which is plotted as a distance radially from the pole. The independent variable is the angular value and covers the range from 0 to 360°. The location of the zero point is not always the same; the graph designer chooses the location that best portrays the relationship between the variables.

Problems

3.1 Table 3.6 is data from a trial run on the Utah salt flats made by an experimental turbine-powered vehicle.

(*a*) Plot the data on rectilinear paper.
(*b*) Determine the equation of the line, using the method of selected points.
(*c*) Interpret the slope of the line.

3.2 Table 3.7 lists the values of velocity recorded on a ski jump in Colorado this past winter.

(*a*) Plot the data on rectilinear paper.
(*b*) Using the method of selected points, determine the equation of the line.
(*c*) Give the average acceleration.

3.3 Table 3.8 is a collection of data for an iron-constantan thermocouple. Temperature is in degrees Celsius and the electromotive force (emf) is in millivolts.

(*a*) Plot a graph, using rectilinear paper, showing the relation of temperature to voltage, with voltage as the independent variable.
(*b*) Using the method of selected points, find the equation of the line.

Table 3.6

Time *t*, s	Velocity *V*, m/s
10.0	15.1
20.0	32.2
30.0	63.4
40.0	84.5
50.0	118
60.0	139

Table 3.7

Time *t*, s	Velocity *V*, m/s
1.0	5.3
4.0	18.1
7.0	26.9
10.0	37.0
14.0	55.2

Table 3.8

Temperature *t*, °C	Voltage (emf), mV
50.0	2.6
100.0	6.7
150.0	8.8
200.0	11.2
300.0	17.0
400.0	22.5
500.0	26.0
600.0	32.5
700.0	37.7
800.0	41.0
900.0	48.0
1 000.0	55.2

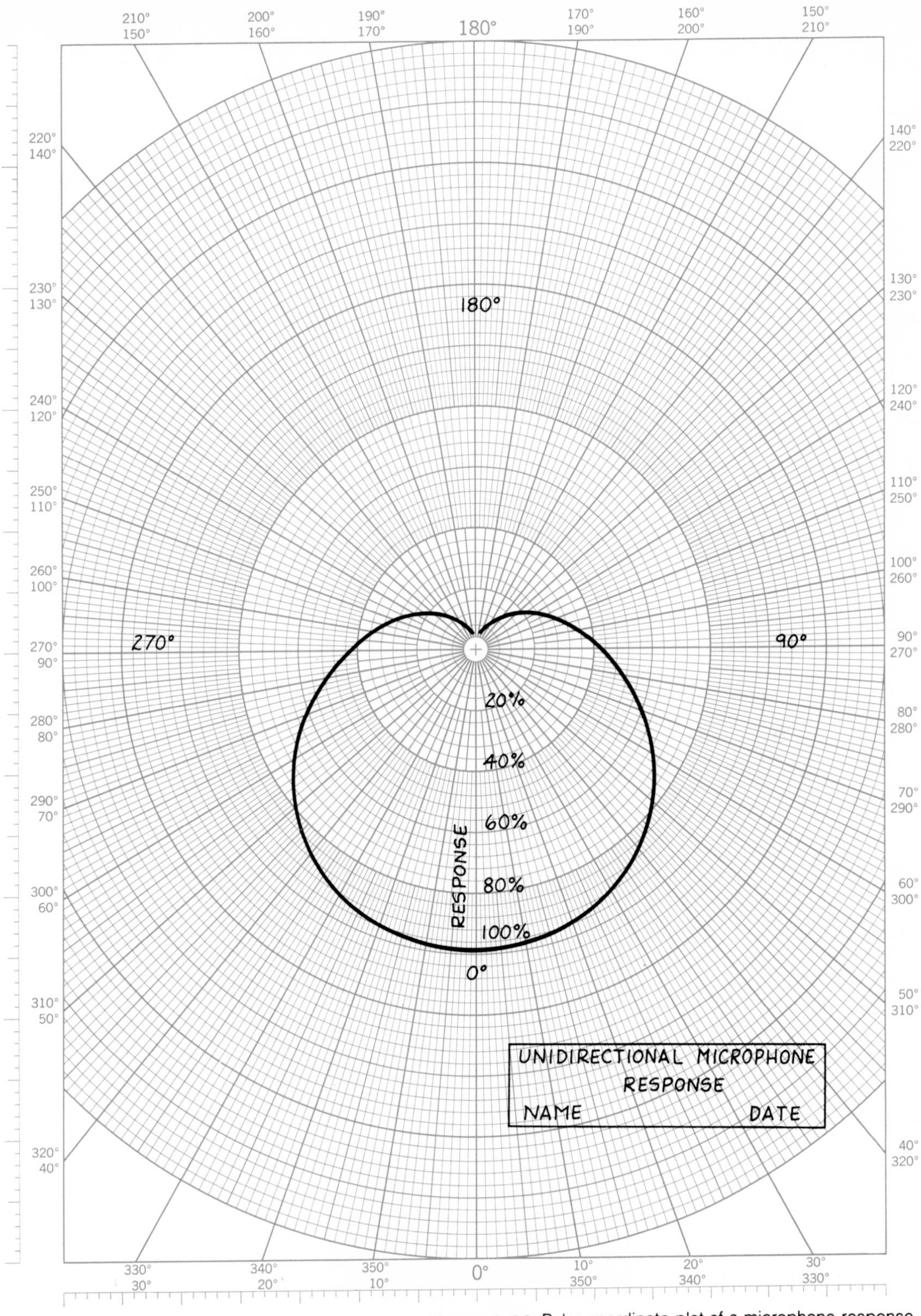

Figure 3.28 Polar coordinate plot of a microphone response.

Table 3.9

Discharge Q, L/s	Power P, kW
3.00	28.5
7.00	33.8
10.00	39.1
13.50	43.2
17.00	48.0
20.00	51.8
25.00	60.0

Table 3.10

Deflection D, mm	Load L, N
2.25	35.0
12.0	80.0
20.0	120.0
28.0	160.0
35.0	200.0
45.0	250.0
55.0	300.0

3.4 A Sessions pump was tested to determine the power required to produce a range of discharges. The test was performed in Oak Park on July 1, 1984. The results of the test are shown in Table 3.9.

(*a*) Plot a graph showing the relation of power required to discharge.
(*b*) Determine the equation of the relationship.
(*c*) Predict the power required to produce a discharge of 37 L/s.

3.5 A spring was tested in Des Moines last Thursday. The test of spring ZX-15 produced the data shown in Table 3.10.

(*a*) Plot the data on rectangular graph paper and determine the equation that expresses the deflection to be expected under a given load.
(*b*) Predict the load required to produce a deflection of 75 mm.
(*c*) What load would be expected to produce a deflection of 120 mm?

3.6 A Johnson furnace was tested 45 days ago in your hometown to determine the heat generated, expressed in thousands of British thermal units per cubic foot of furnace volume, at varying temperatures. The results are shown in Table 3.11.

(*a*) Plot the data on log-log graph paper, with temperature the independent variable.
(*b*) Using the method of selected points, determine the equation that best fits the data.

Table 3.11

Heat released H, 10^3 Btu/ft^3	Temperature T, °F
0.200	172
0.600	241
2.00	392
4.00	483
8.00	608
20.00	812
40.00	959
80.00	1 305

3.7 The capacity of a 20-cm screw conveyor that is moving dry ground corn is expressed in liters per second and the conveyor speed in revolutions per minute. A test was conducted in Cleveland on conveyor model JD172 last week. The results of the test are given in Table 3.12.

(*a*) Plot the data on log-log graph paper.
(*b*) Determine the equation that expresses velocity in terms of capacity.

Table 3.12

Capacity C, L/s	Angular velocity V, r/min
3.01	10.0
6.07	21.0
15.0	58.2
30.0	140.6
50.0	245
80.0	410
110.0	521

3.8 The resistance of a class and shape of electrical conductor was tested over a wide range of sizes at constant temperature. The test was performed in Madison, Wisconsin, on April 4, 1984, at the Acme Electrical Labs. The test results are shown in Table 3.13. The resistance is expressed in milliohms per meter of conductor length.

(*a*) Plot the data on log-log paper.

(*b*) Using the method of selected points, find the equation that expresses resistance as a function of the area of the conductors.

Table 3.13

Area A, mm^2	Resistance R, $m\Omega/m$
0.021	505
0.062	182
0.202	55.3
0.523	22.2
1.008	11.3
3.32	4.17
7.29	1.75

3.9 The area of a circle can be expressed by the formula $A = \pi R^2$. If the radius varies from 0.5 to 5 cm, perform the following:

(*a*) Construct a table of radius versus area mathematically. Use radius increments of 0.5 cm.

(*b*) Construct a second table of log R versus log A.

(*c*) Plot the values from (*a*) on log-log paper and determine the equation of the line.

(*d*) Plot the values from (*b*) on rectilinear paper and determine the equation of the line.

3.10 The volume of a sphere is $V = 4/3\pi r^3$.

(*a*) Prepare a table of volume versus radius, allowing the radius to vary from 2.0 to 10.0 m in 1-m increments.

(*b*) Plot a graph on log-log paper showing the relation of volume to radius using the values from the table in part *a*.

(*c*) Verify the equation given above by the method of selected points.

3.11 A 90° triangular weir is commonly used to measure flow rate in a stream. Data on the discharge through the weir were collected and recorded as shown in Table 3.14.

(*a*) Plot the data on log-log paper, with height as the independent variable.

(*b*) Determine the equation of the line using the method of selected points.

Table 3.14

Discharge, Q, m^3/s	1.5	8	22	45	78	124	182	254
Height, h, m	1	2	3	4	5	6	7	8

3.12 A pitot tube is a device for measuring the velocity of flow of a fluid (see Fig. 3.29). A stagnation point occurs at point 2; by recording the height differential h, the velocity at point 1 can be calculated. Assume for this problem that the velocity at point 1 is known corresponding to the height differential h. Table 3.15 records these values.

(*a*) Plot the data on log-log paper using height as the independent variable.

(*b*) Determine the equation of the line using the method of selected points.

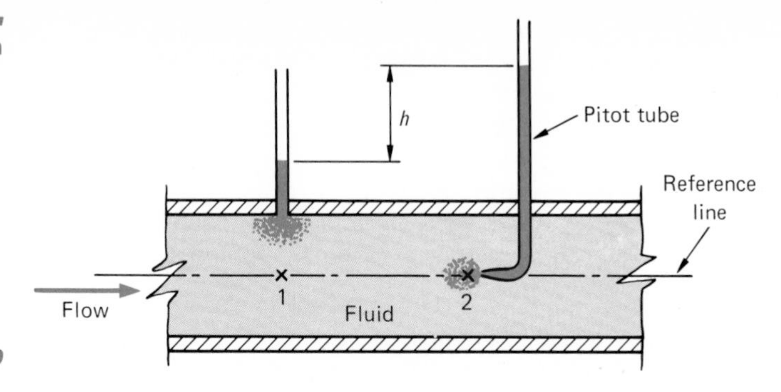

Figure 3.29

Table 3.15

Velocity V, m/s	1.4	2.0	2.8	3.4	4.0	4.4
Height h, m	0.1	0.2	0.4	0.6	0.8	1.0

3.13 A new production facility manufactured 29 parts the first month, but then increased production, as shown in Table 3.16.

(*a*) Plot the data on semilog paper.

(*b*) Using a time variable that defines January to be 1, February to be 2, etc., determine the equation.

Table 3.16

Month	Jan	Feb	Mar	Apr	May	Jun	Jul	Aug
Number	29	40	48	58	85	115	125	180

3.14 The voltage across a capacitor during discharge was recorded as a function of time. (See Table 3.17.)

(*a*) Plot the data on semilog paper, with time as the independent variable.

(*b*) Determine the equation of the line using the method of selected points.

Table 3.17

Time t, s	Voltage V, V
6	98
10	62
17	23
25	9.5
32	3.5
38	1.9
42	1.33

3.15 When a capacitor is to be discharged, the current flows until the voltage across the capacitor is zero. This current flow when measured as a function of time resulted in the data given in Table 3.18.

(*a*) Plot the data points on semilog paper, with time as the independent variable.

(*b*) Determine the equation of the line using the method of selected points.

Table 3.18

Current I, A	1.81	1.64	1.48	1.34	1.21	0.73
Time t, s	0.1	0.2	0.3	0.4	0.5	1.0

3.16 When fluid is flowing in the line, it is relatively easy to begin closing a valve that is wide open. But as the valve approaches a more nearly closed position, it becomes considerably more difficult to force movement. Visualize a circular pipe with a simple flap hinged at one edge being closed over the end of the pipe. The fully open position is $\theta = 0°$, and the fully closed condition is $\theta = 90°$.

A test was conducted on such a valve by applying a constant torque at the hinge position and measuring the angular movement of the valve. The test data are shown in Table 3.19.

(*a*) Plot the data, with angle as the independent variable.
(*b*) By the method of selected points, find the equation relating torque to angular movement.

Table 3.19

Torque T, N · m	Movement θ, degrees
3.0	5.2
6.0	29.3
10.0	40.9
20.0	56.3
35.0	71.0
50.0	84.8

3.17 All materials are elastic to some extent. It is desirable that certain parts of some designs compress when a load is applied to assist in making the part airtight or watertight (e.g., a jar lid). The test results shown in Table 3.20 resulted from a test made at the Herndon Test Labs in Houston on a material known as SILON Q-177.

(*a*) Plot the data on semilog graph paper.
(*b*) Using the method of selected points, find the equation of the relationship.
(*c*) What pressure would cause a 10 percent compression?

Table 3.20

Pressure P, MPa	Relative compression R, %
1.12	27.3
3.08	37.6
5.25	46.0
8.75	50.6
12.3	56.1
16.1	59.2
30.2	65.0

3.18 The rate of absorption of radiation by metal plates varies with the plate thickness and the nature of the source of radiation. A test was made at Ames Labs on October 11, 1982, using a Geiger counter and a constant source of radiation; the results are shown in Table 3.21.

(*a*) Plot the data on semilog graph paper.
(*b*) Find the equation of the relationship between the parameters.
(*c*) What level of radiation would you estimate to pass a 2-in-thick plate of the metal used in the test described above?

Table 3.21

Plate thickness W, mm	Geiger counter C, counts per second
0.20	5 500
5.00	3 720
10.00	2 550
20.00	1 320
27.5	720
32.5	480

3.19 A test of a directional microphone resulted in the data shown in Table 3.22. Output in volts was measured as a sound source was moved around the microphone. Plot the results on polar graph paper. Assume that the plot is symmetrical about the 0 to 180° line.

Table 3.22

Angle, degrees	Output, V
0	1
15	0.98
30	0.97
45	0.94
60	0.91
75	0.87
90	0.82
105	0.78
120	0.72
135	0.65
150	0.58
165	0.52
180	0.50

Table 3.23

Angles, degrees	Directivity, dB
0	0
±5	−2
±10	−4
±15	−7
±20	−11
±25	−18
±30	−10
±35	−6
±40	−4
±45	−3
±50	−4
±55	−7
±60	−9
±65	−11
±70	−15
±75	−20
±80	−22
±85	−20
±90	−17

Table 3.24

Angle, degrees	Relative gain, %
0	100
15	99
30	95
45	75
60	43
75	15
90	2
105	12
120	16
135	12
150	4
165	6
180	10

Table 3.25

Distance from light pole, ft	Angles, degrees
60	270
61	255, 285
65	240, 300
73	225, 315
82	210, 330
92	195, 345
95	180, 0
60	165, 15
37	150, 30
27	135, 45
22	120, 60
21	105, 75
20	90

3.20 The directivity pattern for a theoretical straight-sided conical loudspeaker is presented in Table 3.23. The data are for a frequency of 6.2 kHz. Plot the data on polar graph paper for angles from −90 to 90°.

3.21 The field pattern for a UHF receiving antenna at 540 MHz is shown in Table 3.24. Plot on polar graph paper, assuming symmetry about the 0 to 180° line.

3.22 Lighting fixtures can be designed to produce desired patterns of light. For street lights, it is usually preferred to have patterns that will light the street from sidewalk to sidewalk. If the fixtures are placed near the east curbline, you want the pattern to extend longer to the west and shorter to the east. (Homeowners do not wish to have the strong light in their house windows.) Table 3.25 shows the design contour of one such street light.

(*a*) Plot the data on polar coordinate paper.

(*b*) Using this lighting fixture, how many street lights should be installed in 1 000 ft of street length?

Manipulating Engineering Data

PART TWO

4

Engineering Estimations and Approximations

4.1

Introduction

Much is said, and rightly so, about the great diversity among the many branches of engineering. While it is true that modern engineering has spawned a myriad of specialties, there are likewise many things that engineers have in common, one of which is the need to acquire physical measurements. The nineteenth century physicist Lord Kelvin stated that man's knowledge and understanding are not of high quality unless the information can be expressed in numbers. We all have made or heard statements such as "The water is too hot." This statement may or may not give us an indication of the temperature of the water. At a given temperature water may be too hot for taking a bath but not hot enough for making instant coffee or tea.

The truth is that pronouncements such as "hot," "too hot," "not very hot," etc., are relative to a standard selected by the speaker and have meaning only to those who know what that standard is.

Figure 4.1
How many cubic meters of overburden must be removed to allow mining of the coal under 1 km^2 of surface? (*Ames Laboratory, U.S. Department of Energy.*)

We make measurements of a vast array of physical quantities that allow addition to our knowledge and permit us to transfer this knowledge and understanding to others. Skill in making and interpreting measurements is an essential element in our practice of engineering.

4.2 Errors

To measure is to err! Any time a measurement is taken, some physical object is being compared with a standard. If we measure the distance between two points on the surface of the earth, why doesn't a repetition of the measurement produce identical results? The obvious answer is that errors occur in each attempt; only coincidence will produce exactly the same result. Errors are normally classified in two categories: systematic and accidental (random).

4.2.1 Systematic Errors

Our task is to measure the distance between two fixed points. Assume that the distance is in the range of 1 200 m and that we are experienced and competent and have equipment of high quality to do the measurement. Some of the errors that occur will always have the same sign (+ or −) and are said to be systematic. Assume that a 25-m steel tape is to be used, one that has been compared with the standard at the U.S. Bureau of Standards in Washington, D.C. If the tape is not exactly 25.000 m long, then there will be a systematic error each of the 48 times that we use the tape to measure out the 1 200 m. However, the error can be removed by applying a correction. A second source of error can stem from a difference between the temperature at the time of use and at the time when the tape was compared with the standard. Such an error can be removed if we measure the temperature of the tape and apply a mathematical correction. (The coefficient of thermal expansion for steel is 11.7×10^{-6} per kelvin.) The accuracy of such a correction depends on the accuracy of the thermometer and on our ability to measure the temperature of the tape instead of the temperature of the surrounding air. Another source of systematic error can be found in the difference in the tension applied to the tape while in use and the tension employed during standardization. Again, scales can be used; but as before, their accuracy will be suspect. In all probability, the tape was standardized by laying it on a smooth surface and supporting it throughout. But such surfaces are seldom available in the field. The tape is suspended at times, at least partially. But, knowing the weight of the tape, the tension is applied, and the length of the suspended tape, we can calculate a correction and apply it.

The sources of systematic error just discussed are not all the possible sources, but they illustrate an important problem even encountered in taking comparatively simple measurements. Similar problems occur in all types of measurements: mechanical quantities, electrical quantities, mass, sound, odors, etc. We must be aware of the presence of systematic errors, eliminate those that we can, and quantify and correct for those remaining.

4.2.2 Accidental Errors

In reading Sec. 4.2.1 you may have realized that even if it had been possible to eliminate all the systematic errors, the measurement is still not exact. To elaborate on this point, we will continue with the example of the task of measuring the 1 200-m distance. Several accidental errors can creep in, as follows. When reading the thermometer, we must estimate the reading when the indicator falls between graduations. Moreover, it may appear that the reading is exactly on a graduation when it is actually slightly above or below the graduation. Furthermore, the thermometer may not be accurately measuring the tape temperature but be influenced instead by the temperature of the ambient air. These errors can thus produce measurements that are either too large or too small. Regarding sign and magnitude, the error is therefore random.

Errors can also result from our correcting for the sag in a suspended tape. In such a correction, it is necessary to determine the weight of the tape, its cross-sectional area, its modulus of elasticity, and the applied tension. In all such cases, the construction of the instruments used for acquiring these quantities can be a source of both systematic and accidental errors.

The major difficulty we encounter with respect to accidental errors is that, although their presence is obvious by the scatter in the data, it is impossible to predict the magnitude and sign of the accidental error that is present in any one measurement. Repeating measurements and averaging the results will reduce the random error in the average. However, repeating measurements will not reduce the systematic error in the average result.

Refinement of the apparatus and care in its use can reduce the magnitude of the error; indeed, many engineers have devoted their careers to this task. Likewise, awareness of the problem, knowledge about the degree of precision of the equipment, skill with measurement procedures, and proficiency in the use of statistics allow us to determine the approximate magnitude of the error remaining in measurements. This knowledge, in turn, allows us to accept the error or develop different apparatus and/or methods in our work. It is beyond the scope of this text to discuss quantifying accidental errors. However, Chap. 6 includes a brief discussion of central tendency and standard deviation, which are part of the analysis of accidental errors.

4.3 Significant Digits

As we have just pointed out, measurements cannot be assumed to be exact. Errors are always present regardless of the precautions we take. Keep in mind that we are not speaking of mistakes in data collection or processing; we are referring to data that are correctly collected and processed but are still not exact. Quantities determined by analytical means are not always exact either. Often assumptions are necessary in order to derive methods that can be practically applied. Sometimes we don't understand the phenomenon well enough to get exact answers. Thus, it is clear that we need a way of expressing our results so that our readers will know how "good" we believe these answers to be. Use of significant digits gives us this capability to a limited degree without resorting to the more rigorous approach of stating the estimated percentage error in the result.

A *significant digit*, or *figure*, is defined as any digit used in writing a number, *except* those zeros that are used only for location of the decimal point or those zeros that do not have any nonzero digit on their left. When you read the number 0.001 5, only the digits 1 and 5 are significant, since the three zeros have no nonzero digit to their left. We would say then that this number has two significant figures. If the number is written 0.001 50, it contains three significant figures; the rightmost zero is significant.

Numbers 10 or larger that are not written in scientific notation and that are not counts (exact values) can cause difficulties in interpretation when zeros are present. For example, 2 000 could contain one, two, three, or four significant digits; it is not clear which. If you write the number in scientific notation as 2.000×10^3, then clearly four significant digits are intended. If you want to show only two significant digits, you would write 2.0×10^3. It is our recommendation that, if uncertainty results from using standard decimal notation, you switch to scientific notation so your reader can clearly understand your intent.

You may find yourself as the user of values where the writer was not careful to properly show significant figures. What then? Assuming that the number is not a count or a known exact value, about all you can do is establish a reasonable number of significant figures based on the context of the value and on your experience. Once you have decided on a reasonable number of significant digits, you can then use the number in any calculations that are required.

As a guide to deciding how many significant figures to use, remember that a number containing three significant figures implies a maximum error range of about 1 percent. To explain, the quantity 101 means a number between 100.5 and 101.5. Thus the error range is 1 (± 0.5) which represents about 1 percent of 101. The quantity 999, also containing three significant figures, has an error range of ± 0.5, which is about 0.1 percent of 999. Therefore, we use the guideline that a number with three significant figures has a maximum error range of 1 percent. Likewise, a number containing four significant figures has a maximum error range of 0.1 percent. Only in exceptional cases will precision better than 0.1 percent be necessary. (Parameters used to guide a spacecraft from the earth to a landing on the moon present an example where greater precision might be needed.)

Numerical values obtained from an instrument like a voltmeter, micrometer, or thermometer (assuming an analog display rather than a digital display) normally are recorded to contain one estimated or doubtful digit. That is, the instrument is read by estimating between the smallest graduations on the scale to get the final digit. Thus, the final digit reported is said to be "doubtful." It is normal practice to consider one doubtful digit as significant. For example, if a voltmeter reading is 110.3 V, where the 3 is estimated, the value has four significant digits, counting the single doubtful digit.

Calculators and computers commonly work with numbers having as few as 7 digits or as many as 16 or 17 digits. This is true no matter how many significant digits an input value or calculated value should have. Therefore, you will need to exercise care in reporting values from a calculator display or from a computer output. Most high-level computer languages allow you to control the number of digits that are to be displayed or printed. If a computer output is to be a part of your final solution presentation, you will need to carefully control the output form. If the output is only an intermediate step, you can round the results to a reasonable number of significant figures in your presentation.

As you perform arithmetic operations, it is important that you not lose the significance of your measurements or, conversely, imply precision that does not exist. Rules for determining the number of significant figures that should be reported following computations have been developed by engineering associations. The following rules customarily apply.

1. **Multiplication and division.** The product or quotient should contain the same number of significant digits as are contained in the number with the fewest significant digits.

Examples

a. (2.43)(17.675) = 42.950 25

If each number in the product is exact, the answer should be reported as 42.950 25. If the numbers are not exact, as is normally the case, 2.43 has three significant figures and 17.675 has five. Applying the rule, the answer should contain three significant figures and be reported as 43.0 or 4.30×10^1.

b. (2.479 h)(60 min/h) = 148.74 min

In this case, the conversion factor is exact (a definition) and could be thought of as having an infinite number of significant figures. Thus, 2.479, which has four significant figures, controls the precision, and the answer is 148.7 min, or 1.487×10^2 min.

c. (4.00×10^2 kg)(2.204 6 lbm/kg) = 881.84 lbm

Here, the conversion factor is not exact, but you should not let the conversion factor dictate the precision of the answer if it can be avoided. You should attempt to maintain the precision of the value being converted; you cannot improve its precision. Therefore, you should use a conversion factor that has one or two more significant figures than will be reported in the answer. In this situation, three significant figures should be reported, yielding 882 lbm.

d. 589.62/1.246 = 473.210 27

The answer, to four significant figures, is 473.2.

2. **Addition and subtraction.** The answer should show significant digits only as far to the right as is seen in the least precise number in the calculation.

Example

a.
```
1 725.463
  189.2
   16.73
---------
1 931.393
```

The least precise number in this group is 189.2 so, according to the rule, the answer should be reported as 1 931.4. Using alternative reasoning, suppose these numbers are instrument readings, which means the last reported number in each is a doubtful digit. A column addition that contains a doubtful digit will result in a doubtful digit in the sum. So all three digits to the right of the decimal in the answer are doubtful. Normally we report only one; thus the answer is 1 931.4 after rounding.

b.
```
897.0
- 0.092 2
---------
896.907 8
```

Application of the rule results in an answer of 896.9.

3. **Combined operations.** If products or quotients are to be added or subtracted, perform the multiplication or division first, establish the correct number of significant figures in the subanswer, perform the addition or subtraction, and round to proper significant figures. Note, however, that in calculator or computer applications it is not practical to perform intermediate rounding. It is normal practice to perform the entire calculation and then report a reasonable number of significant figures.

If results from additions or subtractions are to be multiplied or divided, an intermediate determination of significant figures can be made when the calculations are performed manually. Use the suggestion above for calculator or computer answers.

Subtractions that occur in the denominator of a quotient can be a particular problem when the numbers to be subtracted are very nearly the same. For example, 39.7/(772.3 − 772.26) gives 992.5 if intermediate roundoff is not done. If, however, the subtraction in the denominator is reported with one digit to the right of the decimal, the denominator becomes zero and the result becomes undefined. Commonsense application of the rules is necessary to avoid problems.

4. **Rounding.** In rounding a value to the proper number of significant figures, *increase the last digit retained by 1 if the first figure dropped is 5 or greater.* This is the rule normally built into a calculator display control or a computer language.

Examples

a. 827.48 rounds to 827.5 or 827 for four and three significant digits, respectively.

b. 23.650 rounds to 23.7 for three significant figures.

c. 0.014 3 rounds to 0.014 for two significant figures.

4.4 Accuracy and Precision

"Accuracy" and "precision" have different meanings and should not be used interchangeably. *Accuracy* is a measure of the nearness of a value to the correct or true value. *Precision* refers to the repeatability of a measurement. Consider again the example of measuring the 1 200-m distance discussed in Sec. 4.2.1. It is quite possible that we might measure the distance between the two fixed points four times and record 1 202.96, 1 203.13, 1 203.04, and 1 202.91 m. The mean of these measurements is 1 203.01 m. Our measurements stray from the mean by 0.05, 0.12, 0.03, and 0.10 m, respectively. It can reasonably be assumed that because of the closeness of the first four measurements a fifth measurement would not differ greatly. Our quantities appear to be repeatable. But these results have nothing to do with the accuracy of the measurements. If the tape that we used was too short, then we obtained a measurement that is too large a value. Unless we account for the error in tape length, we will continue to get inaccurate, though precise, readings. Conversely, we could have used sloppy procedures with an accurately calibrated tape and obtained values that were in considerable variance with each other, averaged them, and (by chance) obtained a very accurate result.

Methods of analyzing measured data are briefly discussed in Chap. 6.

4.5 Approximations

Even though engineers try for a high level of precision, there are many times when only a close approximation is needed. For years, we used the slide rule as our computational device. We were aware that slide-rule answers had only three significant figures, so we knew

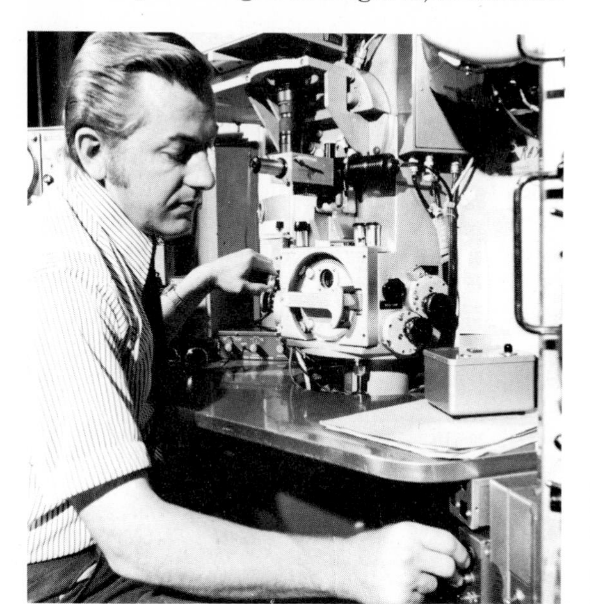

Figure 4.2
Highly accurate surface properties of a metal sample are being determined by an electron microprobe. (*Ames Laboratory, U.S Department of Energy.*)

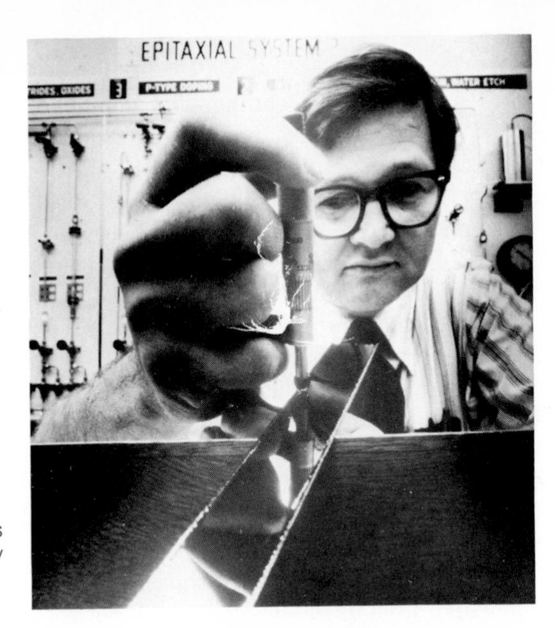

Figure 4.3
The approximate thickness of this metal plate is being measured by the use of a micrometer. (*Westinghouse.*)

when to use a machine, logarithms, or even long-hand computations to improve the precision of the answer.

When we do not have tables and references with us and must estimate a calculation with reasonable accuracy, we have to rely on our basic understanding of the problem under discussion, coupled with our previous experience. If greater accuracy is needed, we will refine our first answer when we have more time and the necessary reference materials.

In the area of our highest competency, we are expected to be able to make rough estimates to provide figures that can be used for tentative decisions. These estimates may be in error by perhaps 10 to 20 percent or even more. The accuracy of these estimates depends strongly on what reference materials we have available, how much time is allotted for the estimate, and, of course, how experienced we are with similar problems. The first example we present will attempt to illustrate what a professional engineer might be called upon to do in a few minutes with no references. It is not the type of problem you, as a beginning student, would be expected to do because you have not yet gained the necessary experience.

Example problem 4.1 A civil engineer is asked to meet with a city council committee to discuss their needs with respect to the disposal of solid wastes (garbage or refuse). The community, a city of 12 000 persons, must begin supplying refuse collection and disposal for its citizens for the first time. In reviewing various alternatives for disposal, a sanitary landfill is suggested. One of the council members is concerned about how much land is going to be needed, so he asks the engineer how many acres will be required within the next 10 years.

Discussion The engineer quickly estimates as follows:

The national average solid-waste production is 2.75 kg/(capita)(d). We can determine that each citizen will thus produce 1 Mg of refuse per year by the following calculation:

$$(2.75 \text{ kg/d})(365 \text{ d/year}) \left(\frac{1 \text{ Mg}}{1\,000 \text{ kg}}\right) \cong 1 \text{ Mg/year}$$

Experience indicates that refuse will probably be compacted to a density of 400 to 600 kg/m^3. On this basis, the per capita landfill volume will be 2 m^3 each year; and 1 acre filled 1 m deep will contain the collected refuse of 2 000 people for a year (1 acre = 4 047 m^2). Therefore, the requirement for 12 000 people will be 1 acre filled 6 m deep. However, knowledge of the geology of the particular area indicates that bedrock occurs at approximately 6 m below the ground surface. The completed landfill should therefore have an average depth of 4 m; consequently, 1.5 acres a year, or 15 acres in 10 years, will be required. The patterns of the recent past indicate that some growth in population and solid-waste generation should be expected. It is finally suggested that the city should plan to use about 20 acres in the next 10 years.

This calculation took only minutes and required no computational device other than pencil and paper. The engineer's experience, rapid calculations, sound basic assumptions, and sensible rounding of figures were the main requirements. And a usable estimate, designed to neither mislead nor to sell a point of view, was provided. If this project proceeds to the actual development of a sanitary landfill, the civil engineer will then gather actual data, refine the calculations, and prepare estimates upon which one would risk a professional reputation.

Example Prob. 4.2 is an illustration of a problem you might be assigned. Here you have the necessary experience to perform the estimation. Not counting the final written presentation, you should be able to do a similar problem in $\frac{1}{2}$ to 1 h.

Example problem 4.2 Suppose that your instructor assigns the following problem: Determine the number of pieces of lumber 5 cm × 10 cm × 2.40 m that can be sawn from the tree nearest to the southeast corner of the building in which you are now meeting. How would you proceed? See Fig. 4.4 for one student's response.

Discussion The assumptions that Jim made seem to be reasonable. Although he did not allow for the width of the saw cut nor for the thickness of the bark, the omissions may very well be consistent with the degree of accuracy involved in his calculations. Most Boy and Girl Scouts have learned to measure heights by the method he used; and all freshman engineering students should be familiar with similar triangles. After determining the height and diameter of the

	9-24-84	Fr E 160 SPECIAL PROBLEM	JAMES JOHNSON	1/2

PROBLEM

ESTIMATE THE NUMBER OF 5cm × 10cm × 2.40m BOARDS THAT CAN BE SAWN FROM THE FIR TREE NEAR THE S.E. CORNER OF THE ENGINEERING BUILDING.

ASSUMPTIONS

1. THE TREE TRUNK IS CONICAL.
2. THE LIMBS CANNOT BE USED - TOO SMALL.
3. ALL PIECES THAT ARE TOO SMALL WILL BE DISCARDED - NO PARTIALS OR PARTICLE BOARDS.
4. THE TREE WILL BE CUT 0.3m ABOVE THE GROUND.

COLLECTED DATA

1. I AM 180cm TALL AND MY SHADOW WAS 135cm.
2. THE TREES SHADOW WAS 14m.
3. THE BASE OF THE TREE HAS A CIRCUMFERENCE OF 120cm.

SOLUTION

HEIGHT OF TREE $\left(\frac{180}{135}\right)$ 14 m = 18.67m – APPROX ~ 19m

DIAMETER OF TREE AT GROUND $\frac{120}{\pi}$ = 38.2cm

APPROX ~ 38 cm

DIAMETER REDUCTION = $\frac{38\text{cm}}{19\text{m}} = 2.0\ \frac{\text{cm}}{\text{m}}$

FIRST SECTION (0.3m – 2.7m)

EFFECTIVE CROSS SECTION @ 2.7m MEASURING FROM GROUND = 38cm – (2.7 × 2.0) = 32.6 cm

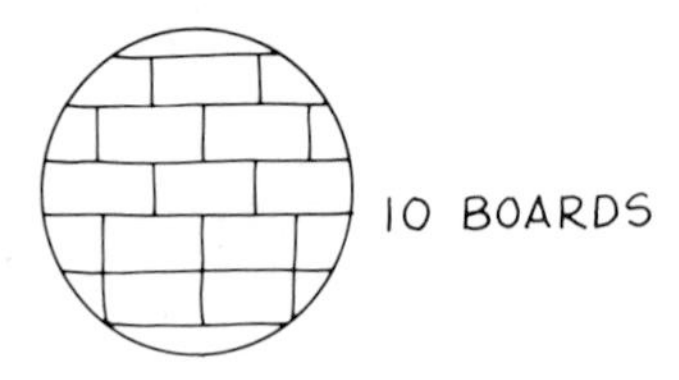

10 BOARDS

SECOND SECTION (2.7m – 5.1m)

38 – (5.1)(2.0) = 27.8 cm

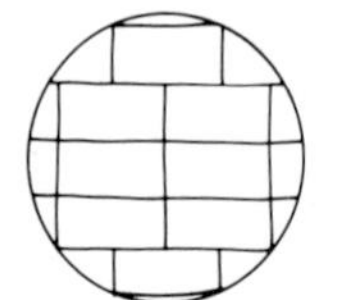

8 BOARDS

Figure 4.4
Student presentation of Example Prob. 4.2:

	9-24-84	Fr E 160 SPECIAL PROBLEM	JAMES JOHNSON	2/2

THIRD SECTION (5.1m – 7.5m) 38 – (7.5)(2.0) = 23.0 cm

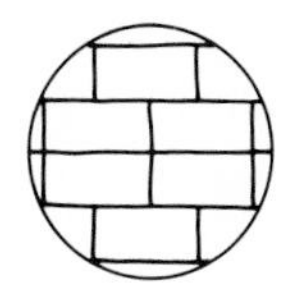

6 BOARDS

FOURTH SECTION (7.5m – 9.9m) 38 – 19.8 = 18.2 cm

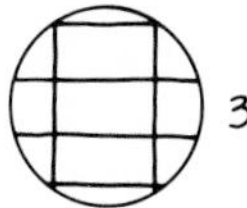

3 BOARDS

FIFTH SECTION (9.9m – 12.3m) 38 – 24.6 = 13.4 cm

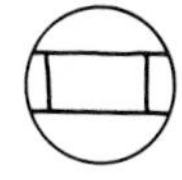

1 BOARD

SIXTH SECTION (12.3m – 14.7m) 3.8 – 29.4 = 9.6 cm TOO SMALL

TOTAL = 10 + 8 + 6 + 3 + 1 = 28 BOARDS

COMMENT

VERY INEFFICIENT TO USE ONLY ONE SIZE BOARD BECAUSE OF TOO LARGE PERCENTAGE OF DISCARD.

tree, Jim applied a graphical technique for determining how many boards could be cut from each 2.4-m section of the tree. He correctly used the upper (smaller) diameter of the section. His task was then reduced to a simple counting of the boards.

We have mentioned previously that both time allotted and physical circumstances (reference material at hand, access to a phone, etc.) are major factors that influence the type of estimate you can produce. Example Probs. 4.3 and 4.4 will show reasonable estimates of the same quantity (paper used) under two different sets of circumstances. In the first case, the student had about 10 min to do the estimate under exam conditions; that is, the student could not leave the classroom seat and had no reference material. The second case resulted from a homework assignment where 1 to 2 h were available and the student could obtain data needed for the estimate.

Example problem 4.3 Estimate the amount of paper used by the students of this college for homework, quizzes, and examinations during this academic year. Express your answer in kilograms. You may use no reference materials.

Discussion The problem requires that data be known or assumed. The calculation is simple and straightforward once the data are established (see Fig. 4.5). Because much data must be assumed without verification, the estimate cannot be expected to be very precise, so only one, or at most two, significant figures should be reported. (Recall that two significant figures suggest a possible error of 10 percent.)

Example problem 4.4 Estimate the amount of paper used by the students of this college for homework, quizzes, and examinations during this academic year. Express your answer in kilograms. Plan to use 1 to 2 h as a homework assignment.

Discussion The results of this estimation are found in Fig. 4.6. Many of the assumptions that had to be made in Example Prob. 4.3 are no longer necessary because time is available to obtain information. The small sample in the survey still produces some uncertainty. In each case, your reader should be reasonably convinced by your techniques; for example, anyone would be willing to accept student population obtained from the registrar.

Another type of estimate that engineers are called upon to make is one where a choice is involved. Estimates are prepared of the various alternatives available, and a decision as to which alternative to follow is then made; perhaps the one that is least expensive is chosen. Example Prob. 4.5 is designed to show estimating for decision purposes. This problem requires experience that a student might have.

3-16-84	FR.E. 160 PROBLEM 4.11	SARA KENILWORTH	1/1

PROBLEM 4.11

ESTIMATE THE MASS (kg) OF PAPER USED BY STUDENTS IN THIS COLLEGE DURING THIS ACADEMIC YEAR FOR HOMEWORK, QUIZZES, AND EXAMS.

ASSUMPTIONS

1. COLLEGE HAS 4200 STUDENTS.
2. 2 SHEETS ARE USED PER CREDIT HOUR PER WEEK.
3. 32 WEEKS PER ACADEMIC YEAR.
4. TABLET OF 50 SHEETS OF PAPER HAS A MASS OF ABOUT 1/2 lbm.
5. STUDENTS AVERAGE 15 CREDITS PER TERM.

CALCULATIONS

$$\text{MASS OF SHEET} = \left(\frac{1/2 \text{ lbm}}{50 \text{ SHEETS}}\right)\left(\frac{454 \text{ g}}{\text{lbm}}\right) = 4.54 \text{ g/SHEET}$$

$$\text{MASS OF PAPER} = (4200 \text{ STUDENTS})\left(\frac{15 \text{ CREDITS}}{\text{STUDENT}}\right)\left(\frac{2 \text{ SHEETS}}{(\text{CREDIT})(\text{WEEK})}\right)\left(\frac{32 \text{ WEEKS}}{\text{YEAR}}\right)\left(\frac{4.54 \text{ g}}{\text{SHEET}}\right)\left(\frac{1 \text{ Kg}}{1000 \text{ g}}\right)$$

$$= 18305.28 \text{ kg/YEAR}$$

$$\cong 1.8 \times 10^4 \text{ kg/YEAR}$$

Figure 4.5
Student presentation of Example Prob. 4.3.

	3-16-84	FR E 160 PROBLEM 4.11	SARA KENILWORTH	1/2

PROBLEM 4.11

ESTIMATE THE MASS (kg) OF PAPER USED BY STUDENTS IN THIS COLLEGE DURING THIS ACADEMIC YEAR FOR HOMEWORK, QUIZZES, AND EXAMS.

ASSUMPTIONS

1. THIS COLLEGE REFERS TO THE COLLEGE OF ENGINEERING, STATE UNIVERSITY
2. ONLY UNDERGRADUATE STUDENTS WILL BE CONSIDERED.
3. THE COMPUTATION IS FOR THE 1983-84 ACADEMIC YEAR.
4. ENGINEERING-PROBLEMS PAPER IS THE STANDARD FOR THE COLLEGE.
5. THE RESULTS OF THE SURVEY BELOW REASONABLY REPRESENT THE ENGINEERING STUDENTS AS A WHOLE.

COLLECTED DATA

1. ACCORDING TO THE UNIVERSITY REGISTRAR, 4 256 ENGINEERING UNDERGRADUATES WERE REGISTERED FOR FALL SEMESTER AND 4 028 FOR SPRING SEMESTER.
2. THE GENERAL CATALOG SHOWS 15 WEEKS OF CLASSES EACH SEMESTER PLUS FINAL EXAM WEEK EACH TERM.
3. A SURVEY OF 8 ENGINEERING STUDENTS, 2 FROM EACH CLASS, FRESHMAN THROUGH SENIOR, PROVIDED THE FOLLOWING DATA:

STUDENTS	SHEETS/WEEK	SHEETS/EXAM WEEK	COMMENTS
1	35	15	
2	18	8	PART TIME
3	45	12	
4	52	20	20 CREDIT HOURS
5	28	10	
6	25	14	
7	38	15	
8	42	18	
AVE	35.4	14	

4. A REAM (500 SHEETS) OF ENGINEERING-PROBLEMS PAPER HAS A MASS OF 3.75 lbm AS DETERMINED WITH A POSTAL SCALE.

THEORY

MASS PER SHEET = (MASS PER REAM) / (500 SHEETS PER REAM)

MASS OF PAPER = MASS FOR FALL SEMESTER + MASS FOR SPRING SEMESTER.

WHERE

MASS FOR FALL = (NUMBER OF STUDENTS) [(SHEETS/WEEK) (CLASS WEEKS) + (SHEETS/EXAM WEEK)] (MASS/SHEET)

MASS FOR SPRING = (NUMBER OF STUDENTS) [(SHEETS/WEEK) (CLASS WEEKS) + (SHEETS/EXAM WEEK)] (MASS/SHEET)

Figure 4.6
Student presentation of Example Prob. 4.4

	3-16-84	FR. E. 160 PROBLEM 4.11	SARA KENILWORTH	2/2

CALCULATIONS

MASS PER SHEET = [(3.75 lbm PER REAM)/(500 SHEETS)](0.453 59 kg/lbm)
= 0.003 402 kg/SHEET

MASS FOR FALL = (4 256 STUDENTS)[(35.4 SHEETS/WEEK)(15 WEEKS) + 14 SHEETS](0.003 402 kg/SHEET)
= 7 891 kg

MASS FOR SPRING = (4 028 STUDENTS)[(35.4 SHEETS/WEEK)(15 WEEKS) + 14 SHEETS](0.003 402 kg/SHEET)
= 7 468 kg

TOTAL MASS OF PAPER = 7 891 kg + 7 468 kg
= 15 359 kg

ESTIMATE OF TOTAL MASS = 1.5×10^4 kg

Example problem 4.5 Determine on the basis of an estimate whether you should paint your room at home or whether you should hire a painter to do the job for you. Your supervisor at the fast-food restaurant where you work has indicated that you could work the hours needed for painting at \$3.50 per hour.

Discussion Figure 4.7 is the result of the estimate. A local painting contractor has provided a guaranteed price for doing the work and will provide the needed materials. You must serve as your own contractor and obtain the materials you need at your expense if you do the work yourself.

Problems

4.1 How many significant figures are contained in each of the following quantities?

(*a*) 256
(*b*) 2.56
(*c*) 0.025 6
(*d*) 4.970
(*e*) 1 072.0
(*f*) 4 000
(*g*) 7.440×10^2
(*h*) 4 001
(*i*) 0.000 1
(*j*) 0.100 0

4.2 How many significant digits are there in each figure?

(*a*) 7 942
(*b*) 0.622
(*c*) 622.0
(*d*) 900
(*e*) $3.660\ 1 \times 10^5$
(*f*) 0.020
(*g*) 0.200
(*h*) 60 min/h
(*i*) 1.609 km/mi
(*j*) 6 290

4.3 Perform the following computations and report the answers with the proper number of significant figures.

(*a*) (2.97) (3.62)
(*b*) 764.99/144
(*c*) [(491) (32.2)]/7 232
(*d*) $[(83.2 \times 10^2)(2.6 \times 10^{-2})(701)]/[(47)(7.83 \times 10^{-4})]$
(*e*) $(0.446\ 39 \times 10^2) + (0.472 \times 10^1)$
(*f*) $(9.24 \times 10^{-3}) + (1.23 \times 10^{-2})$
(*g*) $(-0.664\ 3 + 0.004\ 97)/1\ 792$
(*h*) $47.9/(0.227\ 9 - 0.228)$

4.4 Compute the results of the following expressions. Clearly record the answers with proper significant figures.

(*a*) 292.4/144
(*b*) $292.4\ \text{in}^2/(144\ \text{in}^2/\text{ft}^2)$
(*c*) $[(5.692 \times 10^3) - (7.22 \times 10^2)]/412.4$
(*d*) $[(2.778 \times 10^6) - (8.054 \times 10^7)]/[3.684 + (8.43 \times 10^{-2})]$
(*e*) (\$49.95) (0.109 7)

4.5 What are the results of the following computations expressed with proper significant figures?

(*a*) 4 962 s/(3 600 s/h)
(*b*) $y = 3x^2 + 35x - 60$ for $x = 10.4$
(*c*) $\pi\,(32.924)^2/4$
(*d*) $0.103\ \{[(4.023 - 2.94)/2] + 3.944\}$

	2-17-84	FR.E 160 PROBLEM 4.2	KEVIN NORTH	1/2

PROBLEM 4.2

DETERMINE ON THE BASIS OF AN ESTIMATE WHETHER YOU SHOULD PAINT YOUR ROOM AT HOME OR WHETHER YOU SHOULD HIRE A PAINTER TO DO THE JOB FOR YOU.

ASSUMPTIONS

1. I CAN WORK AT MY REGULAR JOB AT $3.50 PER HOUR, FOR EACH HOUR REQUIRED TO DO THE PAINTING.
2. THE COLOR OF THE ROOM WILL NOT CHANGE, SO ONE COAT OF PAINT IS SUFFICIENT.
3. THE CEILING IS WHITE WHILE THE WALLS ARE NOT, SO TWO COLORS OF PAINT ARE NEEDED.
4. EQUIPMENT SUCH AS A LADDER, ROLLER, PAN, DROP CLOTH, ETC., IS AVAILABLE AT HOME, SO IT WON'T HAVE TO BE PURCHASED.
5. THE LAST TIME THE ROOM WAS PAINTED, IT TOOK 8 HOURS FOR PREPARATION AND PAINTING. TIME REQUIRED SHOULD BE SIMILAR THIS TIME.
6. IT WILL REQUIRE 1 HOUR TO PURCHASE THE SUPPLIES. NEGLECT AUTO EXPENSE.
7. THE INSIDE OF THE CLOSET IS IN GOOD CONDITION AND WILL NOT REQUIRE PAINTING THIS TIME.

COLLECTED DATA

1. A LOCAL PAINTING CONTRACTOR WILL DO THE JOB AND PROVIDE ALL MATERIALS FOR $150.
2. THE ROOM IS 15 × 18 ft, WITH AN 8-ft CEILING.
3. THE DOOR IS 36 × 78 in EACH OF 2 WINDOWS IS 36 × 42 in. THE FOLDING CLOSET DOOR IS 72 × 78 in.
4. THE LATEX PAINT SHOULD COVER 400 ft²/gal.
5. PAINT COSTS $14.95/gal AND $6.95/qt.
6. MASKING TAPE COSTS $3 FOR THE JOB.

CALCULATIONS

CEILING AREA = (15 ft)(18 ft) = 270 ft²

AT 400 ft²/gal, ONE GALLON WILL BE NEEDED (WITH SOME LEFT OVER)

TOTAL WALL AREA = (8 ft)[2(15 ft) + 2(18 ft)] = 528 ft²

UNPAINTED AREA = (36 in)(78 in) + 2(36 in)(42 in) + (72 in)(78 in)
= 11 448 in² (ft²/144 in²) = 79.5 ft²

Figure 4.7
Student presentation of Example Prob. 4.5

	2-17-84	FR.E. 160 PROBLEM 4.2	KEVIN NORTH	2/2

THEREFORE,

PAINTED AREA = $528 ft^2 - 79.5 ft^2 = 448.5 ft^2$

THUS AT $400 ft^2/gal$, 1 gal + 1 qt ARE NEEDED FOR THE WALLS.

MATERIAL COST = CEILING PAINT + WALL PAINT + TAPE

= (1 gal)($14.95/gal) + (1 gal)($14.95/gal) + (1 qt)($6.95/qt) + $3

= $39.85

COST OF LOST WAGES = (8h + 1h)($3.50/h) = $31.50

TOTAL COST = MATERIAL COST + LOST WAGES

= $39.85 + $31.50 = $71.35

CONCLUSION: I SHOULD PAINT THE ROOM AT A COST TO ME OF $71.35 RATHER THAN HIRE A CONTRACTOR AT $150.

Figure 4.7 (cont.)

(e) $[(6)(5.068\,325 \times 10^9) - (6\,685)(4.535 \times 10^6)]/$
$[(6)(7\,470\,125) - 6\,685^2]$

4.6 Estimate the volume of all mathematics texts that will be purchased throughout their college careers by the students enrolled in this introductory engineering class.

4.7 How much gasoline will be used by students of your university during the next recess between terms for the purpose of visiting their homes and returning? Express your answer in gallons and liters. What will be the total cost of this gasoline?

4.8 Estimate the total volume and mass of the concrete and asphalt paving in the largest parking lot on your campus.

4.9 Estimate the volume (capacity) in cubic meters of a water tower located on your campus or nearby (as specified by your instructor).

4.10 How much paint (primer plus two coats) is needed to paint a water tower on your campus or nearby (as specified by your instructor)? Consider only the tank portion, not the supporting structure, if any.

4.11 In preparation of the budget for her next school year, Sally Student needed to estimate the money necessary for her clothing for the year, including all aspects (its purchase price, cleaning costs, etc.). Describe how this could be done by using your own personal situation as an example.

4.12 Compute the surface area (in square meters) of external glass in your classroom building or any other structure designated by your instructor. What is the mass of this glass in kilograms?

4.13 Suppose your school's basketball team plans to play in a 3-day tournament in Kansas City. Your job is to estimate the total cost of transporting the team, food and housing for the team, etc. Consider the essential personnel (team, coaches, and managers) only.

4.14 Your school has plans to require all entering engineering students to purchase their own microcomputers. Estimate the total cost to students for the next fall term based on a system specification worked out by the class or provided by the instructor.

4.15 Based on your local electric energy rates, estimate the cost of lighting for your classroom for this academic year.

4.16 For a local building constructed with a brick facade, estimate the number and total mass of the bricks required.

4.17 Estimate the total number of cards in your school's library card catalog. Do this without consultation with any library personnel.

4.18 By your personal observation, estimate the total number of computer terminals available for general student use on your campus. Determine the maximum number of students that could be served for 2 h each week, assuming the terminals are available 12 h per day, 7 days per week. How does this compare with the number of students on campus?

4.19 Estimate the average number of audio tapes and records owned by a typical freshman engineering student. How much money does each student have invested?

4.20 Each year rain and/or snow falls on your campus. For a specified parking lot or rooftop, estimate the total volume (in cubic meters) and mass (in kilograms) of water that must be carried away during the entire year.

4.21 By utilizing a reasonable survey technique, compute the amount of money annually spent by all engineering students enrolled at your school for long-distance telephone calls.

4.22 Most first-year engineering students own some electronic sound equipment (receivers, radios, speakers, tape decks, etc.). Estimate the average monetary investment by each student. Exclude video equipment (TVs, VCRs, video games, etc.).

4.23 If your campus has a body of water (lake, pond, fountain, etc.), determine the volume of water contained in it (in cubic feet) and the surface area (in square feet). Do the estimate without getting wet or getting in trouble with campus security.

4.24 Estimate the mass of all textbooks you will purchase to complete your first engineering degree. If you have not selected a major, use one that is a likely candidate.

4.25 What is the total floor area (in square feet) and volume (in cubic feet) of the classrooms in the building where this course meets?

4.26 Assuming that the interior of the classroom where this class meets is painted, estimate the amount of paint necessary to refurbish it and make a significant change in wall color. (Alternative: Do the estimate for another interior space designated by your instructor.)

4.27 Following an accident where a student was cut by broken glass in a door, your school has decided to replace all door glass by plastic (Plexiglas or some other brand) in the building where this class meets. Considering interior and exterior doors, estimate the amount (in square feet) of material needed and its approximate cost. Exclude labor costs in this estimate.

4.28 Estimate the total mass (in kilograms) of the exterior walls of a building specified by your instructor. Do not include interior walls, floors, roof, etc. Include exterior doors and windows in your calculation.

4.29 Many students have access to video games, either family- or personally owned. For this class, approximate the total investment in electronic games and game equipment.

4.30 For a building (or portion of a building) on your campus that is carpeted (the Student Union Building, perhaps), estimate the amount (in square yards) of material required to recarpet the space. What is the expected cost of the materials?

5

Dimensions, SI Units, and Conversions

5.1 Introduction

In the past when countries were more isolated from one another than they are now, individual countries tended to develop and use their own set of measures. As the rapid increase in global communication and travel brought countries closer together and the world advanced in technology, the need for a universal system of measurement became clearly apparent. There was such a proliferation of information among all nations that a standard set of dimensions, units, and measurements was vital if this wealth of knowledge was to be of benefit to all.

We will deal in this chapter with the difference between dimensions and units and at the same time explain how there can be an orderly transition from many systems of units to one system, i.e., an international standard.

The standard currently accepted in most industrial nations (it is optional in the United States) is the international metric system, or Système International d'Unites, abbreviated SI. The SI units are a modification and refinement of an earlier system (MKS) that designated the meter, kilogram, and second as fundamental units.

France was the first country, as early as 1840, to officially legislate adoption of the metric system and decree that its use be mandatory.

The United States almost adopted the metric system 150 years ago. In fact, the metric system was made legal in the United States in 1866, but its use was not made compulsory. In spite of many attempts since that time, full conversion to the metric system has not yet been realized in the United States, but significant steps in that direction are presently underway.

5.2 Physical Quantities

Engineers are constantly concerned with the measurements of fundamental physical quantities such as length, time, temperature, force, etc. In order to specify a physical quantity fully, it is not sufficient to indicate merely a numerical value. The magnitude of physical quantities can be understood only when they are compared

From September 1977 Canadian road speed limits will be posted in kilometres per hour (km/h).

Kilometre is pronounced kill-o-metre.

Look for the **km/h** tab below the maximum speed limit sign, indicating that this is the new speed in metric.

100 km/h This speed limit will likely be the most common on freeways. On most rural two-lane roadways, **80 km/h** will be typical.

50 km/h A **50 km/h** speed limit will apply in most cities.

Actual speed limits will be established in accordance with local regulations.

Metric Commission Canada   Commission du système métrique Canada

À compter de septembre 1977, les vitesses maximales seront indiquées en kilomètres par heure (km/h).

Septembre 1977

Surveillez l'indication de l'unité de vitesse **km/h**; ce symbole signifie que la vitesse est mesurée selon le système métrique.

100 km/h Sur les autoroutes, la vitesse maximale la plus courante sera de **100 km/h** tandis que sur les routes à grande circulation, elle sera de **80 km/h**.

50 km/h Dans la plupart des grands centres, la vitesse maximale sera de **50 km/h**.

Les vitesses maximales en vigueur dans votre société seront établies selon les règlements municipaux.

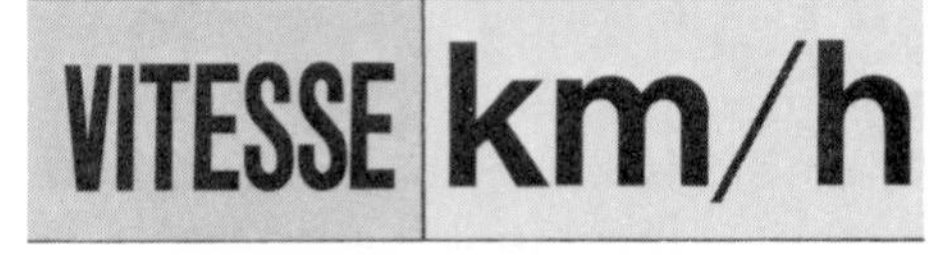

Commission du système métrique Canada   Metric Commission Canada

Figure 5.1
Highway signs in Canada. (*Metric Commission of Canada.*)

with predetermined reference amounts, called *units*. Any measurement is, in effect, a comparison of how many units are contained within the physical quantity. For example, if we call the physical quantity Q, the numerical value V, and the unit U, then

$$Q = VU$$

For this relationship to be valid, the exact reproduction of a unit must be theoretically possible at any time. Therefore standards must be established. They are a set of fundamental unit quantities kept under normalized conditions in order to preserve their values as accurately as possible. We shall speak more about them later.

5.3 Dimensions

Dimensions are used to describe physical quantities. An important element to remember is that dimensions are independent of units. A physical quantity such as length can be represented by the dimension L, for which there are a large number of possibilities available when selecting a unit. For example, in ancient Egypt, the cubit was related to the length of the arm from the tip of the middle finger to the elbow. Measurements were thus a function of physical stature, with variation from one individual to another. Much later, in Britain, the inch was specified as the distance covered by three barley corns, round and dry, laid end to end.

Today we require more precision. For example, the meter is defined in terms of the distance traveled by light in a vacuum in a specified amount of time. We can draw two important points from this discussion: (1) Physical quantities can be accurately measured, and (2) each of these units (cubit, inch, and meter), although distinctly different, has in common the quality of being a length and not an area or a volume.

A technique used to distinguish between units and dimensions is to call all quantities of length simply L. In this way, each new physical quantity gives rise to a new dimension, such as T for time, F for force, M for mass, etc. (Note that there are as many dimensions as there are kinds of physical quantities.)

Moreover, dimensions can be divided into two areas—derived and fundamental. *Derived dimensions* are a combination of *fundamental dimensions*. Velocity, for example, can be defined as a fundamental dimension V. But it is more customary as well as more convenient to consider velocity as a combination of fundamental dimensions, so that it becomes a derived dimension, $V = (L)(T)^{-1}$.

It is advantageous to use as few fundamental dimensions as possible, but the selection of what is to be fundamental and what is to be derived is not fixed. In actuality, any dimension can be selected as a fundamental dimension in a particular field of engineering or science; and for reasons of convenience, it may be a derived dimension in another field.

A *set of fundamental dimensions* is simply a group of dimensions that can be conveniently and usefully manipulated when expressing all physical quantities of a particular field.

A *dimensional system* can be defined as the smallest number of fundamental dimensions which will form a consistent and complete set for a field of science. For example, three fundamental dimensions are necessary to form a complete mechanical dimensional system. These dimensions may be either length (L), time (T), and mass (M) or length (L), time (T), and force (F), depending on the specific application. If temperature is important to the application, a fourth dimension (θ) must be added.

The *absolute system* (so called because dimensions within it are not affected by gravity) has as its fundamental dimensions L, T, and M. An advantage of this system is that comparisons of masses at various locations can be made with an ordinary balance, because the local acceleration of gravity has no influence upon the results.

The *gravitational system* has as its fundamental dimensions L, T, and F. It is widely used in many engineering branches because

Table 5.1 Two basic dimensional systems

Quantity	Absolute	Gravitational
Length	L	L
Time	T	T
Mass	M	$FL^{-1}T^2$
Force	MLT^{-2}	F
Velocity	LT^{-1}	LT^{-1}
Pressure	$ML^{-1}T^{-2}$	FL^{-2}
Momentum	MLT^{-1}	FT
Energy	ML^2T^{-2}	FL
Power	ML^2T^{-3}	FLT^{-1}
Torque	ML^2T^{-2}	FL

it simplifies computations when weight is a fundamental quantity in the computations. Table 5.1 illustrates two of the more basic systems in terms of dimensions. A number of dimensional systems other than those strictly used in mechanics are for systems of heat, electromagnetism, electrical dimensions, etc.

5.4 Units

Once a consistent dimensional system has been selected, one can develop a unit system by choosing a specific unit for each fundamental dimension. The problem one encounters when working with units is that there can be a large number of unit systems to choose from for each complete dimension system, as we have already suggested. It is obviously desirable to limit the number of systems and combinations of systems. The SI previously alluded to is intended to serve as an international standard that will provide worldwide consistency.

There are two fundamental systems of units commonly used in mechanics today. One system used in almost every industrial country of the world is called the *metric system*. It is a decimal-absolute system based on the meter, kilogram, and second (MKS) as the units of length, mass, and time, respectively. The United States has used the other system, normally referred to as the British gravitational system. It is based on the foot, pound-force, and second.

Numerous international conferences on weights and measures over the past 25 years have gradually modified the MKS system to the point that all countries previously using various forms of the metric system are beginning to standardize. The SI is now considered the new international system of units. The United States has adopted the system, but full use will be preceded by a long and expensive period of change. During the transition period, engineers will have to be familiar with SI as well as with other systems and their corresponding conversion factors.

SI Units and Symbols

The International System of Units (SI), developed and maintained by the General Conference on Weights and Measures, is intended as a basis for worldwide standardization of measurements. The name and abbreviation were set forth in 1960. SI at the present time is a complete system that is being universally adopted.

This new international system is divided into three classes of units:

1. Base units
2. Supplementary units
3. Derived units

There are seven base units in the SI. The units (except the kilogram) are defined in such a way that they can be reproduced anywhere in the world.

Table 5.2 lists each base unit along with its name and proper symbol.

Each of the base units is defined below as established at the international General Conference on Weights and Measures (CGPM).*

1. Length: The meter is a length equal to the distance traveled by light in a vacuum during 1/299 792 458 s. The meter was defined by the CGPM that met in 1983.

2. Time: The second is the duration of 9 192 631 770 periods of radiation corresponding to the transition between the two hyperfine levels of the ground state of the cesium-133 atom. The second was adopted by the thirteenth CGPM in 1967.

3. Mass: The standard for the unit of mass, the kilogram, is a cylinder of platinum-iridium alloy kept by the International Bureau of Weights and Measures in France. A duplicate copy is maintained in the United States. The unit of mass was adopted by the First and Third GCPMs in 1889 and 1901. It is the only base unit nonreproducible in a properly equipped lab.

4. Electric current: The ampere is a constant current which, if maintained in two straight parallel conductors of infinite length and of negligible circular

*The initials stand for Conférence Générale des Poids et Mesures.

Table 5.2 Base units

Quantity	Name	Symbol
Length	meter	m
Mass	kilogram	kg
Time	second	s
Electric current	ampere	A
Thermodynamic temp	kelvin	K
Amount of substance	mole	mol
Luminous intensity	candela	cd

cross section and placed one meter apart in vacuum, would produce between these conductors a force equal to 2×10^{-7} newton per meter of length. The ampere was adopted by the Ninth CGPM in 1948.

5. Temperature: The kelvin, a unit of thermodynamic temperature, is the fraction 1/273.16 of the thermodynamic temperature of the triple point of water. The kelvin was adopted by the Thirteenth CGPM in 1967.

6. Amount of substance: The mole is the amount of substance of a system that contains as many elementary entities as there are atoms in 0.012 kilogram of carbon-12. The mole was defined by the Fourteenth CGPM in 1971.

7. Luminous intensity: The base unit candela is the luminous intensity in a given direction of a source that emits monochromatic radiation of frequency 540×10^{12} hertz and that has a radiant intensity in that direction of 1/683 watts per steradian.

The units listed in Table 5.3 are called *supplementary units* and may be regarded as either base units or as derived units.

The unit for a plane angle is the radian, a unit that is used frequently in engineering. The steradian is not as commonly used. These units can be defined in the following way:

1. Plane angle: The radian is the plane angle between two radii of a circle that cut off on the circumference an arc equal in length to the radius.

2. Solid angle: The steradian is the solid angle which, having its vertex in the center of a sphere, cuts off an area of the sphere equal to that of a square with sides of length equal to the radius of the sphere.

As indicated in previous discussion, derived units are formed by combining base, supplementary, or other derived units. Symbols for them are carefully selected to avoid confusion. Those which have special names and symbols, as interpreted for the United States by the National Bureau of Standards, are listed in Table 5.4 together with their definitions in terms of base units.

At first glance, Figure 5.2 may appear complex, even confusing. However, if you study the examples below, you will no doubt agree that a considerable amount of information is presented in a concise flowchart. To get the point of it quickly, be aware that the solid lines denote multiplication and the broken lines indicate divison. The arrows pointing toward the units (circled) are significant and arrows going away have no meaning for that particular unit. Con-

Table 5.3 Supplementary units

Quantity	Name	Symbol
Plane angle	radian	rad
Solid angle	steradian	sr

Table 5.4 Derived units

Quantity	SI unit symbol	Name	Base units
Frequency	Hz	hertz	s^{-1}
Force	N	newton	$kg \cdot m \cdot s^{-2}$
Pressure, stress	Pa	pascal	$kg \cdot m^{-1} \cdot s^{-2}$
Energy or work	J	joule	$kg \cdot m^2 \cdot s^{-2}$
Quantity of heat	J	joule	$kg \cdot m^2 \cdot s^{-2}$
Power, radiant flux	W	watt	$kg \cdot m^2 \cdot s^{-3}$
Electric charge	C	coulomb	$A \cdot s$
Electric potential	V	volt	$kg \cdot m^2 \cdot s^{-3} \cdot A^{-1}$
Potential difference	V	volt	$kg \cdot m^2 \cdot s^{-3} \cdot A^{-1}$
Electromotive force	V	volt	$kg \cdot m^2 \cdot s^{-3} \cdot A^{-1}$
Capacitance	F	farad	$A^2 \cdot s^4 \cdot kg^{-1} \cdot m^{-2}$
Electric resistance	Ω	ohm	$kg \cdot m^2 \cdot s^{-3} \cdot A^{-2}$
Conductance	S	siemens	$kg^{-1} \cdot m^{-2} \cdot s^3 \cdot A^2$
Magnetic flux	Wb	weber	$kg \cdot m^2 \cdot s^{-2} \cdot A^{-1}$
Magnetic flux density	T	tesla	$kg \cdot s^{-2} \cdot A^{-1}$
Inductance	H	henry	$kg \cdot m^2 \cdot s^{-2} \cdot A^{-2}$
Luminous flux	lm	lumen	$cd \cdot sr$
Illuminance	lx	lux	$cd \cdot sr \cdot m^{-2}$
Celsius temperature*	°C	degree Celsius	K
Activity (radionuclides)	Bq	becquerel	s^{-1}
Absorbed dose	Gy	gray	$m^2 \cdot s^{-2}$
Dose equivalent	S_v	sievert	$m^2 \cdot s^{-2}$

*The thermodynamic temperature (T_K) expressed in kelvins is related to Celsius temperature ($t_{°C}$) expressed in degrees Celsius by the equation $t_{°C} = T_K - 273.15$.

Table 5.5 Common derived units

Quantity	Units	Quantity	Units
Acceleration	$m \cdot s^{-2}$	Molar entropy	$J \cdot mol^{-1} \cdot K^{-1}$
Angular acceleration	$rad \cdot s^{-2}$	Molar heat capacity	$J \cdot mol^{-1} \cdot K^{-1}$
Angular velocity	$rad \cdot s^{-1}$	Moment of force	$N \cdot m$
Area	m^2	Permeability	$H \cdot m^{-1}$
Concentration	$mol \cdot m^{-3}$	Permittivity	$F \cdot m^{-1}$
Current density	$A \cdot m^{-2}$	Radiance	$W \cdot m^{-2} \cdot sr^{-1}$
Density, mass	$kg \cdot m^{-3}$	Radiant intensity	$W \cdot sr^{-1}$
Electric charge density	$C \cdot m^{-3}$	Specific heat capacity	$J \cdot kg^{-1} \cdot K^{-1}$
Electric field strength	$V \cdot m^{-1}$	Specific energy	$J \cdot kg^{-1}$
Electric flux density	$C \cdot m^{-2}$	Specific entropy	$J \cdot kg^{-1} \cdot K^{-1}$
Energy density	$J \cdot m^{-3}$	Specific volume	$m^3 \cdot kg^{-1}$
Entropy	$J \cdot K^{-1}$	Surface tension	$N \cdot m^{-1}$
Heat capacity	$J \cdot K^{-1}$	Thermal conductivity	$W \cdot m^{-1} \cdot K^{-1}$
Heat flux density	$W \cdot m^{-2}$	Velocity	$m \cdot s^{-1}$
Irradiance	$W \cdot m^{-2}$	Viscosity, dynamic	$Pa \cdot s$
Luminance	$cd \cdot m^{-2}$	Viscosity, kinematic	$m^2 \cdot s^{-1}$
Magnetic field strength	$A \cdot m^{-1}$	Volume	m^3
Molar energy	$J \cdot mol^{-1}$	Wavelength	m

sider the pascal, for example: Two arrows point toward the circle—one solid and one broken. This means that the unit pascal is formed from the newton and meter squared, or N/m^2.

Other derived units, such as those included in Table 5.5, have no special names but are combinations of base units and units with special names.

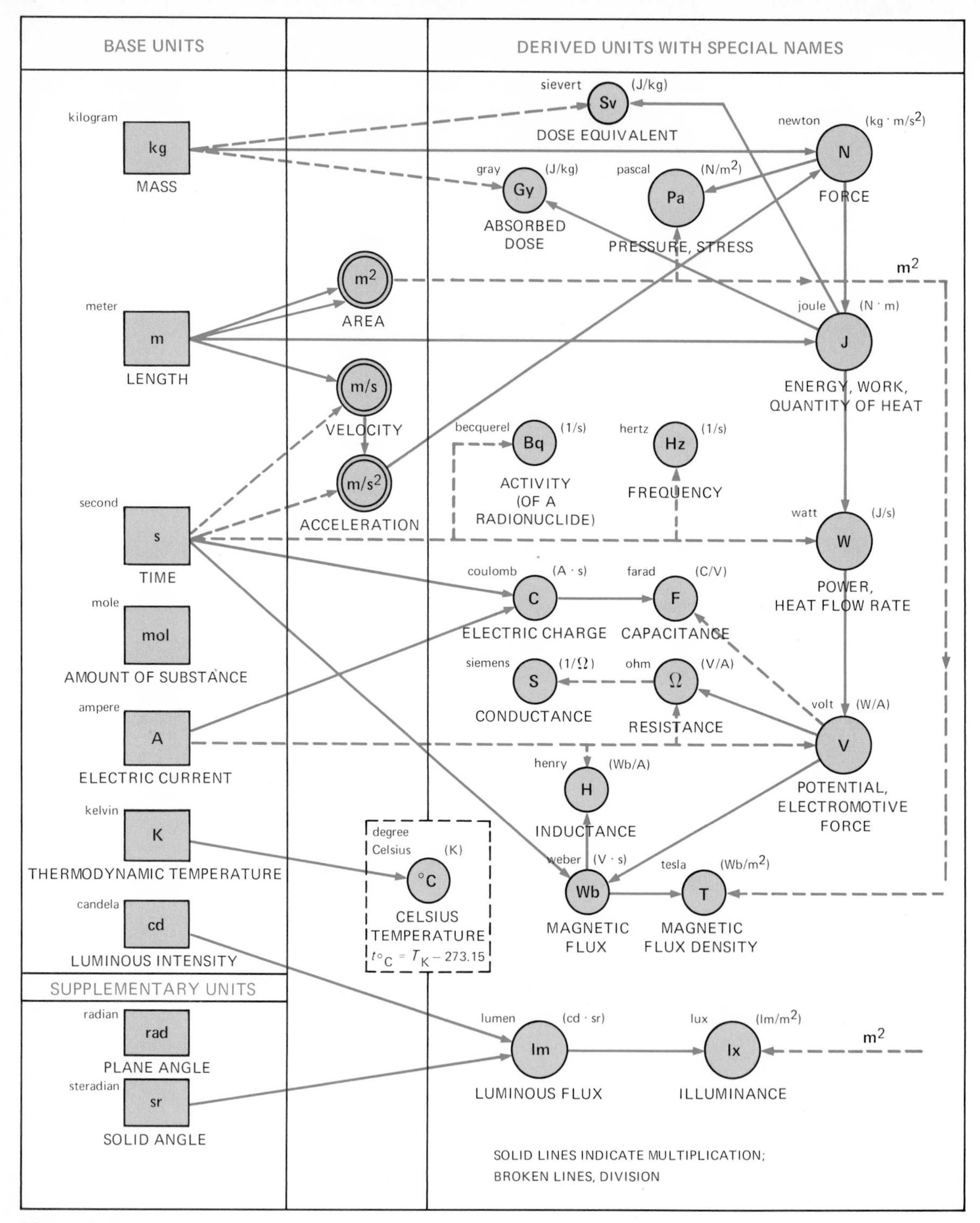

Figure 5.2
Graphical illustration of how certain SI units are derived in a coherent fashion from base and supplementary units. (*National Bureau of Standards.*)

Table 5.6 Decimal multiples

Multiplier	Prefix name	Symbol
10^{18}	exa	E
10^{15}	peta	P
10^{12}	tera	T
10^{9}	giga	G
10^{6}	*mega	M
10^{3}	*kilo	k
10^{2}	hecto	h
10^{1}	deka	da
10^{-1}	deci	d
10^{-2}	centi	c
10^{-3}	*milli	m
10^{-6}	*micro	μ
10^{-9}	nano	n
10^{-12}	pico	p
10^{-15}	femto	f
10^{-18}	atto	a

*Most often used.

Being a decimal system, the SI is convenient to use, because by simply affixing a prefix to the base, a quantity can be increased or decreased by factors of 10 and the numerical quantity can be kept within manageable limits. Table 5.6 lists the multiplication factors along with their prefix names and symbols.

The proper selection of prefixes will also help eliminate nonsignificant zeros and leading zeros in decimal fractions. The numerical value of any measurement should be recorded between 0.1 and 1 000. This rule is suggested because it is easier to make realistic judgments when working with numbers between 0.1 and 1 000. For example, suppose that you are asked the distance to a nearby town. It would be more understandable to respond in kilometers than meters. That is, it is easier to visualize 10 km than 10 000 m.

Moreover, the use of certain prefixes is preferred over that of others. Those representing powers of 1 000, such as kilo, mega, milli, and micro, will reduce the number you must remember. These preferred prefixes should be used, with the following three exceptions which are still common because of convention.

1. When expressing area and volume, the prefixes hecto-, deka-, deci-, and centi- may be used, for example, cubic centimeter.
2. When discussing different values of the same quantity or expressing them in a table, calculations are simpler to perform when you use the same unit multiple throughout.
3. Sometimes a particular multiple is recommended as a consistent unit even though its use violates the 0.1 to 1 000 rule. For example, many companies use the millimeter for linear dimensions even when the values lie far outside this suggested range. The cubic decimeter (commonly called liter) is also used.

Recalling the discussion of significant figures in Chap. 4, we see that the SI prefix notations can be used to a definite advantage.

Consider the previous example of 10 km. When giving an estimate of distance to the nearest town there is certainly an implied approximation in the use of a round number. Suppose that we were talking about a 10 000-m Olympic track and field event. The accuracy of such a distance must certainly be greater than something between 5 000 and 15 000 m. This example is intended to illustrate the significance of the four zeros (10 000). If all four zeros are in fact significant, then the race is accurate within 1 m (9 999.5 to 10 000.5). If only three zeros are significant, then the race is accurate to within 10 m (9 995 to 10 005).

There are two logical and acceptable methods available of eliminating confusion concerning zeros:

1. Use proper prefixes to denote intended significance.

Distance	Precision
10 000 m	5 000 to 15 000 m
10.000 km	9 999.5 to 10 000.5 m
10.00 km	9 995 to 10 005 m
10.0 km	9 950 to 10 050 m

2. Use scientific notation to indicate significance.

Distance	Precision
10 000 m	5 000 to 15 000 m
10.000×10^3 m	9 999.5 to 10 000.5 m
10.00×10^3 m	9 995 to 10 005 m
10.0×10^3 m	9 950 to 10 050 m

Selection of a proper prefix is customarily the logical way to handle problems of significant figures; however, there are conventions that do not lend themselves to the prefix notation. An example would be temperature in degrees Celsius; that is, $4.00(10^3)$°C is the conventional way to handle it, not 4.00 k°C.

5.6 Rules for Using SI Units

Along with the adoption of SI comes the responsibility to thoroughly understand and properly apply the new system. Obsolete practices involving English and metric units are widespread. This section provides rules that should be followed when working with SI units.

5.6.1 Unit Symbols and Names

1. Periods are never used after symbols unless the symbol is at the end of a sentence (i.e., SI unit symbols are not abbreviations).

2. Unit symbols are written in lowercase letters unless the symbol derives from a proper name, in which case the first letter is capitalized.

Lowercase	Uppercase
m, kg, s, mol, cd	A, K, Hz, Pa, C

3. Symbols rather than self-styled abbreviations should always be used to represent units.

Correct	Not correct
A	amp
s	sec

4. An s is never added to the symbol to denote plural.
5. A space is always left between the numerical value and the unit symbol.

Correct	Not correct
43.7 km	43.7km
0.25 Pa	0.25Pa

Exception: No space should be left between numerical values and the symbols for degree, minute, and second of angles and for degree Celsius.

6. There should be no space between the prefix and the unit symbols.

Correct	Not correct
mm, MΩ	k m, μ F

7. When writing unit names, all letters are lowercase except at the beginning of a sentence, even if the unit name is derived from a proper name.

8. Plurals are used as required when writing unit names. For example, henries is plural for henry, etc. The following exceptions are recommended.

Singular	Plural
lux	lux
hertz	hertz
siemens	siemens

With these exceptions, unit names form their plurals in the usual manner.

9. No hyphen or space should be left between a prefix and the unit name. There are three cases where the final vowel in the prefix is omitted, but these are the only exceptions: megohm, kilohm, and hectare.

10. The symbol should be used in preference to the unit name because unit symbols are standardized. An exception to this is made when a number is written in words preceding the unit; e.g., we would write ten meters, not ten m. The same is true the other way, e.g., 10 m, not 10 meters.

5.6.2 Multiplication and Division

1. When writing unit names as a product, always use a space (preferred) or a hyphen.

Correct usage
newton meter or newton-meter

2. When expressing a quotient using unit names, always use the word "per" and not a solidus (/). The solidus, or slash mark, is reserved for use with symbols.

Correct	Not correct
meter per second	meter/second

3. When writing a unit name that requires a power, use a modifier, e.g., squared or cubed, after the unit name. For area or volume, the modifier can be placed before the unit name.

Correct	Correct
meter per second squared	square millimeter

4. When expressing products using unit symbols, the center dot is preferred.

Correct
N · m for newton meter

5. When denoting a quotient by unit symbols, any of the following methods are accepted form:

Correct
m/s or $m \cdot s^{-1}$ or $\frac{m}{s}$

In more complicated cases, negative powers or parentheses should be considered. Use m/s^2 or $m \cdot s^{-2}$ but not m/s/s for acceleration; use $kg \cdot m^2/(s^3 \cdot A)$ or $kg \cdot m^2 \cdot s^{-3} \cdot A^{-1}$ but not $kg \cdot m^2/s^3/A$ for electric potential.

5.6.3 Numbers

1. To denote a decimal point, use a period on the line. When expressing numbers less than 1, a zero should be written before the decimal marker.

Example
15.6
0.93

2. Since a comma is used in many countries to denote a decimal point, its use is to be avoided in grouping digits. Where it is desired to avoid this confusion, recommended practice calls for separating the digits into groups of three, counting from the decimal to the left or right, and using a small space to separate the groups.

Correct and recommended procedure			
6.513 824	76 851	7 434	0.187 62

5.6.4 Calculating with SI Units

Before we look at some suggested procedures that will simplify calculations in SI, the following positive characteristics of the system should be reviewed.

Only one unit is used to represent each physical quantity, e.g., the meter for length, the second for time, etc. The SI metric units are *coherent*; that is, each new derived unit is a product or quotient of the fundamental and supplementary units without any numerical factors. Since coherency is a strength of the SI system, it would be worthwhile to demonstrate this characteristic by using two examples. Consider the use of the newton as the unit of force instead of pound-force (lbf). It is defined by Newton's second law, $F = ma$. It is the force that imparts an acceleration of one meter per second squared to a mass of one kilogram. Thus,

$$1\ \mathrm{N} = (1\ \mathrm{kg})(1\ \mathrm{m/s^2})$$

Consider also the joule, a unit that replaces the British thermal unit, calorie, foot-pound-force, electronvolt, and horsepower-hour to stand for any form of energy. It is defined as the amount of work done when an applied force of one newton acts through a distance of one meter in the direction of the force. Thus,

$$1\ \mathrm{J} = (1\ \mathrm{N})(1\ \mathrm{m})$$

To maintain the coherency of units, however, time must be expressed in seconds rather than minutes or hours, since the second is the base unit. Once coherency is violated, then a conversion factor must be included and the advantage of the system is diminished.

But there are certain units *outside* SI that are accepted for use in the United States, even though they diminish the system's coherence. These exceptions are listed in Table 5.7.

Calculations using SI can be simplified if you

1. Remember that all fundamental relationships such as the following still apply, since they are independent of units.

$$F = ma \qquad \mathrm{KE} = \tfrac{1}{2}mv^2 \qquad E = RI$$

2. Recognize how to manipulate units and gain a proficiency in doing so. Since watt = J/s = N · m/s, you should realize that N · m/s = $(\mathrm{N/m^2})(\mathrm{m^3/s})$ = (pressure)(volume flow rate).

3. Understand the advantage of occasionally adjusting all variables to base units. Replace N with kg · $\mathrm{m/s^2}$, Pa with kg · $\mathrm{m^{-1} \cdot s^{-2}}$, etc.

4. Develop a proficiency with exponential notation of numbers to be used in conjunction with unit prefixes.

$$1\ \mathrm{mm^3} = (10^{-3}\ \mathrm{m})^3 = 10^{-9}\ \mathrm{m^3}$$

$$1\ \mathrm{ns^{-1}} = (10^{-9}\ \mathrm{s})^{-1} = 10^{9}\ \mathrm{s^{-1}}$$

Table 5.7 Non-SI units accepted for use in the United States

Quantity	Name	Symbol	SI equivalent
Time	minute	min	60 s
	hour	h	3 600 s
	day	d	86 400 s
Plane angle	degree	°	π/180 rad
	minute	′	π/10 800 rad
	second	″	π/648 000 rad
Volume	liter	L*	10^{-3} m^3
Mass	metric ton	t	10^3 kg
	unified atomic mass unit	u	$1.660\ 57 \times 10^{-27}$ kg (approx)
Land area	hectare	ha	10^4 m^2
Energy	electronvolt	eV	1.602×10^{-19} J (approx)

*Both "L" and "l" are acceptable international symbols for liter. The uppercase letter is recommended for use in the United States because the lowercase "l" can be confused with the numeral 1.

5.7 Special Characteristics

A term that should be avoided when using SI is "weight." Frequently we hear statements such as "The man weighs 100 kg." A better statement would be "The man has a mass of 100 kg." To clarify any confusion, let's look at some basic definitions.

First, the term "mass" should be used to indicate only a quantity of matter. Mass, as we know, is measured in kilograms against an international standard.

Force, as defined previously, is measured in newtons. It denotes an acceleration of one meter per second squared to a mass of one kilogram.

The acceleration of gravity varies at different points on the surface of the earth as well as with distance from the earth's surface. The accepted standard value of gravitational acceleration is 9.806 650 m/s^2.

Gravity is instrumental in measuring mass with a balance or scale. If you use a beam balance to compare an unknown quantity against a standard mass, the effect of gravity on the two masses cancels out. If you use a spring scale, mass is measured indirectly, since the instrument responds to the local force of gravity. Such a scale can be calibrated in mass units and be reasonably accurate when used where the variation in the acceleration of gravity is not significant.

In the English gravitational system, the unit pound is sometimes used to denote both mass and force. We will use the convention that pound-mass (lbm) is a unit of mass and pound-force (lbf) is a unit of force. Thus, pound-mass can be directly converted to the SI unit kilogram, and pound-force units convert to newtons. Another English unit describing mass is the slug. The slug is 32.174 times the size of the pound-mass (1 slug = 32.174 lbm).

A word of caution when using English gravitational units in Newton's second law ($F = ma$). The combination of units, lbf, lbm, and ft/s^2, is not a coherent (consistent) set (that is, 1 lbf imparts an acceleration of 32.147 ft/s^2 to 1 lbm rather than 1 ft/s^2 required for coherency). A coherent set of English units is lbf, slug, and ft/s^2. Thus we suggest that you convert mass quantities from lbm to slugs before substituting into $F = ma$.

The following example problem clarifies the confusion that exists in the use of the term "weight" to mean either force or mass. In everyday use, the term "weight" nearly always means mass; thus, when a person's weight is discussed, the quantity referred to is mass.

Example problem 5.1 A "weight" of 100.0 kg* is suspended by a rope (see Fig. 5.3). Calculate the tension in the rope in newtons when the mass is lifted vertically at constant velocity and the local gravitational acceleration is (*a*) 9.807 m/s^2 and (*b*) 1.63 m/s^2 (approximate value for the surface of the moon).

Theory Tension in the rope when the mass is at rest or moving at constant velocity is

$$F = mg$$

where g is the local acceleration of gravity and m is the mass of object.

Assumption Neglect the mass of the rope.

*The unit itself indicates mass.

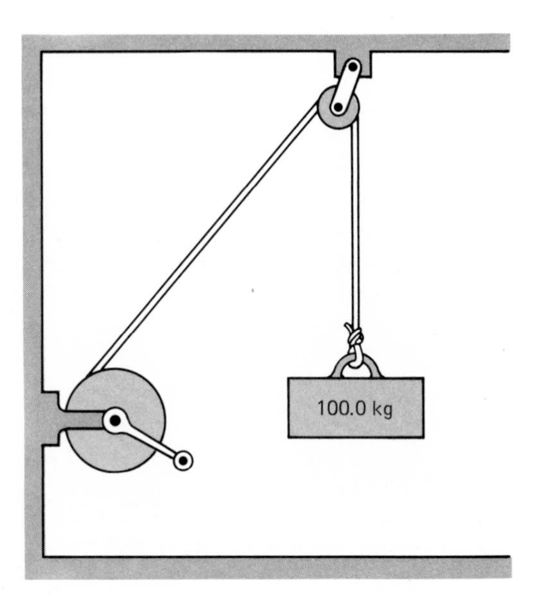

Figure 5.3

Solution

(*a*) For $g = 9.807\ \text{m/s}^2$ (given to four significant figures)

$$F = (100.0\ \text{kg})(9.807\ \text{m/s}^2)$$
$$= 0.980\ 7\ \text{kN}$$

(*b*) For $g = 1.63\ \text{m/s}^2$

$$F = (100.0\ \text{kg})(1.63\ \text{m/s}^2)$$
$$= 0.163\ \text{kN}$$

5.8 Conversion of Units

Although the SI system is the international standard, there are many other systems in use today. It would be fair to say that most of the current work force of graduate engineers has been schooled using terminology such as slugs, pound-mass, pound-force, etc., and a very high percentage of the total United States population is more familiar with degrees Fahrenheit than degrees Celsius.

For this reason and because it will be some time before the SI total system becomes the single standard in this country, you must be able to convert between unit systems.

Four typical systems of mechanical units presently being used in the United States are listed in Table 5.8. The table does not provide a complete list of all possible quantities; it is presented to demonstrate the different terminology that is associated with each unique system.

If a physical quantity is expressed in any system, it is a simple matter to convert the units from that system to another. To do this, the basic unit conversion must be known and a logical unit analysis must be followed.

Table 5.8 Mechanical units

Quantity	Absolute system		Gravitational system	
	MKS	CGS	Type I	Type II
Length	m	cm	ft	ft
Mass	kg	g	slug	lbm
Time	s	s	s	s
Force	N	dyne	lbf	lbf
Velocity	$\text{m} \cdot \text{s}^{-1}$	$\text{cm} \cdot \text{s}^{-1}$	$\text{ft} \cdot \text{s}^{-1}$	$\text{ft} \cdot \text{s}^{-1}$
Acceleration	$\text{m} \cdot \text{s}^{-2}$	$\text{cm} \cdot \text{s}^{-2}$	$\text{ft} \cdot \text{s}^{-2}$	$\text{ft} \cdot \text{s}^{-2}$
Torque	$\text{N} \cdot \text{m}$	$\text{dyne} \cdot \text{cm}$	$\text{lbf} \cdot \text{ft}$	$\text{lbf} \cdot \text{ft}$
Moment of Inertia	$\text{kg} \cdot \text{m}^2$	$\text{g} \cdot \text{cm}^2$	$\text{slug} \cdot \text{ft}^2$	$\text{lbm} \cdot \text{ft}^2$
Pressure	$\text{N} \cdot \text{m}^{-2}$	$\text{dyne} \cdot \text{cm}^{-2}$	$\text{lbf} \cdot \text{ft}^{-2}$	$\text{lbf} \cdot \text{ft}^{-2}$
Energy	J	erg	$\text{ft} \cdot \text{lbf}$	$\text{ft} \cdot \text{lbf}$
Power	W	$\text{erg} \cdot \text{s}^{-1}$	$\text{ft} \cdot \text{lbf} \cdot \text{s}^{-1}$	$\text{ft} \cdot \text{lbf} \cdot \text{s}^{-1}$
Momentum	$\text{kg} \cdot \text{m} \cdot \text{s}^{-1}$	$\text{g} \cdot \text{cm} \cdot \text{s}^{-1}$	$\text{slug} \cdot \text{ft} \cdot \text{s}^{-1}$	$\text{lbm} \cdot \text{ft} \cdot \text{s}^{-1}$
Impulse	$\text{N} \cdot \text{s}$	$\text{dyne} \cdot \text{s}$	$\text{lbf} \cdot \text{s}$	$\text{lbf} \cdot \text{s}$

6 Statistics

6.1 Introduction

Statistics, as used by the engineer, can most logically be called a branch of applied mathematics. It constitutes what some call the science of decision making in a world full of uncertainty. In fact, a degree of uncertainty exists in most day-to-day activities, from something as simple as the tossing of a coin to the results of an election, the outcome of a ball game, or the comparison of the efficiency of two production processes.

There can be little doubt that it would be virtually impossible to understand a great deal of the work done in engineering without having a thorough knowledge of statistics. Numerical data derived from surveys and experiments constitute the raw material upon which interpretations, analyses, and decisions are based; and it is essential that engineers learn how to use properly the information derived from such data.

Everything concerned even remotely with the collection, processing, analysis, interpretation, and presentation of numerical data belongs to the domain of statistics.

Figure 6.1
A researcher enters data into a personal computer that displays the data as a scatter diagram. General conclusions about the data can be made rapidly from the plot. (*International Business Machines Corporation.*)

There exist today a number of different and interesting stories about the origin of statistics, but most historians believe it can be traced to two dissimilar areas: games of chance and political science.

During the eighteenth century, various games of chance involving the mathematical treatment of errors led to the study of probability and, ultimately, to the foundation of statistics. At approximately the same time, an interest in the description and analysis of the voting of political parties led to the development of methods that today fall under the category of *descriptive statistics*, which is basically designed to summarize or describe important features of a set of data without attempting to infer conclusions that go beyond the data.

Descriptive statistics is an important part of the entire subject area; it is still used whenever a person wishes to represent data derived from observation.

In more recent years, however, statisticians have shifted their emphasis from methods that merely describe the data to methods that make generalizations about the data, called *statistical inference*.

To understand the distinction between descriptive statistics and statistical inference, consider the following example.

Suppose that two freshman engineering students are enrolled in mathematics and each completes five quizzes. Student A receives grades of 94, 89, 92, 80, and 85; student B receives grades of 82, 61, 88, 78, and 81. On the basis of this information, it can be said that student A has an average of $(94 + 89 + 92 + 80 + 85)/5$, or 88, and that student B has an average of $(82 + 61 + 88 + 78 + 81)/5$, or 78.

Manipulating numbers belongs to the domain of descriptive statistics. Concluding that A is a better student than B is a generalization, or a statistical inference.

From the information alone, it does not follow that student A is better than student B. Student B may have had an off day on the second quiz, may have been ill, or may have studied the wrong material for the quiz. On the other hand, student A may have studied the correct material for the quiz. There are always uncertainties, so that in this example, it may or may not be correct to conclude that one student is better than the other. The careful evaluation and analysis of all elements involving chance or risk that are normally taken when making such generalizations is an integral part of any statistical inference.

An important step to take when considering the generalization of data is that of carefully examining how the variables were controlled. For instance, if student A had been told which pages to study and student B had not, it is obvious that no reasonable or meaningful comparison can be made.

This last point is mentioned to stress the fact that in statistics, it is not enough to consider only sets of data and calculated results when arriving at conclusions. Items such as control and authenticity of collected data and how the experiment or survey was planned are of major importance. Unless proper care is taken in the planning and execution stages, it may be impossible to arrive at valid results or conclusions.

6.2 Frequency Distribution

Various ways of describing measurements and observations, such as the grouping and classifying of data, are a fundamental part of statistics. In fact, when dealing with a large set of collected numbers, a good overall picture of the data can often be conveyed by proper grouping into classes. The following example will serve to illustrate this point.

Consider the individual test scores received by students on the first major exam in their freshman computations course (see Table 6.1).

Table 6.2 is a type of numerical arrangement showing scores distributed among selected classes. Some information such as the highest and lowest values will be lost once the raw data have been sorted and grouped.

The construction of numerical distributions as in this example normally consists of the following steps: select classes into which the data are to be grouped; distribute data into appropriate classes; and count the number of items in each class.

Since the last two steps are essentially mechanical processes, our attention will be directed primarily toward the *classification* of data.

Two things must be considered when arranging data into classes: the number of classes into which the data are to be grouped and the range of values each class is to cover. Both these areas are somewhat arbitrary, but they do depend on the nature of the data and the ultimate purpose the distribution is to serve.

The following are guidelines that should be followed when constructing a frequency distribution.

1. Use no less than 6 and no more than 15 classes.
2. Select classes that will accommodate all the data points.
3. Make sure that each data point fits into only one class.
4. Whenever possible, make the class intervals of equal length.

Table 6.1

92	71	89	91	53	93	90	96	95
98	76	96	94	68	91	82	82	44
88	87	93	78	85	98	82	90	70
78	70	87	88	89	95	99	88	88
77	65	85	64	79	50	81	80	76

Table 6.2

Test scores	Tally	Frequency
41–50	||	2
51–60	|	1
61–70	卌	5
71–80	卌 |||	8
81–90	卌 卌 卌 |	16
91–100	卌 卌 |||	13
	Total	45

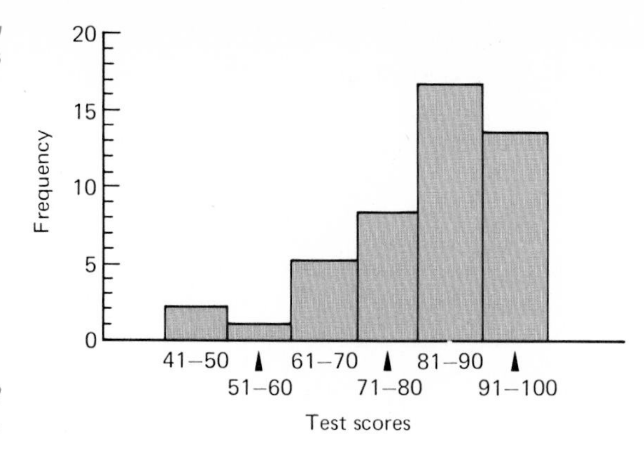

Figure 6.2
Test scores.

The numbers in the right-hand column of Table 6.2 are called the *class frequencies*, which denote the number of items that are in each class.

Since frequency distributions are constructed primarily to condense large sets of data into more easily understood forms, it is logical to display or present the data graphically. The most common form of graphical presentation is called the *histogram.*

It is constructed by representing measurements or grouped observations on the horizontal axis and class frequencies along the graduated and calibrated vertical axis. This representation affords a graphical picture of the distribution with vertical bars whose bases equal the class intervals and whose heights are determined by the corresponding class frequencies. Figure 6.2 demonstrates a histogram of the test scores tabulated in Table 6.2.

6.3 Measures of Central Tendency

The solution of many engineering problems in which a large set of data is collected can be somewhat facilitated by the determination of single numbers that describe unique characteristics about the data. The most popular measure of this type is called the arithmetic mean.

The *arithmetic mean*, or mean of a set of n numbers, is defined as the sum of the numbers divided by n. In order to develop a notation and a simple formula for arithmetic mean, it is helpful to use an example.

Suppose that the average, or mean, height of a starting basketball team is to be determined. Let the height in general be represented by the letter x and the height of each individual player be represented by x_1, x_2, x_3, x_4, and x_5. More generally, there are n measurements that are designated $x_1, x_2, \ldots, x_n$. From this notation, the mean can be written as follows:

$$\text{Mean} = \frac{x_1 + x_2 + x_3 + \cdots + x_n}{n}$$

A mathematical notation that indicates the summation of a series of numbers is normally written

$$\sum_{i=1}^{n} x_i$$

which represents $x_1 + x_2 + x_3 + \cdots + x_n$. This notation will be written in the remainder of the chapter as Σx_i, but the intended summation will be from 1 to n.

When a set of all possible observations is used, it is referred to as the *population;* when a portion or subset of that population is used, it is referred to as a *sample*. The notation for arithmetic mean will be $\bar{x}$ when the x values are representative of a random sample and not an entire population. For a population the symbol μ is used to represent the mean.

The standard notations discussed above provide the following common expression for the arithmetic mean of a sample:

$$\bar{x} = \frac{\Sigma x_i}{n} \qquad 6.1$$

The mean is a popular measure of central tendency because (1) it is familiar to most people, (2) it takes into account every item, (3) it always exists, (4) it is always unique, and (5) it lends itself to further statistical manipulations.

One disadvantage of the arithmetic mean, however, is that any gross error in a number can have a pronounced effect on the value of the mean. To avoid this difficulty, it is possible to describe the "center" of a set of data with other kinds of statistical descriptions. One of these is called the *median*, which can be defined as the value of the middle item of data arranged in increasing or decreasing order of magnitude. For example, the median of the five numbers 15, 27, 10, 18, and 22 can be determined by first arranging them in increasing order: 10, 15, 18, 22, and 27. The median is 18.

If there are an even number of items, there is never a specific middle item, so the median is defined as the mean of the values of the two middle items. For example, the median of six numbers—5, 9, 11, 14, 16, and 19—is (11 + 14)/2, or 12.5.

The mean and median of a set of data rarely coincide. Both terms describe the center of a set of data, but in different ways. The median divides the data so that half of all items is greater than or equal to the median; the mean may be thought of as the center of gravity of the data.

The median, like the mean, has certain desirable properties. It always exists and is always unique. Unlike the mean, the median is not affected by extreme values. If the exclusion of the highest and lowest values causes a significant change in the mean, then the median should be considered as the indicator of central tendency of that data.

In addition to the mean and the median, there is one other average, or center, of a set of data that we call the *mode*. It is simply the value that occurs with the highest frequency. In the following set of numbers—18, 19, 15, 17, 18, 14, 17, 18, 20, 19, 21, and 14—the number 18 is the mode because it appears more often than any of the other values.

An important point for a practicing engineer to remember is that there are any number of ways to suggest the middle, center, or average value of a data set. If comparisons are to be made, it is essential that similar methods be compared. It is only logical to compare the mean of brand A with the mean of brand B, not the mean of one with the median of the other. If one particular item, brand, process, etc., is to be compared with another, the same measures must be used. If the average grade in one section of college calculus is to be compared with the average grade in other sections, the mean of each section would be one important statistic.

6.4 Measures of Variation

It is not likely that the mean values of the course grades of different sections of college calculus will be of equal magnitude. And the extent to which the means are dissimilar is of fundamental importance.

Measures of variation indicate the degree to which data are dispersed, spread out, or bunched together. Suppose that by coincidence two sections of a college calculus course have exactly the same mean grade values on the first-hour exam. It would be of interest to know how far individual scores varied from the mean. Perhaps one class was bunched very closely around the mean, while the other class demonstrated a wide variation, with some very high scores and some very low scores. This situation is typical and is often of interest to the engineer.

It is reasonable to define this variation in terms of how much each number in the sample deviates from the mean value of the sample, that is, $x_1 - \overline{x}$, $x_2 - \overline{x}$, . . . , $x_n - \overline{x}$. If you wanted an average deviation from the mean you might try adding $x_1 - \overline{x}$ through $x_n - \overline{x}$ and dividing by n. But this does not give a useful result, since the sum of the deviations is always zero. The procedure generally followed is to square each deviation, sum the resulting squares, divide the sum by n, and finally take the square root. The formula obtained by this technique gives the *standard deviation* s for a sample as

$$s = \left[\frac{\Sigma(x_i - \overline{x})^2}{n}\right]^{0.5} \tag{6.2}$$

The equivalent formula for the standard deviation σ of the entire population is

$$\sigma = \left[\frac{\Sigma(x_i - \mu)^2}{n}\right]^{0.5} \tag{6.3}$$

If you wish to use the standard deviation s of a sample to estimate the standard deviation σ of the population from which the sample is taken, you can obtain a good estimate from Eq. (6.2) as long as the sample size is 30 or larger.

It has been found, however, that for small samples ($n < 30$) Eq. (6.2) underestimates the magnitude of the population standard deviation. Statisticians have shown that if n in Eq. (6.2) is replaced by $n - 1$, the resulting equation is useful for estimating the population standard deviation from the standard deviation of the small sample. Thus, for small samples, s is given by

$$s = \left[\frac{(x_i - \bar{x})^2}{n - 1}\right]^{0.5} \qquad 6.4$$

An alternate form of Eq. (6.4) which is sometimes easier to use, is derived by expanding $(x_i - \bar{x})^2$, substituting for $\bar{x}$ from Eq. (6.1), and reducing terms. It is

$$s = \left[\frac{n(\Sigma x_i^2) - (\Sigma x_i)^2}{n(n - 1)}\right]^{0.5} \qquad 6.5$$

Because Eq. (6.5) was obtained from Eq. (6.4), it is useful for small samples. For a sample size of 30, the value of s calculated by Eq. (6.5) differs from one calculated by Eq. (6.2) by less than 2 percent. As sample sizes grow, this difference becomes even less. Thus for large samples, either the n or $n - 1$ form of the standard deviation equations could be used with good results.

Another common measure of variation is called the *variance;* it is the square of the standard deviation. Therefore, for small samples, the sample variance is given by

$$s^2 = \frac{\Sigma(x_i - \bar{x})^2}{n - 1} \qquad 6.6$$

Formulas giving the variance for large samples and for populations are obtained by squaring Eqs. (6.2) and (6.3), respectively.

Example problem 6.1 A midwestern university campus has 10 540 male students. Using a random selection process, 50 of these students were chosen and were weighed to the nearest pound (pound-mass); the raw data were as recorded in Table 6.3. The data were then grouped (Table 6.4), and the histogram in Fig. 6.3 was constructed. Calculate the sample mean, sample standard deviation, and sample variance of the data.

Table 6.3

164	171	154	160	158	150	159	185	168	158
143	159	162	165	160	167	166	164	152	172
177	165	170	155	155	163	180	157	145	160
149	153	137	173	157	175	163	147	156	156
162	167	165	166	162	136	158	170	162	159

Table 6.4

Range	Frequency
136–140	2
141–145	2
146–150	3
151–155	5
156–160	13
161–165	11
166–170	7
171–175	4
176–180	2
181–185	1
	50

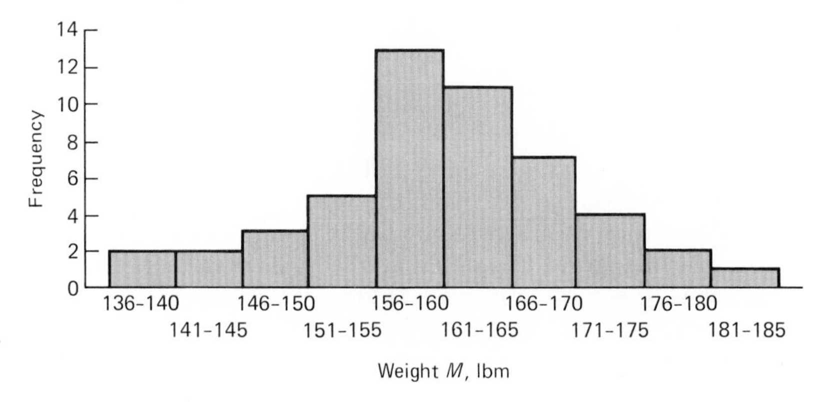

Figure 6.3
Histogram.

Table 6.5

Mass x_i, lbm	x_i^2	$x_i - \bar{x}$	$(x_i - \bar{x})^2$
164	26 896	3.26	10.63
143	20 449	−17.74	314.71
177	31 329	16.26	264.39
149	22 201	−11.74	137.83
·	·	·	·
·	·	·	·
·	·	·	·
156	24 336	−4.74	22.47
159	25 281	−1.74	3.03
8 037	1 296 661	0.00	4 793.74

Solution From Eq. (6.1) the sample mean can be calculated:

$$\bar{x} = \frac{\Sigma x_i}{n}$$

$$= \frac{8\,037}{50}$$

$$= 160.74 \text{ lbm}$$

From Eq. (6.4) the sample standard deviation can be determined [you can also use Eq. (6.2)]:

$$s = \left[\frac{\Sigma(x_i - \bar{x})^2}{n - 1}\right]^{0.5}$$

$$= \left(\frac{4\,793.74}{49}\right)^{0.5}$$

$$= 9.89 \text{ lbm}$$

The sample standard deviation can also be determined from Eq. (6.5):

$$s = \left[\frac{n(\Sigma x_i^2) - (\Sigma x_i)^2}{n(n - 1)}\right]^{0.5}$$

$$= \left[\frac{50(1\,296\,661) - 8\,037^2}{50(49)}\right]^{0.5}$$

$$= 9.89 \text{ lbm}$$

The sample variance can be calculated from Eq. (6.6):

$$s^2 = \frac{\Sigma(x_i - \bar{x})^2}{n - 1}$$

$$= \frac{4\,793.74}{49}$$

$$= 97.83 \text{ lbm}^2$$

By examining the raw data in Example Prob. 6.1 we can see the range in variation of values that occurs from a random sample. Certainly we would expect to find both larger and smaller values if all males at the university, i.e., the entire population, were measured and recorded. If we were to select additional random values and develop a second sample from the population, we would expect to find a different sample mean and a different sample standard deviation. We would not expect, however, the differences in these measures of central tendency and variation to be significant if the two samples were truly random in nature.

6.5 Continuous Distribution

Random variables are classified according to the values that the variable can assume. Discrete random variables may only take on a finite set of values. The flipping of a coin (two outcomes) or the number of automobiles that pass a certain location in a fixed time are examples of discrete random variables.

In contrast to discrete variables, a random variable is continuous when it can assume values on a continuous scale. Quantities such as time and temperature are examples of continuous variables.

Histograms, which were discussed earlier, can be used to deter-

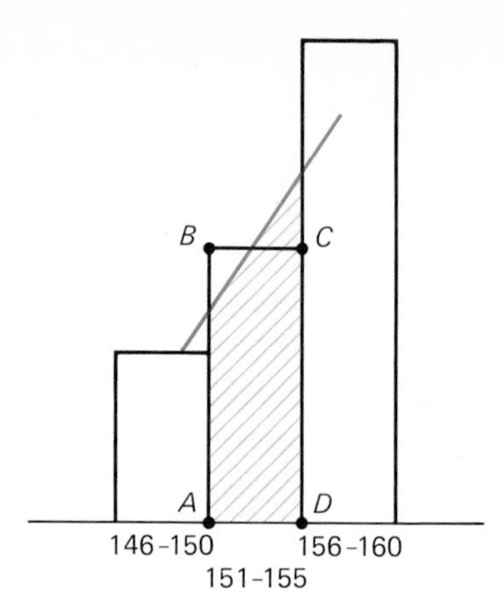

Figure 6.4
Segment of histogram.

mine the probability of a value falling into a given classification. Histograms permit examination of the area of the rectangle representing that classification. For example, one portion of the histogram from Fig. 6.3 is enlarged and shown in Fig. 6.4. The area of rectangle $ABCD$ as a portion of the entire area of the histogram represents the probability that a male weighs between 151 and 155 lbm. It should also be apparent that the area of the rectangle $ABCD$ is nearly equal to the shaded area under the continuous curve that could be constructed to represent the histogram.

More generally, if a histogram is approximated by means of a smooth curve (sometimes called a frequency distribution), the probability associated with any interval is related to the area under the curve bounded by the interval.

6.6 Normal Distribution

Among the many continuous distributions used in statistics, the normal distribution is by far the most useful.

The normal distribution is a theoretical frequency distribution for a specific type of data set. Its graphical representation is a bell-shaped curve that extends indefinitely in both directions. As can be seen in Fig. 6.5, the curve comes closer and closer to the horizontal axis without ever reaching it, no matter how far the axis is extended from the mean (μ).

The location and shape of the normal curve can be specified by

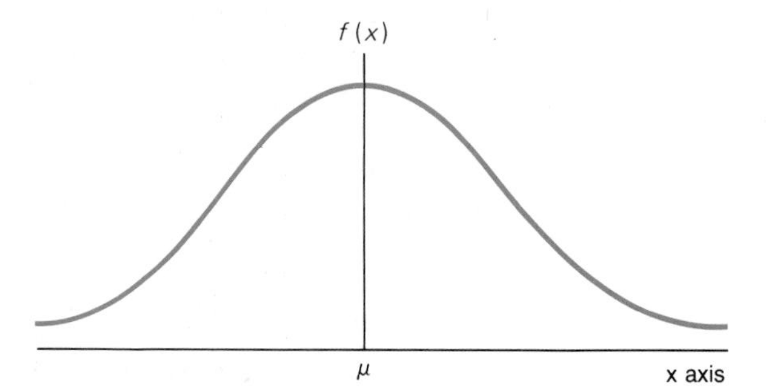

Figure 6.5
Normal distribution.

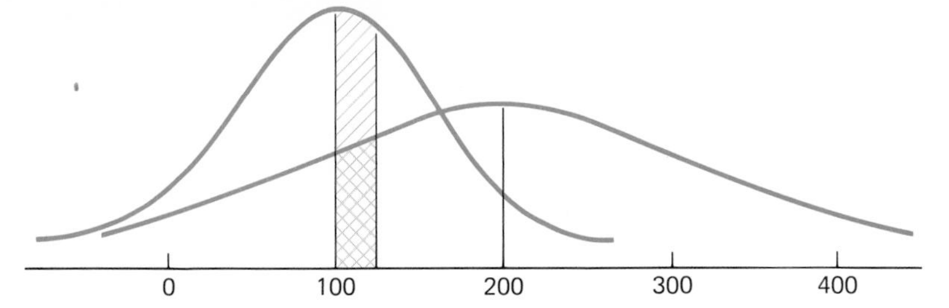

Figure 6.6
Normal curves having standard deviations and means of different magnitudes.

two parameters: (1) the population mean (μ), which locates the center of the distribution, and (2) the population standard deviation (σ), which describes the amount of variability or dispersion of the data.

Mathematically the normal distribution is represented by Eq. (6.7):

$$f(x) = \frac{1}{\sigma\sqrt{2\pi}} e^{-(1/2)[(x-\mu)/\sigma]^2} \tag{6.7}$$

This expression can be used to determine the area under the curve between any two locations on the x axis as long as we know the mean and standard deviation of the data:

$$\text{Area} = \int_{x_1}^{x_2} f(x)\, dx \tag{6.8}$$

Since the evaluation of this expression is difficult, in practice we obtain areas under the curve from a special table of values that was developed from this equation.

As indicated previously, a normal curve is symmetrical about the mean; however, the specific shape of the distribution depends on the deviation of the data about the mean. As can be seen in Fig. 6.6, when the data are bunched around the mean, the curve drops rapidly toward the x axis. However, when the data have a wide deviation about the mean, the curve approaches the x axis more slowly. This presents a problem because if you examine the two curves in Fig. 6.6, the area under the curves between x values of 100 and 125 is not the same for the two distributions.

Thus with each different mean and standard deviation we would have to construct a separate table of normal-curve areas. To avoid having to use many tables, a transformation can be applied converting all curves to a standard form that has $\mu = 0$ and $\sigma = 1$ (see Fig. 6.7). Thus we can normalize the distribution by performing a change of scale that converts the units of measurement into standard units by means of the following equation:

$$z = \frac{x - \mu}{\sigma} \tag{6.9}$$

In order to determine areas under a standardized normal curve,

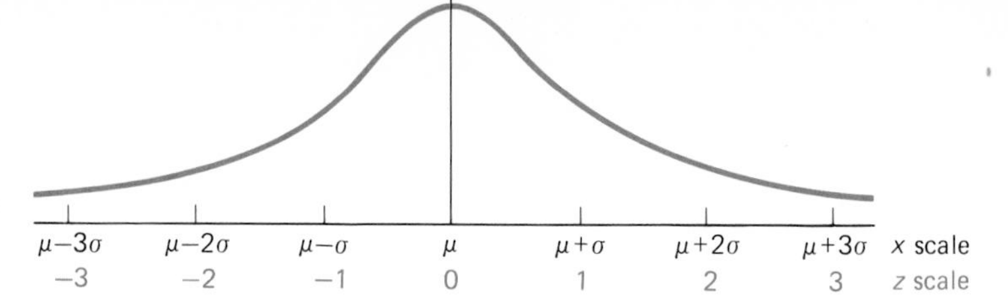

Figure 6.7 Normal curve with normalized distribution.

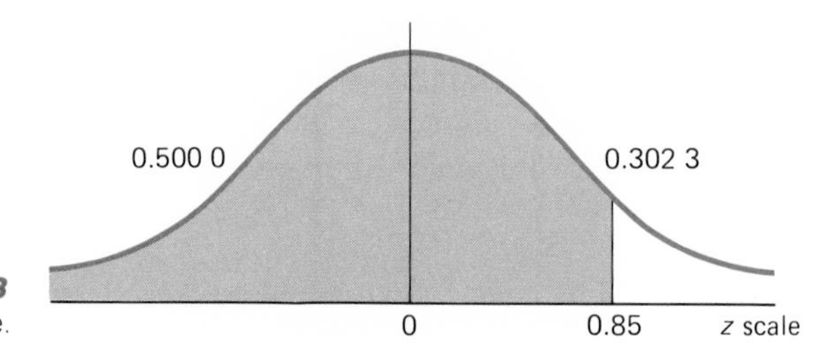

Figure 6.8 Normal curve, z scale.

we must convert x values into z values and then use the table in App. D. Note that numbers in the table have no negative values. Because of the symmetry of the normal curve about the mean, this does not reduce the utility of the table.

Before looking at example problems that demonstrate the practical application of this concept, we will demonstrate the use of the normal-curve table. The total area under the curve to the left of $z = 0$ as well as the area to the right of $z = 0$ is equal to 0.500 0 because for this standardized normal curve, the area beneath it equals 1.000.

Referring to Fig. 6.8, we shall calculate the probability of getting a z value less than 0.85.

The area under the curve for $z < 0.85$ is determined from 0.500 0 (area left of $z = 0$) plus a table value of 0.302 3 for $z = 0.85$ (area between $z = 0$ and $z = 0.85$). Adding these values gives us a probability of 0.802 3.

When working problems with this table, we must remember that the data should closely approximate a normal distribution, and we must know the population mean (μ) and the standard deviation (σ).

Example problem 6.2 A random variable has a normal distribution with $\mu = 50$ and $\sigma = 10$. What is the probability of the variable assuming a value between 40 and 80? (See Fig. 6.9.)

Solution Normalize the values of μ and σ:

$$z_1 = \frac{40 - 50}{10} = -1.0$$

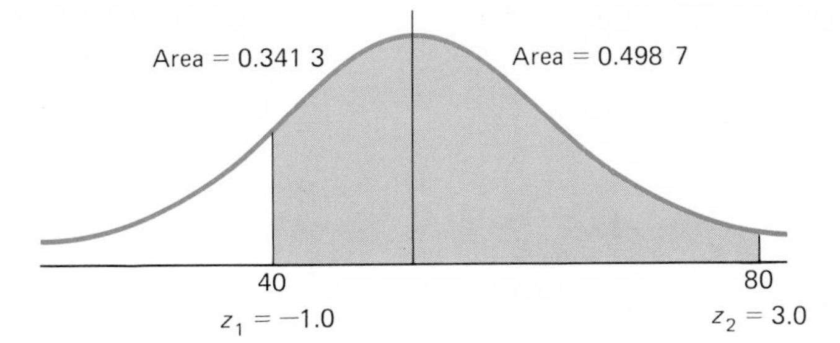

Figure 6.9

$$z_2 = \frac{80 - 50}{10} = +3.0$$

From the z table in App. D,

Area = probability = 0.341 3 + 0.498 7 = 0.840 0

In most statistical applications we do not know the population parameters; instead we collect data in the form of a random sample from that population. It is possible to substitute sample mean and sample standard deviation, provided the sample size is sufficiently large ($n \geqq 30$). A different theory called the *t distribution* is applicable to smaller sample sizes, but this distribution is not covered in our introduction to statistics.

Example problem 6.3 Assume that a normal distribution is a good representation of the data provided in Table 6.3, Example Prob. 6.1.

(*a*) Determine the probability of a male student weighing more than 170 lbm.
(*b*) Determine the percentage of students who weigh between 140 and 150 lbm.

Solution

(*a*) $$z_1 = \frac{170 - 160.74}{9.89}$$

$$= 0.94$$

From the normal-curve table, area = 0.326 4. Since the area under the curve to the right of $z = 0$ is 0.500 0,

Probability = 0.500 0 − 0.326 4 = 0.173 6

(*b*) $$z_1 = \frac{140 - 160.74}{9.89}$$

$$= -2.10$$

$$z_2 = \frac{150 - 160.74}{9.89}$$

$$= -1.09$$

For $z_1 = 2.10$,

Area = 0.482 1 (area between $z = -2.10$ and $z = 0$)

For $z_2 = 1.09$,

Area = 0.362 1 (area between $z = -1.09$ and $z = 0$)

The desired area is the difference, that is,

Probability = area = 0.482 1 − 0.362 1 = 0.120 0

Therefore, we would expect 12 percent of the males to weigh between 140 and 150 lbm.

6.7 Linear Regression

There are many occasions in engineering analysis when the ability to predict or forecast the outcome of a certain event is extremely valuable. The difficulty with most practical applications is the large number of variables that may influence the analysis process. Regression analysis is a study of the relationships among variables. If the situation results in a relationship among three or more variables, the study is called *multiple regression.* There are many problems, however, that can be reduced to a relationship between an independent and a dependent variable. This introduction will limit the subject and treat only two-variable regression analyses.

Of the many equations that can be used for the purposes of prediction, the simplest and most widely used is a linear equation of the form $y = mx + b$, where m and b are constants. Once the constants have been determined, it is possible to calculate a predicted value of y (dependent variable) for any value of x (independent variable).

Before investigating the regression concept in more detail, we must examine how the regression equation is established.

If there is reason to believe that a relationship exists between two variables, the first step is to collect data. For example, suppose x denotes the age of an automobile in years and y denotes the annual maintenance cost. Thus, a sample of n cars would reveal the age $x_1, x_2, x_3, \ldots, x_n$ and the corresponding annual maintenance cost $y_1, y_2, y_3, \ldots, y_n$.

The next step would be to plot the data on rectangular coordinate paper. The resulting graph is called a *scatter diagram.*

Techniques used to correctly plot data were presented in Chap. 3. Determination of the best line through these points was also introduced in Chap. 3 and is normally referred to as *curve fitting.*

From the scatter diagram shown in Fig. 6.10, it may be possible to construct a straight line that adequately represents the data, in

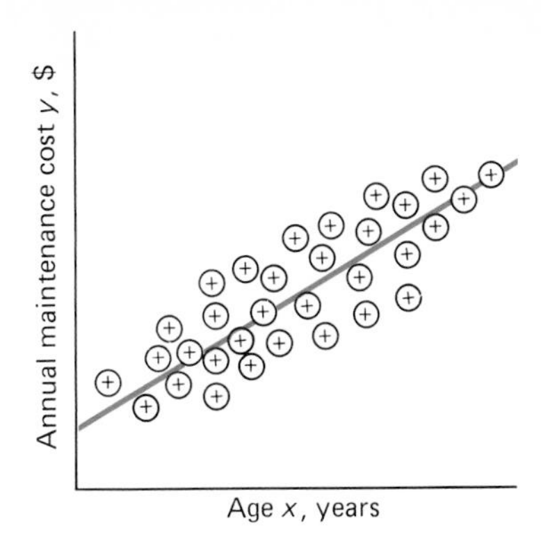

Figure 6.10
Scatter diagram.

which case a linear relationship exists between the variables. In other cases, the line may be curved, and the relationship between variables is nonlinear in nature.

Ideally, we would hope to determine the best possible line (straight or curved) through the points. A standard approach to this problem is called the *method of least squares* and is explained below.

To demonstrate how the process works, as well as to explain the concept of the method of least squares, consider the following situation. A class of 20 students is given a math test and the resulting scores are recorded. Each student's IQ score is also available. Both scores for the 20 students are shown in Table 6.6.

First, the data must be plotted on rectangular coordinate paper (see Fig. 6.11). As you can see by observing the plotted data, there is no limit to the number of straight lines that could be drawn through the points. In order to find the line of best fit, it is necessary to state what is meant by "best." The method of least squares requires that the sum of the squares of the vertical deviations from the data points to the straight line be a minimum.

To demonstrate how a least-squared line is fit to data, let us consider this problem further. There are n pairs of numbers (x_1, y_1), (x_2, y_2), . . . , (x_n, y_n), where $n = 20$, with x and y being IQ and math scores, respectively.

Table 6.6

Student #	Math score	IQ	Student #	Math score	IQ
1	85	120	11	100	130
2	62	115	12	85	130
3	60	100	13	77	118
4	95	140	14	63	112
5	80	130	15	70	122
6	75	120	16	90	128
7	90	130	17	80	125
8	60	108	18	100	140
9	70	115	19	95	135
10	80	118	20	75	130

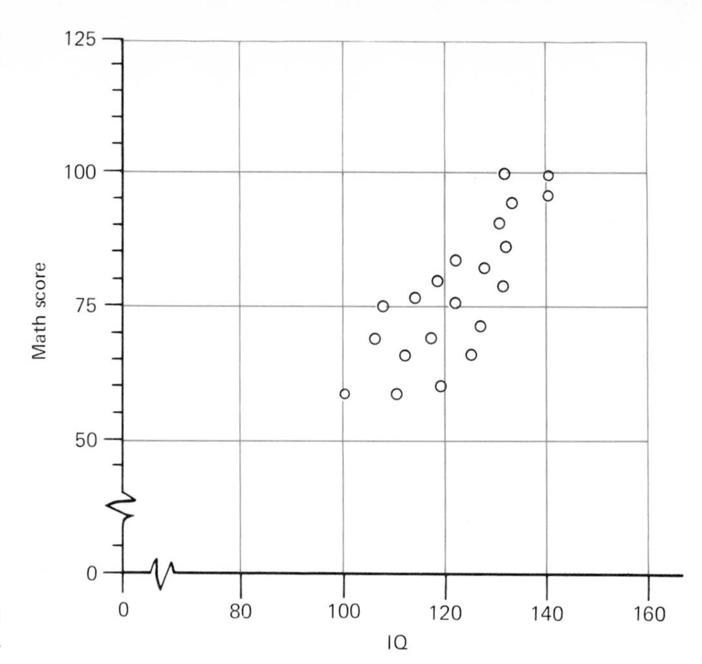

Figure 6.11
Math and IQ scores.

Suppose that the equation of the line that best fits the data is of the form

$$y' = mx + b \tag{6.10}$$

where the symbol y' (y prime) is used to differentiate between the observed values of y and the corresponding values calculated by means of the equation of the line. In other words, for each value of x, there exist an observed value (y) and a calculated value (y') obtained by substituting x into the equation $y' = mx + b$.

The least-squares criterion requires that the sum of all $(y - y')^2$ terms, as illustrated in Fig. 6.12, be the smallest possible. One must determine the constants m and b so that the differences between the observed and the predicted values of y will be minimized.

When this analysis is applied to the linear equation $y = mx + b$, it follows that we wish to minimize the summation of all deviations:

$$\text{SUM} = \sum_{i=1}^{n} [y_i - (mx_i + b)]^2 \tag{6.11}$$

The minimization process involves differential calculus and is not presented here. The resulting expressions are

$$nb + m\Sigma x_i = \Sigma y_i \tag{6.12}$$

$$b\Sigma x_i + m\Sigma x_i^2 = \Sigma x_i y_i \tag{6.13}$$

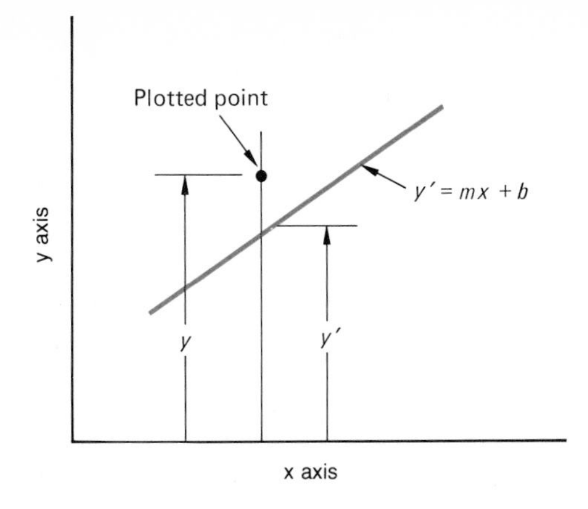

Figure 6.12
y-axis deviation.

Solving these two equations simultaneously for m and b gives

$$m = \frac{n(\Sigma x_i y_i) - (\Sigma x_i)(\Sigma y_i)}{n(\Sigma x_i^2) - (\Sigma x_i)^2} \tag{6.14}$$

$$b = \frac{\Sigma y_i - m(\Sigma x_i)}{n} \tag{6.15}$$

Table 6.7 is a tabulation of the values necessary to determine the constants m and b for the math score–IQ problem. The independent variable is the IQ, and the dependent variable is the math score.

Table 6.7

Independent variable		Dependent variable		
IQ	IQ²	Math score	Math score²	(IQ)(Math score)
120	14 400	85	7 225	10 200
115	13 225	62	3 844	7 130
100	10 000	60	3 600	6 000
140	19 600	95	9 025	13 300
130	16 900	80	6 400	10 400
120	14 400	75	5 625	9 000
130	16 900	90	8 100	11 700
108	11 664	60	3 600	6 480
115	13 225	70	4 900	8 050
118	13 924	80	6 400	9 440
130	16 900	100	10 000	13 000
130	16 900	85	7 225	11 050
118	13 924	77	5 929	9 086
112	12 544	63	3 969	7 056
122	14 884	70	4 900	8 540
128	16 384	90	8 100	11 520
125	15 625	80	6 400	10 000
140	19 600	100	10 000	14 000
135	18 225	95	9 025	12 825
130	16 900	75	5 625	9 750
2 466	306 124	1 592	129 892	198 527

Substituting the values from Table 6.7 into Eqs. (6.14) and (6.15), we get the following values for the two constants:

$$m = \frac{20(198\,527) - (2\,466)(1\,592)}{20(306\,124) - 2\,466^2}$$

$$= 1.081$$

$$b = \frac{1\,592 - (1.081)(2\,466)}{20}$$

$$= -53.7$$

The equation of the line relating math score and IQ using the method of least squares becomes

$$\text{Math score} = -53.7 + 1.08(\text{IQ})$$

Interesting questions arise from this problem. Can IQ be used to predict success on a math exam and, if so, how well? Regression analysis or estimation of one variable (dependent) from one or more related variables (independent) does not provide information about the strength of the relationship. Section 6.8 will provide a method to determine how well an equation developed from the method of least squares describes the strength of the relationship between variables.

The method of least squares as explained above is a most appropriate technique for determination of the best-fit line. You should clearly understand that this method as presented is *linear regression* and is valid only for *linear* relationships. You should recall that in Chap. 3 we used the method of selected points to determine three types of empirical equations: linear, power, and exponential. Recall also that the power and exponential curves are not straight lines on rectangular graph paper but *are* straight lines on log-log (power) and semilog (exponential) plots. Because these data plot straight on the proper graph paper, we can apply the method of least squares to the linear relationship between log x and log y (power) or between log y and x (exponential). Example Probs. 6.4 and 6.5 will demonstrate the methods used for power and exponential curves, respectively.

Example problem 6.4 Find the equation of the line of best fit for the data given in Example Prob. 3.2.

Solution This data is plotted as a straight line in Figs. 3.25 and 3.26. In Fig. 3.25 the logarithms of the data, both time and distance, are plotted. Therefore, if we use log t in place of x and log s in place of y [the linear form is $\log s = m(\log t) + \log b$] in Eqs. (6.14) and (6.15), we can solve for m and log b. Note carefully that the constants in the linear form are m and log b. Refer to Table 6.8.

Table 6.8

t	s	Independent variable $\log t$	$(\log t)^2$	Dependent variable $\log s$	$(\log t)(\log s)$
1	4.9	0.000 0	0.000 0	0.609 2	0.000 0
2	19.6	0.301 0	0.090 6	1.292 3	0.389 0
3	44.1	0.477 1	0.227 6	1.644 4	0.784 5
4	78.4	0.602 1	0.362 5	1.894 3	1.140 6
5	122.5	0.699 0	0.488 6	2.088 1	1.459 6
6	176.4	0.778 2	0.605 6	2.246 5	1.748 2
		2.857 4	1.774 9	9.855 8	5.521 9

Substitute into Eq. (6.14):

$$m = \frac{6(5.521\,9) - (2.857\,4)(9.855\,8)}{6(1.774\,9) - (2.857\,4)^2}$$

$$= 2.00$$

Substitute into Eq. (6.15), using log b, *not* b:

$$\log b = \frac{9.855\,8 - 2.0(2.857\,4)}{6}$$

$$= 0.690\,17$$

$$b = 4.9$$

The equation is $s = 4.9t^2$, which checks with the solution obtained in Example Prob. 3.2.

Example problem 6.5 Using the method of least squares, find the equation that best fits the data shown in Table 6.9. (This is the same data as in Table 3.5.)

Solution This data produces a straight line on semilog graph paper; therefore, its equation will be of the form $y = be^{mx}$ or ln $y =$

Table 6.9

Fuel consumption, mm^3/s	Velocity, m/s
25.2	10.0
44.6	20.0
71.7	30.0
115	40.0
202	50.0
367	60.0
608	70.0

Table 6.10

Independent variable		Dependent variable	
v	v^2	$\ln Q$	$v(\ln Q)$
10	100	3.226 8	32.27
20	400	3.797 7	75.95
30	900	4.272 5	128.17
40	1 600	4.744 9	189.80
50	2 500	5.308 3	265.41
60	3 600	5.905 4	354.32
70	4 900	6.410 2	448.71
280	14 000	33.665 8	1 494.63

$mx + \ln b$. An examination of the equation and the graph paper leads us to the following:

1. Since the line is straight, the method of least squares can be used.
2. Since the x axis is a uniform scale, the independent variable values (velocity in this problem) may be used without adjustment.
3. Since the y axis is a log scale, the dependent variable values (fuel consumption) must be the logarithms of the data, not the raw data.

Table 6.10 provides us with the needed values to substitute into Eqs. (6.14) and (6.15).

Substitute into Eq. (6.14):

$$m = \frac{7(1\,494.63) - (280)(33.665\,8)}{7(14\,000) - 280^2}$$

$$= 0.052\,9 \text{ (0.052 3 by method of selected points)}$$

Substitute into Eq. (6.15), using $\ln b$ rather than b:

$$\ln b = \frac{33.665\,8 - (.052\,9)(280)}{7}$$

$$= 2.693\,4$$

$$b = 14.78$$

The equation becomes $Q = 15e^{0.053v}$.

6.8 Coefficient of Correlation

The technique of finding the best possible straight line to fit experimentally collected data is certainly useful, as previously discussed. The next logical and interesting question is how well such a line actually fits. It stands to reason that if the differences between the observed y's and the calculated y''s are small, the sum of squares $\Sigma(y - y')^2$ will be small; and if the differences are large, the sum of squares will tend to be large.

Although $\Sigma(y - y')^2$ provides an indication of how well a least-squares line fits particular data, it has the disadvantage that it depends on the units of y. For example, if the units of y are changed from dollars to cents, it will be like multiplying $\Sigma(y - y')^2$ by a factor of 10 000. To avoid this difficulty, the magnitude of $\Sigma(y - y')^2$ is normally compared with $\Sigma(y - \bar{y})^2$. This allows the sum of the squares of the vertical deviations from the least-squares line to be compared with the sum of squares of the deviations of the y's from the mean.

To illustrate, Fig. 6.13*a* shows the vertical deviation of the y's from the least-squares line, while Fig. 6.13*b* shows the deviations of the y's from their collective mean. It is apparent that where there is a close fit, $\Sigma(y - y')^2$ is smaller than $\Sigma(y - \bar{y})^2$.

In contrast, consider Fig. 6.14. Again, Fig. 6.14*a* shows the vertical deviation of the y's from the least-squares line, and Fig. 6.14*b* shows the deviation of the y's from their mean. In the latter case, $\Sigma(y - y')^2$ is approximately the same as $\Sigma(y - \bar{y})^2$. This would seem to indicate that if the fit is good, as in Fig. 6.13, $\Sigma(y - y')^2$

Figure 6.13
Deviation from y' and $\bar{y}$.

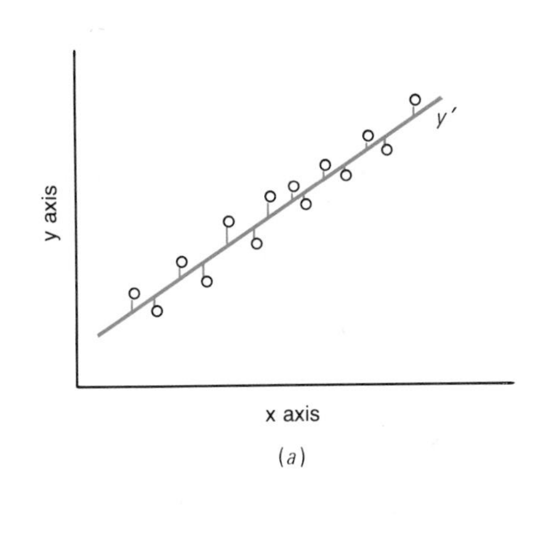

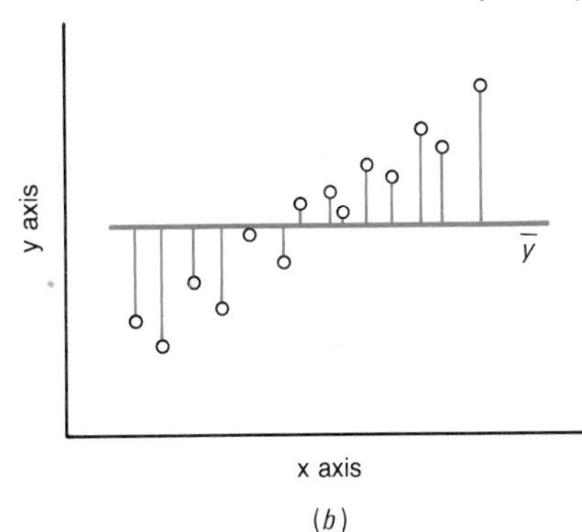

Figure 6.14
Deviation from y' and $\bar{y}$.

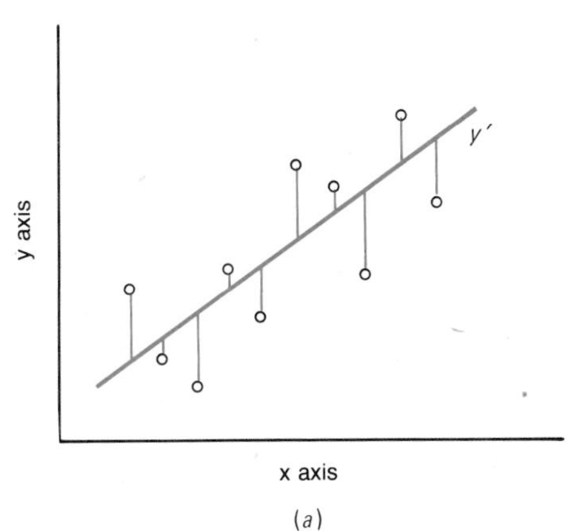

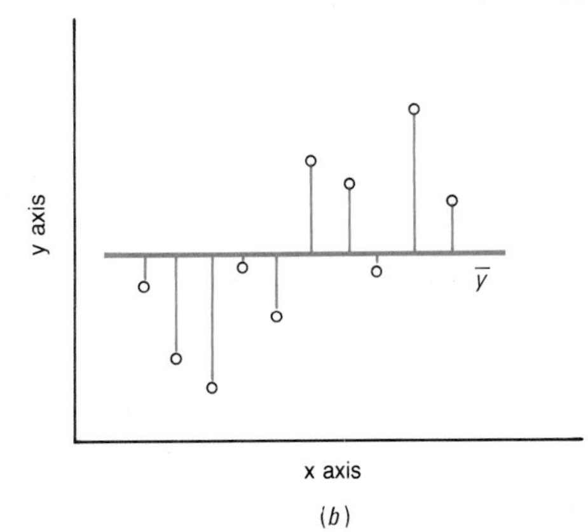

is much less than $\Sigma(y - \bar{y})^2$; and if the fit is as poor as in Fig. 6.14, the two sums of squares are approximately equal.

The coefficient of correlation puts this comparison on a precise basis:

$$r = \pm\sqrt{1 - \frac{\Sigma(y_i - y')^2}{\Sigma(y_i - \bar{y})^2}} \qquad 6.16$$

If the fit is poor, the ratio of the two sums is close to 1 and r is close to zero. However, if the fit is good, the ratio is close to zero and r is close to $+1$ or -1. From the equation, it is obvious that the ratio can never exceed 1. Hence, r cannot be less than -1 or greater than $+1$.

The statistic is undoubtedly the most widely used measure of the strength of a linear relationship between any two variables. It indicates the goodness of fit of a line determined by the method of least squares, and this in turn indicates whether there exists a relationship between x and y.

Although Eq. 6.16 serves to define the coefficient of correlation, it is seldom used in practice. An alternative form of the formula is

$$r = \frac{n(\Sigma x_i y_i) - (\Sigma x_i)(\Sigma y_i)}{\sqrt{n(\Sigma x_i^2) - (x_i)^2}\sqrt{n(\Sigma y_i^2) - (\Sigma y_i)^2}} \qquad 6.17$$

The interpretation of r is not difficult if it is ± 1 or 0: When it is 0, the points are scattered and the fit of the regression line is so poor that a knowledge of x does not help in the prediction of y; when it is $+1$ or -1, all the points actually lie on the straight line, so an excellent prediction of y can be made by using x values. The problem arises when r falls between 0 and $+1$ or 0 and -1.

The simplest physical interpretation of r can be explained in the following manner. If the coefficient of correlation is known for a given set of data, then $100r^2$ percent of the variation of the y's can be attributed to differences in x, namely, to the relationship of y with x. If $r = 0.6$ in a given problem, then 36 percent, that is, $100(0.6^2)$, of the variation of the y's is accounted for (perhaps caused) by differences in x values.

Again consider the problem on IQ and math scores, substituting values from Table 6.7 into the equation for the correlation coefficient.

$$r = \frac{(20)(198\ 527) - (2\ 466)(1\ 592)}{\sqrt{(20)(306\ 124) - 2\ 466^2}\sqrt{(20)(129\ 892) - 1\ 592^2}}$$

$$= 0.87$$

$$100r^2 = 76 \text{ percent}$$

This would indicate that 76 percent of the variations in math scores can be accounted for by differences in IQ.

One word of caution when using or considering results from linear regression and coefficients of correlation: There is a fallacy in interpreting high values of r as implying cause-effect relations. If the increase in television coverage of professional football is plotted against the increase in traffic accidents at a certain intersection over the past 3 years, an almost perfect positive correlation (+1.0) can be shown to exist. This is obviously not a cause-effect relation, so it is wise to interpret the correlation coefficient carefully. The variables must have a measure of association if the results are to be meaningful.

Problems

6.1 If utility bills paid by residents of a small town during the month of October varied from $27.82 to $82.27, construct a table with eight classes into which these amounts may be grouped.

6.2 A state highway department is interested in the variation of two-lane highways. By measuring the width of all two-lane concrete roads, the following data are collected. Each reading is recorded to the nearest tenth of a decimeter.

56.7	64.9	61.4	66.9	61.5
71.2	66.6	68.2	68.9	63.6
67.8	67.5	67.6	71.3	56.2
64.2	55.7	68.8	68.5	65.9
67.5	72.0	63.9	64.3	73.0
61.4	63.4	59.6	68.7	64.4
70.3	68.2	68.2	69.2	65.6

(a) Group these measurements into a frequency-distribution table having five equal classes from 54.0 to 73.9.
(b) Construct a histogram of the distribution.
(c) Determine the median, mode, and mean of the data.
(d) Calculate the standard deviation.

6.3 A farm-implement manufacturing company in the midwest purchases steel castings from a Chicago-area foundry. Thirty castings are selected at random and weighed, and their masses are recorded to the nearest kilogram, as shown below.

235	232	228	228	240	231
225	220	218	230	222	242
207	233	222	211	228	228
238	232	230	226	236	247
227	227	229	229	224	228

(a) Group the measurements into a frequency-distribution table having six equal classes from 206 to 247.
(b) Construct a histogram of the distribution.
(c) Determine the median, mode, and mean of the data.
(d) Calculate the standard deviation.

6.4 An approximation of missile velocities was recorded over a predetermined fixed distance. Each value was rounded to the nearest 10 m/s.

980	660	950	1 010
930	880	870	960
750	1 020	970	890
970	900	1 030	950
1 000	940	970	600

(a) Group these measurements into a frequency distribution table having six equal classes that range from 500 to 1 099.
(b) Construct a histogram of the distribution.
(c) Determine the median, mode, and mean of the data.
(d) Calculate the standard deviation.

6.5 The following test scores were recorded by a class of freshman engineering students on a chemistry test:

59	90	67	47	70	40	70	77	80	85
58	70	67	62	75	87	61	73	88	70
83	63	72	95	62	65	90	58	69	99
58	69	60	72	88	79	80	68	100	75
70	31	93	79	72	64	52	65	77	72

(a) Group these test scores into a frequency-distribution table.
(b) Construct a histogram of the distribution.
(c) Determine the median, mode, and mean of the data.
(d) Calculate the standard deviation.

6.6 Survey at least 30 engineering students to ascertain each student's investment in computation equipment; the investment figure should reflect the value of all equipment, whether bought personally or received as a gift.
(a) Find the mean, mode, and median investment.
(b) Find the standard deviation.
(c) Assuming your sample to be a normal distribution, half of the engineering students at your school probably have invested between $\$$________ and $\$$________ each.

6.7 Find out from members of this class how many credits toward graduation each will have at the end of this term.
(a) Find the mean, mode, and median of the data.
(b) Find the standard deviation.
(c) Assuming that your class is representative of all of those taking this course this term, 12 percent of the students currently taking this course will have completed more than ________ credits at the end of this term.

6.8 The scores on the final exam for 50 students taking engineering computations produced a mean of 78 and a standard deviation of 12. If the distribution of the scores is approximately normal, determine
(a) The expected percentage of students earning 90 or higher
(b) The number of students expected to have scores below 60
(c) The expected number of C's (70–79)
(d) The predicted maximum score of the lowest 15 percent of the class

6.9 The quality control department measured the length of 100 bolts randomly selected from a specified order. The mean length was found to be 9.76 cm, and the standard deviation was 0.01 cm. If the bolt lengths are normally distributed, find:
(a) The percentage of bolts shorter than 9.74 cm
(b) The percentage of bolts longer than 9.78 cm
(c) The percentage of bolts that meet the length specification of 9.75 ± 0.02 cm
(d) The percentage of bolts that are longer than the nominal length of 9.75 cm

6.10 A sample of 40 resistors randomly taken from yesterday's production were tested with the following result: mean resistance = 985Ω and standard deviation = 55Ω. Assuming a normal distribution, compute the following:

(a) The percentage of resistors with resistance greater than 1000Ω
(b) The yield (that is, percentage good) if the acceptable values are 1000Ω ± 10 percent
(c) The yield if acceptable values include only 1000Ω ± 5 percent

6.11 On the right are grades that 20 students obtained on the mid-term and final examination in a freshman graphics course:

(a) Using the method of least squares, determine the best straight line through the data.
(b) Plot the data on linear graph paper.
(c) Represent the equation in (*a*) on the graph from (*b*).
(d) Calculate and interpret the coefficient of correlation.

88	78
75	85
97	91
68	82
86	81
91	75
53	64
84	91
77	78
92	83
62	52
83	73
36	50
51	40
81	83
91	87
82	80
74	70
85	89
96	98

6.12 Table 6.11 is data from a trial run on the Utah salt flats made by an experimental turbine-powered vehicle.

(a) Plot the data on rectilinear paper.
(b) Using the method of least squares, determine the equation of the line of best fit.
(c) Draw the line on the graph.
(d) Calculate and interpret the coefficient of correlation.

Table 6.11

Time *t*, s	Velocity *V*, m/s
10.0	15.1
20.0	32.2
30.0	63.4
40.0	84.5
50.0	118
60.0	139

6.13 Table 6.12 lists values of velocity of ski flyers in Colorado this past winter.

(a) Plot the data on rectilinear paper.
(b) Using the method of least squares, determine the equation of the line of best fit.
(c) Draw the line on the graph.
(d) Calculate and interpret the coefficient of correlation.

Table 6.12

Time *t*, s	Velocity *V*, m/s
1.00	5.30
4.00	18.1
7.00	26.9
10.0	37.0
14.0	55.2

6.14 A Johnson furnace was tested 45 days ago in your hometown to determine the heat generated, expressed in thousands of British thermal units per cubic foot of furnace volume, at varying temperatures.

The results are shown in Table 6.13.

(a) Plot the data on log-log graph paper, with temperature the independent variable.
(b) Using the method of least squares, find the equation of the line of best fit.
(c) Draw the line on the graph.
(d) Calculate the coefficient of correlation.

Table 6.13

Heat released *H*, 10^3 Btu/ft^3	Temperature *T*, °F
0.200	172
0.600	241
2.00	392
4.00	483
8.00	608
20.00	812
40.00	959
80.00	1 305

6.15 The capacity of a screw conveyor that is moving dry ground corn is expressed in liters per second and the conveyor speed in revolutions per minute. A test was conducted in Cleveland on the conveyor model JD172 last week. The results of the test are shown in Table 6.14.

(a) Plot the data on log-log graph paper.
(b) Using the method of least squares, find the equation of the line of best fit.
(c) Draw the line on the graph.
(d) Calculate the coefficient of correlation.

Table 6.14

Capacity C, L/s	Angular velocity V, r/min
3.01	10.0
6.07	21.0
15.0	58.2
30.0	140.6
50.0	245
80.0	410
110.0	521

6.16 The resistance of a size and class of electrical conductor was tested over a wide range of sizes at constant temperature. The test was performed at Madison, Wisconsin, on April 4, 1984, at the Acme Electrical Labs. The test results are shown in Table 6.15. The resistance is expressed in milliohms per meter of conductor length.

(a) Plot the data on log-log graph paper.
(b) Using the method of least squares, find the equation of the line of best fit.
(c) Draw the line on the graph.
(d) Find the coefficient of correlation.

Table 6.15

Area A, mm^2	Resistance R, $m\Omega/m$
0.021	505
0.062	182
0.202	55.3
0.523	22.2
1.008	11.3
3.32	4.17
7.29	1.75

6.17 Table 6.16 is a collection of data for an iron-constantan thermocouple. Temperature is in degrees Celsius and the emf is in millivolts.

(a) Plot the data on rectilinear graph paper.
(b) Using the method of least squares, find the equation of the line of best fit.
(c) Draw the line on the graph.
(d) Calculate and interpret the coefficient of correlation.

Table 6.16

Temperature t, °C	Voltage (emf), mV
50.0	2.6
100.0	6.7
150.0	8.8
200.0	11.2
300.0	17.0
400.0	22.5
500.0	26.0
600.0	32.5
700.0	37.7
800.0	41.0
900.0	48.0
1 000.0	55.2

6.18 A Sessions pump was tested to determine the power required to produce a range of discharge. The test was performed in Oak Park on July 1, 1984. The results of the test are shown in Table 6.17.

(a) Plot the data on rectilinear graph paper.
(b) Using the method of least squares, find the equation of the line of best fit.
(c) Draw the line on the graph.
(d) Calculate and interpret the coefficient of correlation.

Table 6.17

Discharge Q, L/s	Power P, kW
3.00	28.5
7.00	33.8
10.00	39.1
13.50	43.2
17.00	48.0
20.00	51.8
25.00	60.0

6.19 A spring was tested in Des Moines last Thursday. The test of spring ZX-15 produced the data in Table 6.18.

(a) Plot the data on rectilinear graph paper and determine the equation that expresses the deflection to be expected under a given load. Use the method of least squares.
(b) Draw the line on the graph.
(c) Predict the load required to produce a deflection of 75 mm.
(d) Calculate and interpret the coefficient of correlation.

Table 6.18

Deflection D, mm	Load L, N
2.25	35.0
12.0	80.0
20.0	120.0
28.0	160.0
35.0	200.0
45.0	250.0
55.0	300.0

6.20 A 90° triangular weir is commonly used to measure flow rate in a stream. Data on the discharge through the weir were collected and recorded in Table 6.19 for various heights.

(a) Plot the data on log-log graph paper.
(b) Using the method of least squares, find the equation of the line of best fit.
(c) Draw the line on the graph.
(d) Calculate the coefficient of correlation.

Table 6.19

Height h, m	Discharge Q, m^3/s
1.0	1.5
2.0	8.0
3.0	22
4.0	45
5.0	78
6.0	124
7.0	182
8.0	254

6.21 A Pitot tube is a device for measuring the velocity of flow of a fluid (see Fig. 3.29). A stagnation point occurs at point 2; by recording the height differential h, the velocity at point 1 can be calculated. Assume for this problem that the velocity at point 1 is known corresponding to the height differential h. Table 6.20 records these values.

(a) Plot the data on log-log graph paper.
(b) Using the method of least squares, find the equation of the line of best fit.
(c) Draw the line on the graph.
(d) Find the coefficient of correlation.

Table 6.20

Height h, m	Velocity V, m/s
0.10	1.40
0.20	2.00
0.40	2.80
0.60	3.40
0.80	4.00
1.00	4.40

Table 6.21

Time t, s	Voltage V, V
6	98
10	62
17	23
25	9.5
32	3.5
38	1.9
42	1.33

Table 6.22

Time t, s	Current I, A
0.10	1.81
0.20	1.64
0.30	1.48
0.40	1.34
0.50	1.21
1.00	0.73

Table 6.23

Torque T, N·m	Movement θ, degrees
3.0	5.2
6.0	29.3
10.0	40.9
20.0	56.3
35.0	71.0
50.0	84.8

6.22 The voltage across a capacitor during discharge was recorded as a function of time (see Table 6.21).

(a) Plot the data on semilog graph paper.
(b) Using the method of least squares, find the equation of the line of best fit.
(c) Draw the line on the graph.
(d) Calculate the coefficient of correlation.

6.23 When a capacitor is to be discharged, the current flows until the voltage across the capacitor is zero. This current flow when measured as a function of time resulted in the data given in Table 6.22.

(a) Plot the data on semilog graph paper.
(b) Using the method of least squares, find the equation of the line of best fit.
(c) Draw the line on the graph.
(d) Calculate the coefficient of correlation.

6.24 When fluid is flowing in the line, it is relatively easy to begin closing a valve that is wide open. But as the valve approaches a more nearly closed position, it becomes considerably more difficult to force movement. Visualize a circular pipe with a simple flap hinged at one edge being closed over the end of the pipe. The fully open position is $\theta = 0°$, and $\theta = 90°$ is the fully closed condition.

A test was conducted on such a valve by applying constant torque at the hinged position and measuring the angular movement of the valve. The test data are shown in Table 6.23.

(a) Plot the data on semilog graph paper.
(b) Using the method of least squares, find the equation of the line of best fit.
(c) Draw the line on the graph.
(d) Calculate the coefficient of correlation.

6.25 All materials are elastic to some degree. It is desirable that certain parts of some designs compress when a load is applied to assist in making the part air tight or watertight (e.g., a jar lid). The test results shown in Table 6.24 resulted from a test made at the Herndon Test Labs in Houston on a material known as SILON Q-177.

(a) Plot the data on semilog graph paper.
(b) Using the method of least squares, find the equation of the line of best fit.
(c) Draw the line on the graph.
(d) Calculate the coefficient of correlation.

Table 6.24

Pressure P, MPa	Relative compression R, %
1.12	27.3
3.08	37.6
5.25	46.0
8.75	50.6
12.3	56.1
16.1	59.2
30.2	65.0

6.26 The rate of absorption of radiation by metal plates varies with the plate thickness and the nature of the source of radiation. A test was made at Ames Labs on October 11, 1982, using a Geiger counter and a constant source of radiation; the results are shown in Table 6.25.

(a) Plot the data on semilog graph paper.
(b) Using the method of least squares, find the equation of the line of best fit.
(c) Draw the line on the graph.
(d) Calculate the coefficient of correlation.

Table 6.25

Plate thickness W, mm	Geiger counter C, counts per second
0.20	5 500
5.00	3 720
10.00	2 550
20.00	1 320
27.50	720
32.50	480

6.27 Survey at least 15 students (U.S. citizens) in your housing unit regarding the number of long-distance calls per year in which they have been involved, either as the caller or the receiver, and the number of miles from their homes to campus.

(a) Plot the data on the type of graph paper that appears to give the nearest to a straight-line relationship.
(b) Using the method of least squares, calculate the line of best fit.
(c) What is the correlation factor? What does it mean?

Computers in Engineering

7

Introduction to Computing Systems

7.1 Introduction

In the history of any profession, major developments occur from time to time of such importance that the profession and the society it serves are permanently changed. The invention of the steam engine was such a development for the industrial sector in the eighteenth century. Anesthesia was a significant advancement to the medical profession. Now, for the engineering profession, the impact of computers is direct, sweeping, and extreme. Computers are a vital part of all phases of business and industry today; their use has resulted in rapid developments in transportation, health care, business management, education, and national defense. As the technology grows and new applications are discovered, computers are becoming a part of our home lives as well as our professional responsibilities.

Barely 30 years old, the computer industry is growing at a rate greater than 15 percent a year. Couple this with the fact that 20 percent more computing power is available each year because of advances in technology and we have an effective growth rate of nearly 40 percent per year. Another way of stating this is that computing capability is doubling about every 2 years. The dramatic effect on society in general is obvious.

This unparalleled advance in computing capability puts special demands on you as an engineer. You must learn how computers work, what they can do, and what they are used for. You will also have to surmise what their effect on society might be. Otherwise, you may find it difficult to compete in the engineering profession. When used properly, computers can greatly extend your effectiveness. Not only do they save you time in doing routine computations but their speed enables you to examine many more alternatives before making decisions. Even at a cost of several hundred dollars per hour for time on a large computing system, a few minutes of computation can yield substantial results. As a comparison, it has been estimated that for a moderately sized computer system, one

Figure 7.1
Computers are helping to link more closely the design and manufacturing functions in all areas of industry. Here a computer system is assisting the control of parts movement in an assembly operation. (*Allen-Bradley.*)

machine-second is equivalent to one person-month for arithmetic computations.

A computer may be defined broadly as a device capable of accepting information or data, processing it (that is, manipulating it), and providing usable results from it. Because engineers use the computer as a means of computing, they must understand the computer's capabilities and have access to it by knowing how to program data into it. To program is to prepare a detailed sequence of operating instructions for a particular problem. Most engineering problems are mathematical in nature, so engineers must have a complete knowledge of the problem parameters and necessary mathematical analyses before they can determine which type of computing machine would be most efficient and economical. (It may not be feasible to program a computer solution if a rough estimate of the answer is all that is desired at the time.)

The six basic steps for logical problem analysis and solution (the engineering method) given in Chap. 2 are repeated here to show how and where the computer can assist the problem solver.

1. Recognize and understand the problem.
2. Accumulate facts.
3. Select appropriate theory or principle.
4. Make necessary assumptions.
5. Solve the problem.
6. Verify and check results.

The computer enters the solution process at step 5. Selection of the appropriate computational device depends on the complexity and accuracy requirements of the mathematical model developed in step 4. The computer can assist at step 6 by performing calculations and comparing results from alternative methods of solution. In order for the computer to provide the needed computational assistance, it must be supplied a definite procedure.

The procedure prescribed for solving a specified problem by means of well-defined rules or processes is called an *algorithm*. The most difficult part of writing an algorithm is that of stating precisely and fully each step necessary. (It must be remembered that a computer cannot supply missing steps because it is incapable of creative thinking.) To appreciate the challenge of writing an algorithm, let's consider writing one for calculating the square root of a number. We can verify the correctness of our algorithm by applying it to find the square root of 42 to three significant figures.

Steps in the algorithm	Example
1. Divide number by 2 and use the result as first trial root.	$\frac{42}{2} = 21 = \text{trial root}$
2. Add trial root to the fraction formed from the original number divided by the trial root and divide result by 2.	$\frac{21 + \frac{42}{21}}{2} = \frac{23}{2} = 11.5$
3. Square result.	$11.5^2 = 132.25$
4. Compare with original number. (If result is not within prescribed accuracy, let the computed trial root from step 2 be the new trial root and repeat steps 2, 3, and 4.)	$132.25 > 42$
5. Continue the process until desired accuracy is obtained.	Trial root = 11.5

The reader should continue to execute the algorithm, verifying that the successive trial roots are 7.576, 6.560, and 6.481—the last of which, when squared, yields 41.99. The square-root algorithm thus yields 6.48 as the square root of 42 to three significant figures.

The remainder of this chapter is devoted to a brief look at the evolution of computers, a discussion of the digital computer, a per-

Figure 7.2
A computer operator monitors system performance on a large centralized facility located on a university campus.

ception of the impact of digital computers on engineering, and a glance at the future of digital computing. The ensuing chapter will concentrate on the necessary procedures for preparing a problem for computer solution.

7.2 Evolution of Computing Equipment

The first counting devices were undoubtedly based on the 10 fingers of the hands. Pebbles were used to count higher values. Around 1 000 BC the first form of a calculating instrument, the abacus, appeared. It is a rectangular frame with beads strung on parallel wires. It can be used to add and subtract rapidly by manually sliding and thereby grouping the beads.

In 1642, Blaise Pascal, a French scientist, constructed the first mechanical calculator generally considered to be the forerunner of the modern desk calculator. Pascal's device replaced the beads and wires with toothed wheels with 10 cogs (teeth) per wheel, each cog representing the number 1. The wheels were placed side by side to enable the carrying operation to take place. When one wheel completed a revolution, a ratchet caused the adjacent wheel on the left to move one notch, thereby effecting a carrying, or borrowing, operation as in longhand arithmetic. As with the abacus, the Pascal device could perform multiplication and division only by successive additions and subtractions.

Up to this point, computing devices were digital in nature; that is, they dealt only with data in the form of discrete numbers (digits). Next appeared the first forms of another method of computing, the analog. Analog computation is a means of representing numbers by a continuous range of physical quantities such as lengths, rotations, voltages, or currents. For example, a temperature may be represented by a voltage, which is its analog. It is possible then to add two numbers by adding the voltages which represent the numbers.

Consider the measurement of the passage of one minute of time. The passage of the minute can be measured *digitally* by counting 60 seconds, one at a time; or *analogically*, by observing one revolution of the second hand on a clock. An analog is thus a physical variable that remains similar to another variable in that the proportional relationships are the same over some specified range.

The first form of analog computation, the slide rule, used length to represent numbers. In 1614, John Napier described the natural logarithms, which allowed a person to multiply and divide by adding exponents (logarithms). Henry Briggs and Napier jointly published a table of logarithms using 10 as a base. One year later Edmund Gunter constructed the logarithmic line, which enabled the user to add and subtract lengths (representing logarithms of numbers) with a pair of dividers. About 1630, William Oughtred invented the sliding logarithmic scales, thereby helping the user add or subtract lengths rapidly (thus multiplying or dividing numbers). This device is considered the forerunner of the slide rule. By 1650, two types

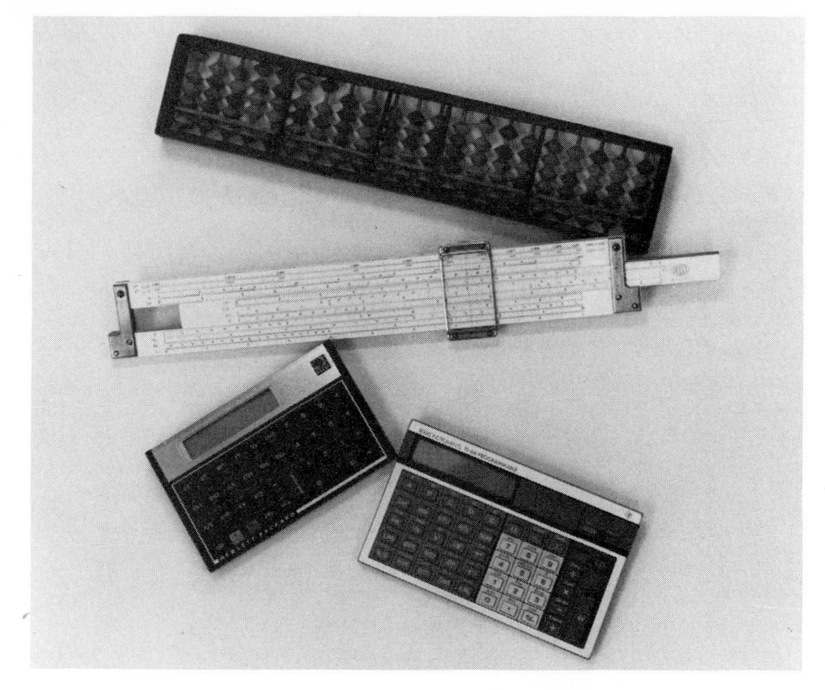

Figure 7.3
The evolution of computational devices used by engineers throughout history. Mechanical means of computing have been replaced with electrical devices.

of computers were in common use: the digital, in the form of the abacus and Pascal's machine and the analog, in the form of the slide rule.

Baron von Leibniz, a German mathematician, provided the next major digital mechanical computing device. His machine, built in 1671, could perform multiplication directly through the use of gears, instead of as a series of additions, which was required by Pascal's machine. Unfortunately, the poor machining techniques used in manufacturing Leibniz's calculator limited its reliability, so its capability was not realized for more than 50 years. Mechanical calculators, which were manufactured as recently as the 1950s, incorporated many of the features of Leibniz's calculating machine.

The first attempt at producing a computer with all the elements of a modern digital computer—namely, memory, control (sequencing of the operations), and calculating unit—was made in 1833 by Charles Babbage, an English mathematician. Babbage worked 20 years on the machine, but manufacturing limitations prevented its successful completion. Babbage attempted to incorporate one important idea that is basic to modern computers: the ability of the computer to modify the course of a calculation according to intermediate results. In the study of computer programming this concept is called *branching*. Because he developed this idea, Babbage is often referred to as the grandfather of the computer.

The first large-scale digital computer using modern principles was the Harvard Mark I, completed in 1944. This machine was electrically powered but performed arithmetic operations mechanically, much like Babbage's machine. The Mark I was extremely slow by today's standards. For example, division of two 10-digit numbers took about a minute, while today it takes a few nanoseconds (1 ns = 10^{-9} s).

The inventor of the first electronic digital computer was John V. Atanasoff of Iowa State College (now Iowa State University). Four major concepts were used in the ABC, the Atanasoff-Berry Computer, which are considered the basis for modern digital computers.

1. Electricity and electronics were the media for the computer.
2. Base 2 numbers (binary) were used.
3. Capacitors were used for memory.
4. Computation took place by direct logical action, not enumeration.

Atanasoff, working with Clifford Berry, a graduate student, built the ABC between 1939 and 1942 for the purpose of solving systems of linear equations. World War II came, and work on the ABC was suspended before it could be used extensively.

John Mauchly of Ursinus College in Pennsylvania had visited with Atanasoff in 1940 about constructing a general-purpose digital computer. Mauchly then joined with J. Presper Eckert to build the ENIAC, or the *e*lectronic *n*umerical *i*ntegrator *a*nd *c*omputer, which was completed in 1946. This machine contained over 18 000 vacuum tubes, required 130 kW of power, and occupied an area of 140 m^2. The significant contribution of this machine was a speed increase of 10 000 over that of the electromechanical computers.

For many years Mauchly and Eckert were credited as inventors of the digital computer and had a patent for the ENIAC. However, after several years of litigation, a federal judge ruled that the idea

Figure 7.4
John V. Atanasoff has recently been recognized as the inventor of the digital computer. His ABC device shown here was developed at Iowa State University. (*Iowa State University Information Service.*)

for the ENIAC was derived from John V. Atanasoff and declared the Mauchly-Eckert patent invalid. Perhaps the words of Dr. Arthur Burks, a computer historian and a member of the team that built the ENIAC, will put the matter into perspective. Dr. Burks stated that the modern digital computer revolution "began with three steps. The first was the Atanasoff-ABC Computer at Iowa State, an electronic digital computer. The second step was the ENIAC, the first general purpose computer. Third was the stored program computer."

As we have seen, it was a short time between the construction of the first large computer and today's widespread use of sophisticated high-speed equipment. New technology in the past 40 years has far surpassed the computing achievements in the 3 000-year period from the abacus to ENIAC.

7.3 Computer Systems

The two classifications of computers are analog and digital. *Analog computers* deal with continuously varying quantities in time. For example, the steering mechanism of an automobile is an analog device which responds to a rotation of the steering wheel with a proportional turn of the front wheels. The major drawback of the analog computer is that it is not easily adaptable to general problems. An analog computer is most useful for solving time-dependent ordinary differential equations which occur in the application of

Figure 7.5
Analog computers, such as the one shown here, can simulate a physical system, thus enabling the engineer to predict performance without the need of a prototype.

Figure 7.6
A large scale computing system and its peripheral devices. This system can support the engineering activities of a large organization. (*Control Data Corporation.*)

Newton's laws to physical phenomena. The electronic analog computer represents problem variables with electric voltages. Thus if the differential equations describing the physical system can be written, then a circuit can be wired in the analog computer to represent the variables and the behavior of the actual system can be simulated. Note that for each physical system to be studied, a new circuit would have to be designed and wired.

Digital computers operate with discrete quantities and are therefore quite adaptable to the processing of numbers. They are more accurate and versatile than the analog machines. In addition, the variety of digital computers available, along with their compactness and portability, make these computers more economical for most computing tasks. Digital computers are classified as "general purpose" or "special purpose." The processor used in a microwave oven is an example of a special-purpose digital computer. General-purpose digital computers will be discussed in the next section.

We will limit further discussion to digital computers because of the wider range of their applications to engineering problems today.

7.4 Digital Computers

The impact of digital computers on all aspects of our lives is due primarily to the technology of microelectronics. In the late 1960s a general-purpose chip (a thin silicon piece upon which electronic circuits are constructed) was developed. Named the Intel 4004, this chip could be programmed in a multitude of ways to perform many functions. The Intel 4004 was commonly called a *microprocessor* because it was a computer in a very small unit. The advantage of having the central processing unit (CPU) of a computer built of one segment rather than of many components as in earlier computers led to greater reliability and feasibility for mass production. Since

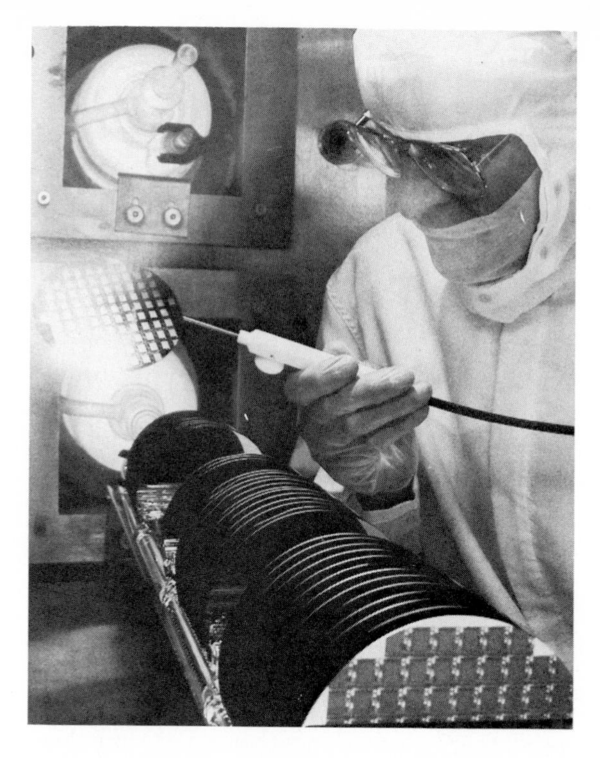

Figure 7.7
Integrated circuit chips on wafers are shown during a portion of the manufacturing process. (*Digital Equipment Corporation.*)

the development of the Intel 4004 chip, advances in integrated-circuit concepts and microfabrication techniques have led to the hundreds of types of digital computers that are available in a wide range of prices.

The operational components of a digital computer are represented in Fig. 7.8. The input and output components, the storage system, and the central processing unit (CPU) are the hardware components. The CPU is made up of the arithmetic unit and a controller. Software in the form of an operating system and programs (instructions) directs the sequence of events and computational tasks required by

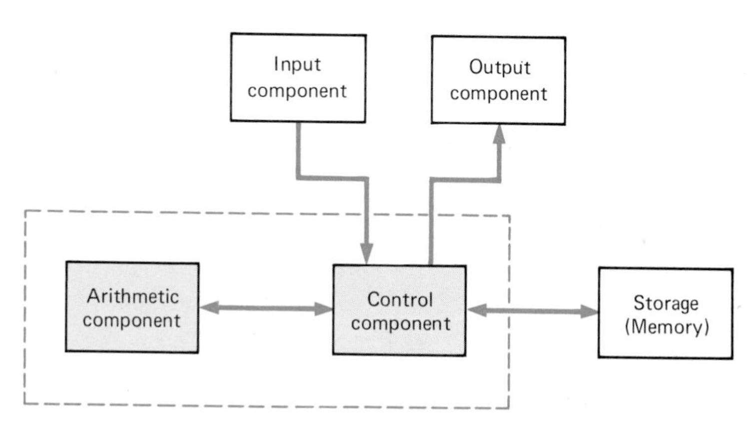

Figure 7.8
Operational components of a digital computer.

quired by the hardware. In effect, the user accesses the hardware with commands understood by the operating system, proceeds to input appropriate data, has the data processed according to programmed instructions, and directs the output.

A digital computer processes instructions at an almost incomprehensible rate. The actual rate is very difficult to predict or measure. One measure of computational speed is millions of instructions per second (MIPS). The following very simple program of instructions will illustrate the process.

```
   K = 25
   DO 10 I = 1, 300 000
   J = K + I
10 CONTINUE
```

This is a FORTRAN program that evaluates the mathematical expression J = K + I for 300 000 values of I. The control of the execution of the instructions is handled by the so-called DO loop, which repeatedly computes the expression for values of I from 1 to 300 000 in increments of 1, that is, 1, 2, 3, . . . , 300 000. If we assume one instruction is the addition, one is the check for the final value of I, and one is incrementing I, we have a total of three instructions per cycle of the loop. As a measure of computational speed, there are machines that can execute this program in 1 s (i.e., at a speed of 1 MIPS) or less. While not a practical program, the example shows how the hardware is controlled by a set of instructions (software).

We will now investigate the hardware components of a computer in greater detail.

7.4.1 Input-Output Components

The input-output (I/O) equipment provides the means for us to communicate with what amounts to a high-speed electronic moron. I/O systems are designed to move information between peripheral devices and main storage, into which it can be inserted, in which it can be retained, and from which it can be retrieved. Once information is placed in the computer in storage, then access and computation speeds are based on electronic circuitry. I/O systems serve as an interface between the user and the machine.

Consider another brief set of FORTRAN instructions to a computer.

```
10 PRINT *, 'ENTER INTEGER VALUES FOR K, I'
   READ *, K, I
   J = K + I
   PRINT *, J
   GO TO 10
```

This program was written for use with an interactive terminal.

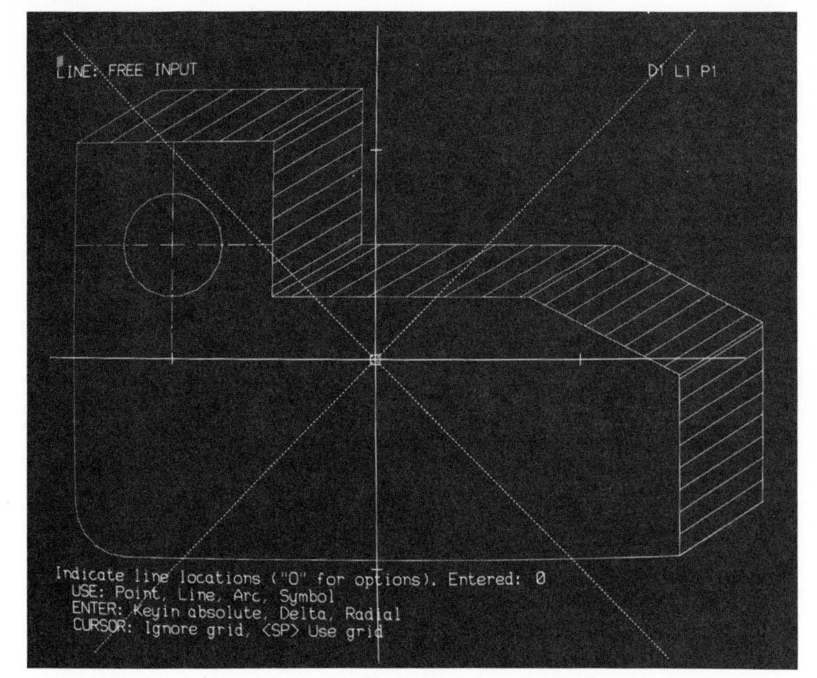

Figure 7.9
Entering graphics interactively at a terminal. The terminal interfaces with a host computer which in turn serves many additional terminals simultaneously. The host may be physically located very close to the terminals or at a remote site where the link is made through a telephone line. (*Tektronix, Inc.*)

When the program is executed, the user is "prompted" by the instruction to enter two integers through the keyboard. Once the numbers are entered, the machine adds the two numbers, prints the sum, and sends control back to the start (statement 10). Since the user is interacting with the machine, we can see that the central processing unit (CPU) will be idle most of the time; the major share of activity is in the I/O. The inefficient use of the CPU may be reduced considerably with the use of timesharing.

The development of the timesharing concept has led to more efficient use of I/O components in conjunction with high-speed CPUs. *Timesharing* means that several I/O channels have access to a single CPU and main storage system. Figure 7.10 illustrates the timesharing concept schematically. Timesharing is accomplished with the use of a communication buffer, or storage device, that compensates for the tremendous difference in the flow rates of data between I/O devices and a CPU by holding the information and feeding it at the rate at which the I/O device operates.

Requests for processing are placed through the appropriate I/O device. The requests are in turn stored in the buffer for processing by the CPU and storage component. The computer can scan all the I/O terminals in a fraction of a second. If one user has a request that may take several minutes to generate on the I/O terminal, this does not tie up the CPU and storage system, since it can process requests, in the order presented to the buffer, from any of the other I/O terminals. The high-speed CPU and storage systems are there-

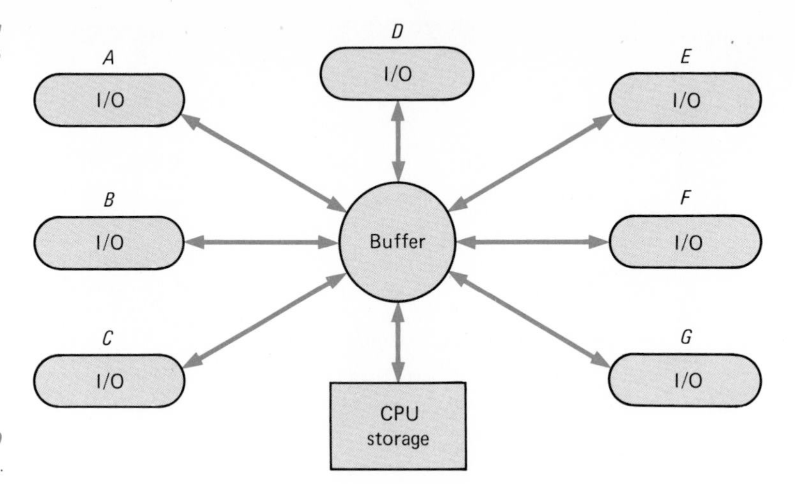

Figure 7.10
The principle of timesharing.

fore never tied up waiting for information from the slower I/O components. If systems are synchronized properly, the user will probably not realize that the computer is being used for other requests simultaneously.

We have discussed the differences in speed between the I/O device and the CPU that must be accounted for in any digital computer system. What is the process by which control, through a set of instructions, is established for a desired function?

First, a *source program* is prepared by the programmer. The source program is a set of instructions that is familiar to us because the language is natural (mathematical expressions are obvious for one thing). The two previous examples were written in FORTRAN, one of the high-level languages commonly used in engineering and science. Other widely used source program languages are BASIC and Pascal. Secondly, the source program is translated into an *object program*, which is generally in machine code. A machine code instruction is one which the central processor can interpret and execute. These codes are generally written using binary, octal, or hexadecimal number bases. In any case, most of us are not able to read or produce machine code. Thirdly, on most systems, the object program must be linked to other object programs to create an *executable program*, which is the actual set of instructions executed by the computer. All of this activity preliminary to the computations requires software of a very sophisticated nature. Once the appropriate instructions (software) are available, communication with a computer depends upon the hardware.

Each of the following devices can serve as both input and output components for a digital computer:

1. Cathode-ray tube (CRT) terminal
2. Card reader/punch
3. Magnetic disk

4. Magnetic tape
5. Video device

Each of these devices is used for input only:

1. Interactive keyboard
2. Human voice
3. Optical reader

Finally, two devices are used for output only:

1. Printer
2. Plotter

A computer terminal, which may have its own intelligence or may be hooked to a remote host computer, has a CRT for displaying input, intermediate results, and output. A keyboard enables output to be sent to the computer. This combination of hardware is called an *interactive system*. In the case of the system having a hookup to a host, there may be many other similar systems operating on a timesharing basis. The type of hookup to the host may also vary. If the connection is a direct line, transmission speeds of 19 200 or more bits per second (baud rate) can be attained. A *bit* is one binary character (0 or 1). If the connection is across a telephone line using a modem, then the baud rate may be as low as 300.

Magnetic disks are commonly used with computer systems requiring rapid access to large data sets (see Fig. 7.11). The large hard disks range from 203.2 to 406.4 mm in diameter and contain

Figure 7.11
A bank of disk drives serves as auxiliary memory for a large computing system.

Figure 7.12
An automatic card reader used as an input device to mainframe system.

thousands of tracks. The smaller floppy disks used with the personal computers range from 76.2 to 203.2 mm in diameter.

The punched card is a widely used input device. Originally developed by Hollerith in 1880, the punched card contains 80 columns, each column representing a single character. Punched cards take a long time to prepare (one to five characters per second) but can be punched without connecting to the computer. Once the cards are punched, checked, and arranged, a card reader can process several hundred cards per minute (see Fig. 7.12). Punched paper tape is essentially the same as punched cards, in that both display a pattern of holes or cuts to represent data.

The most common output medium is the printed page. Printed pages can be generated by a line printer that operates by impact printing (by means of devices that actually strike the paper to produce print), producing one line at a time with a maximum rate today of over 1 000 lines per minute (see Fig. 7.13).

Magnetic tape (Fig. 7.14) is a relatively high-speed I/O medium. It can provide highly reliable data rates of up to 680 000 characters per second. Information is stored on the tape in parallel channels along the length of the tape by magnetized spots. A tape may contain seven or more channels. Character densities of 1 600 per inch have been achieved. The tape may be retained indefinitely or erased and reused. Another form of magnetic I/O media is the magnetic ink character. Common in the banking business, the magnetic ink character is visually as well as magnetically readable. The numbers on the bottom of printed checks are examples of magnetic ink characters.

Optically readable characters are a popular means of representing data that is prepared for the computer manually. Instead of all the information being written out, a character is denoted by marking a

Figure 7.17
This microcomputer system is used in an engineering office for word processing by a secretary. (*Digital Equipment Corporation.*)

4. A printer and possibly a plotter for hard-copy output
5. Capability to link with a computer network containing other microcomputers or minicomputers and mainframes

A minicomputer will, in general, include the following:

1. A CPU and memory on the order of 1M byte or more.
2. Magnetic-disk (hard-disk) capability of essentially unlimited capacity.
3. Several ports that connect with one or more terminals. These terminals are referred to as "dumb" terminals if they do not possess any capability to operate without the host minicomputer support.
4. Additional ports that connect to other peripheral devices such as printers, plotters, and other computer systems.

The mainframe systems have additional CPU capacity, memory on the order of 2M bytes or more, and the capability of supporting from three or four terminals to over a hundred terminals, depending on the application. Today there are mainframe systems that have been classified as *supercomputers*. As the technology expands, the separate classifications of digital computers will become more difficult to define. Today's supercomputers will become tomorrow's mainframes, and today's mainframes will be classified as minicomputers.

7.5 Engineering and the Digital Computer

The techniques for engineering design and problem solving are rapidly changing because of the digital computer. Perhaps the two areas where the impact of the computer on engineering has been the greatest are (1) computer graphics and (2) the integrated-data-base concept for the total design and manufacturing cycle.

7.5.1 Computer graphics

The profound effect of computer graphics on the engineering function lies in the nature of computer graphics: the capability of displaying pictures on a screen. The mind interprets pictures much more quickly than it does words and numbers. When we look at a page containing written material and numerical quantities, we need to study it for a few moments before we are capable of discussing the material. With a picture, we quickly comprehend the image and are able to make appropriate judgments within a matter of seconds. Thus a computer-graphics system allows the engineer to function at a high level of productivity. See Fig. 7.18. The display of pictures enhances the engineer's logic and intuition and permits better and faster decisions in the overall engineering effort.

An engineer and a computer-graphics system blend into a productive working combination because the strengths of one in performing engineering tasks are a drawback of the other. We can see this by listing the characteristics of the engineer and of the computer.

First, the characteristics of a computer include:

1. Large capacity (time-independent) for storing data
2. Excellent performance on repetitive tasks
3. Good numerical analysis capability
4. Poor capability to detect significant information
5. No intuitive analysis capability
6. Rapid production of output (electronic and mechanical means)

Figure 7.18 An engineer is developing solid models at a computer graphics work station. (*Control Data Corporation.*)

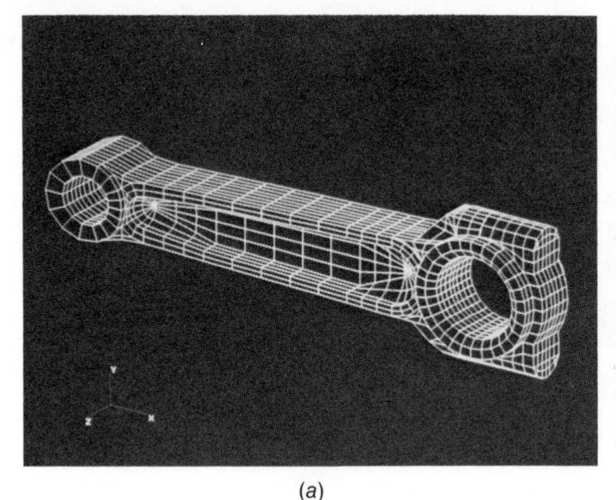

(*a*)

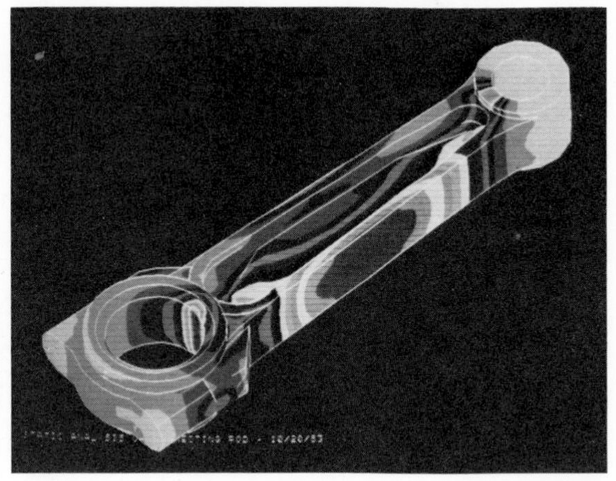

(*b*)

Figure 7.19
(*a*) A wire-frame model of a connecting rod prior to structural analysis. (*b*) The stress patterns in the connecting rod are shown as contrasting colors. Designers use this type of display to modify shapes for more efficient performance. (*Control Data Corporation.*)

On the other hand, the characteristics of the engineer include:

1. Time-dependent capacity for storing data
2. Poor tolerance for repetitive tasks
3. Poor numerical analysis capability
4. Good ability to detect significant information
5. Good intuitive analysis capability
6. Slow production of ouput (manual)

We can see that the characteristics are mutually supportive.

7.5.2 The Integrated Data Base

In modern industry, the computer supports all the design and manufacturing functions. The concept of an integrated (common) data base has brought these functions together and requires the engineer to be aware of ramifications of changes anywhere in the design and manufacturing process. Figure 7.20 is a schematic of the relationship between the various engineering functions and the integrated data base.

To illustrate how the integrated data base works, consider the shell of the space shuttle shown on a computer graphics CRT in Fig. 7.21. The performance requirement of the shuttle led to the unique concept of covering the skin of the shuttle with some 8 000 tiles to dissipate heat during reentry. Once the geometry of the surface of the shuttle had been determined, a data base of the geometry was established. Imagine this data base to be hundreds of thousands of (x, y, z) coordinates representing points on the surface. From the data base each of the 8 000 tiles was designed to conform to a specific section of the geometry of the shuttle's surface. Once designed, each tile shape was verified by checking against the data base.

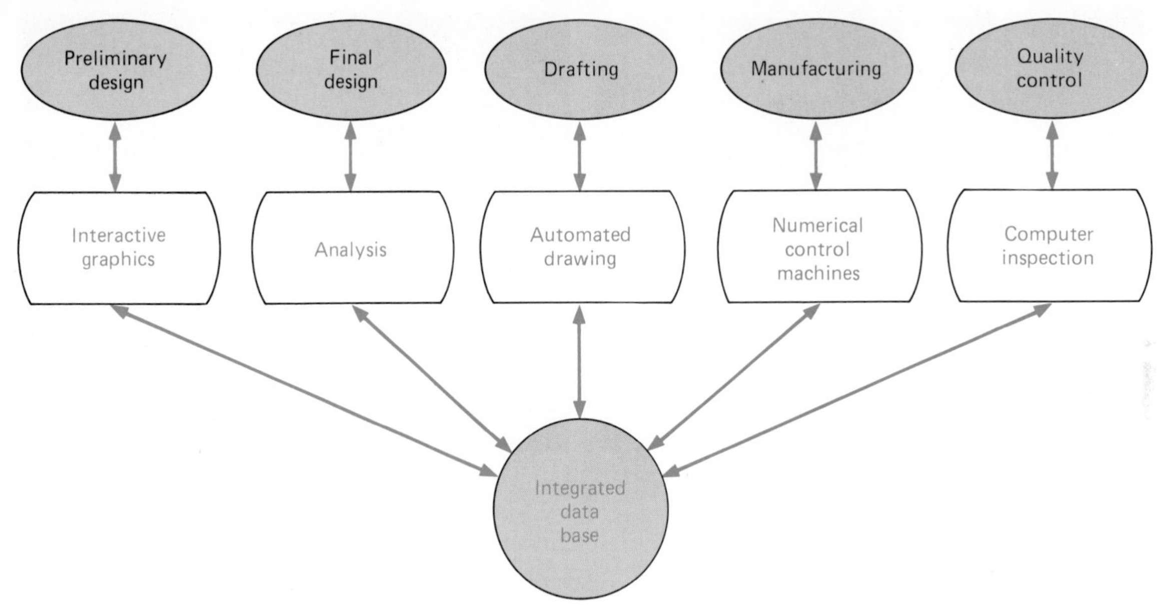

Figure 7.20
Computer graphics and integrated data base support of engineering functions.

Figure 7.21
Computer-aided design utilizing computer graphics provides the engineer with a powerful tool to produce better designs in a shorter time period. (*Control Data Corporation.*)

Aerodynamics and structural engineers worked with this same data base in determining loads and resulting stresses and deformations on the vehicle during the mission. Again using the same data base, the principles of thermodynamics were applied to determine temperature distributions on the surface during flight, particularly during reentry. The results of this analysis were available to

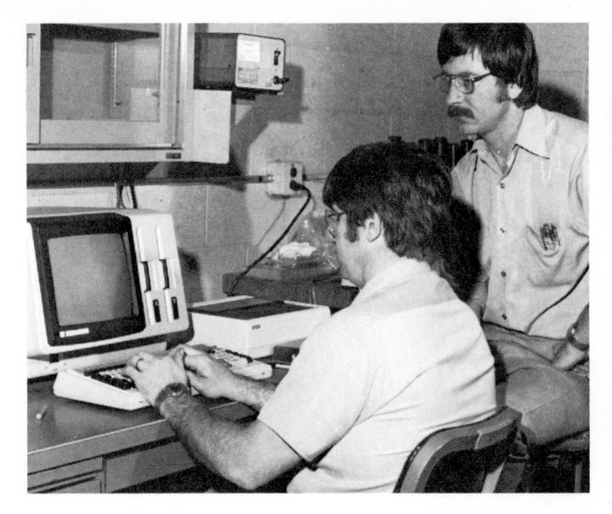

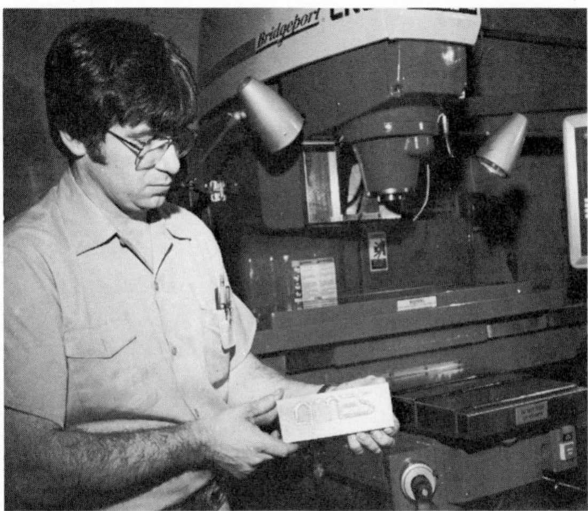

Figure 7.22
The link provided by an integrated data base ensures that the product manufactured is the same as the one designed. Here a numerically controlled (NC) machine produces parts from a computer-generated tape. (*Ames Laboratory, U.S. Department of Energy.*)

the materials engineers to assist in the selection of tile material and fastening substance.

Keep in mind that all of this engineering effort took place with computer assistance. Any changes to the data base suggested from analyses would be instantly available to all groups with access to the computer network. In many cases, the groups having access to the network, and thus to the data base, were subcontractors located in different parts of the country. This method of managing the vast data bases that exist in engineering is far superior to anything we have known in the past. Changes are essentially instantaneous and not subject to misunderstanding because of slow communication caused by the sheer volume of paperwork necessary to keep abreast of the changes.

The use of computer graphics in the design aspect of engineering is commonly called *CAD* (*computer-aided design*). For *computer-aided manufacturing*, the acronym *CAM* is used. Of particular interest to the engineering community today is the design-for-manufacturing concept, CAD/CAM. This implies an integrated data base that includes analysis based on the manufacturing capability and materials availability for the particular industry. The design engineers are thus able to ascertain whether their concepts can be produced exactly as designed. If not, changes can be made at this stage rather than after an expensive prototype is developed or badly designed products are marketed. The time from idea to marketable product is shortened considerably if manufacturing problems are ironed out before the design is finalized.

7.6 A Look Ahead

Exciting times are ahead for those of us involved in technology. Computer graphics, discussed in Sec. 7.5.1, will continue to have a great impact on engineering. We will see improved images on the

Figure 7.23
Solid modeling further enhances the engineer's visualization of three-dimensional objects. The ability to "see" the design before a costly prototype is constructed saves time and money and links to a better design. (*International Business Machines Corporation.*)

screen as solid-modeling techniques become available to more areas of industry and education. We will have access to a wider variety of data bases to increase our capability for wise engineering decisions. The shortened design-to-manufacturing time, made possible through the use of integrated data bases, will improve the quality and availability of consumer products.

From a manufacturing standpoint, the incorporation of flexible manufacturing centers will enhance products. These centers will be able to automatically produce any needed mechanical part. This means that the machines will be automatically set up, the material inserted, and the machining operation completed without any need for human intervention. Robotics will play an important role in manufacturing procedures as well as in many other tasks which may be difficult or dangerous for humans to undertake. Each of the automatic processes, whether with metalworking machinery or robots, requires a computer with appropriate software. The number of applications are enormous in this field.

Perhaps the most significant changes will occur in the actual structure of a computer. How far can we go with electronic miniaturization? The principal advantage is that the smaller the device, the more closely together the several devices that make up a computer system can be placed and the greater the speed becomes. In addition, if more and more transistors can be packed onto a single chip, fewer chips are needed for a given computer capacity. One limitation to the effort at compactness is the capability of manufacturing the super-miniature components. It has been demonstrated in the laboratory that a switching device can react in 13 ps (a picosecond is 10^{-12} second). In that time an electrical signal travels about 1 mm. A computer using these switching components would have to be contained in a package about the size of a baseball in order to have effective communication. Compare this to the ENIAC mentioned earlier.

The packaging of computer components will continue to be a challenge to engineers and scientists. Cost-effectiveness is obtained if the number of interconnectors required continues to decrease. The difficulty of cooling the components in a dense package is one limit to the size of the chips. Perhaps new materials and new methods of cooling will be developed to improve heat dissipation.

Software development will continue to be a challenge. The increased use of personal computers by the general population has created a need for "user-friendly" software in all aspects of work and leisure. This demand for software will continue to grow as we learn more about the potential of the computer.

We must learn more about how we work with computers. What do we expect as a response from the machine when we make errors? How do we "shortcut" some of the responses when we have more experience than others on a particular computer? What about our physical and mental ability to work on a computer for an extended period of time? All of these and many other questions need to be answered for effective computer design in the future.

Some 10 years ago, when computers first began to have impact on the general public, many said the computer would cause unemployment and a depressed economic situation. Today, with the experience of the past 10 years to guide us, we see that the effect is just the opposite. More opportunities are opening in the workplace because of computer technology. It appears that the "information revolution" is well underway, and like so many technological developments in the past, it will improve our lifestyle significantly.

Figure 7.24
Computer use now extends into all levels of education. It is important that one learn both the potential and the limitations of a computer in order to be an effective and productive user. (*Digital Equipment Corporation.*)

Problems

7.1 Write a brief summary of the contribution to the development of the modern computer by each of the following:

(a) Edmund Gunter
(b) Howard Aiken
(c) Herman Hollerith
(d) John Von Neumann
(e) J. P. Eckert
(f) John Mauchly
(g) J. V. Atanasoff

7.2 Write a short report on an application of each of the following classes of computers in the engineering field of your choice.

(a) Large-scale digital computer
(b) Minicomputer
(c) Microcomputer

7.3 Write an effective algorithm for the following:

(a) Cooking a hard-boiled egg
(b) Multiplying two integer numbers of two digits each
(c) Finding the roots of the equation $Ax^2 + Bx = C$
(d) Computing a worker's net pay knowing the hourly rate and hours worked and assuming appropriate deductions

7.4 One of the classic algorithms was developed about 250 B.C. by Eratosthenes. Called the Sieve of Eratosthenes, the algorithm can find all the prime numbers between 1 and some specified integer N. Find documentation of the algorithm and use it to determine the prime numbers

(a) Between 1 and 18
(b) Between 26 and 54

7.5 Using the engineering method of problem solving, develop a procedure to determine whether any three given lengths can constitute a right triangle.

7.6 Using the engineering method of problem solving, determine the solutions to the following equations to four significant figures. Set up an algorithm for the computational procedure.

(a) $x = \sin x + 0.5$
(b) $x \log x - 1 = 0$

7.7 From a manufacturer or supplier, obtain an owner's manual or other descriptive information for one of the following devices. Make a brief oral report to the class emphasizing the functions and operation of the microprocessor or microcomputer contained in the device.

(a) A programmable microwave oven
(b) A video game
(c) An automatic bank teller
(d) An electronic timer for television
(e) An auto fuel-injection control system
(f) A temperature controller for a home or industrial furnace

7.8 For a computer-graphics system on your campus, write a brief report on any or all of the following:

(a) Possible input devices
(b) Type of display device and its characteristics (color, resolution, etc.)
(c) Possible output devices
(d) A list of the software packages that may be used by your discipline in coursework

8 Preparation for Computer Solutions

8.1 Introduction

In using the engineering method of problem solution, you oftentimes reach a point at which a decision must be made about which computational device will be used. In many cases, you simply use a calculator to process the numerical values, or you leave the answer in algebraic form. Problems of a sophisticated nature or ones involving lengthy, repetitive calculations should be considered for computer solution, but the decision to use a computer should not be made hastily. Because the digital computer performs calculations very quickly, a large volume of expensive, unwanted, or incorrect computations may be generated unless a well-thought-out program is developed.

In Chap. 2 the engineering method of problem solving was described in terms of the following six-step procedure:

1. Recognize and understand the problem.
2. Accumulate facts.
3. Select appropriate theory or principle.
4. Make necessary assumptions.
5. Solve the problem.
6. Verify and check results.

The mechanics of computation fits into this procedure at step 5. Before deciding on the method of computation and the equipment to be used, you must proceed through the several preliminary steps: You must define the problem, outline all the knowns and unknowns, and then develop the mathematical statements to be solved or calculated.

At this point—step 5—you must determine how to perform the calculations necessary to solve the problem. That is, you must proceed as though there were a problem within a problem. And you must decide what type of computational equipment to use. Let's suppose for purposes of discussion that the computation is complex

Figure 8.1
A modern digital computing system. (*Digital Equipment Corporation*)

and iterative in nature and that something more than a simple calculator is needed. Computing equipment that could be used includes programmable calculators, personal computers, minicomputers, and mainframe machines. Availability of a particular computer and relative computing speed are a couple of things to consider in making a good choice.

Once the decision has been made, a three-phase procedure within step 5 is recommended for obtaining a successful solution:

5*a*. Construct a flowchart for the solution procedure.

5*b*. Write the program in a computer language appropriate for the machine you will use.

5*c*. Follow the correct procedures for inputting your program and data into the computer and for retrieving and interpreting the output (results).

Computers and calculators perform only operations that they are directed to do. Directions are given to the computer via the program, which is a series of steps, a routine, that must be sequentially followed to solve the problem. Devising the plan is called *programming*. The one who prepares the sequence of instructions, without necessarily converting it into a detailed code is called a *programmer*. In the substeps just given, 5*a* can be completed without regard to the machine to be used, but 5*b* and 5*c* must take into account the computer and peripheral equipment. We will devote the remainder of this chapter therefore to step 5*a*, the construction of flowcharts. The details of the programming language and operation of the computer equipment are beyond the scope of this text. For further information about programming, consult one of the many texts that are available.

Flowcharting

A *flowchart* is a graphical representation of an *algorithm*, which is a procedure for solving a problem. In developing an algorithm and then a flowchart, it is advantageous to think in terms of the big picture before focusing on the details of each subsection. For example, when designing a house, an architect must first plan where the kitchen, bathroom, bedrooms, and other rooms are going to be located before specifying where electric, water, and sewer lines should be placed. Likewise, an algorithm is designed by working out large blocks to assure that the global logic is satisfied before deciding what detailed procedure should be used within each block. Once the algorithm is established, a flowchart can be drawn.

A set of graphical symbols is used to describe each step of the flowchart. Although there are many symbols in general use, we will define a small subset that is somewhat generic in nature; that is, it does not denote any particular device or method for performing the operation. For example, a general input-output symbol is used that does not suggest a computer card, magnetic tape, or specific document. The symbol says only that communication with the computer should occur using whatever device may be available. The next step, that of coding the program in a particular language for a given machine, will require the programmer to be more specific.

The five symbols that we will use are shown in Fig. 8.2. They are the symbols that denote input-output, process instruction, conditional test (decision), start-stop, and connector. These symbols are connected by lines (flow lines) with arrows to indicate the flow direction. This is illustrated in the simple sequence in Fig. 8.3. In this case, the flow begins at the start position and goes sequentially from operation to operation until the stop symbol is reached. No decisions are made and no step or series of steps is repeated. Dashed flow lines in some flowcharts simply mean repeated symbols have been omitted.

Figure 8.2
Flowchart symbols.

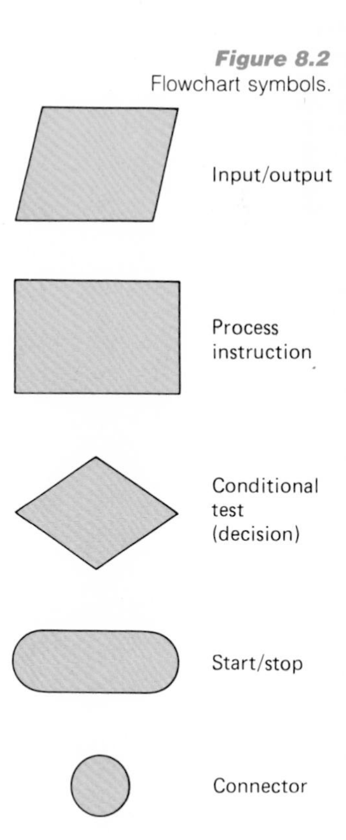

The fundamental decision structure is illustrated in Fig. 8.4. It contains a decision symbol that asks a question with a yes/no (true/false) answer or states a condition with two possible outcomes. Thus, based on the outcome of a decision, one of the two sets of operations will be performed. Each branch of the decision structure may contain as many operations as are necessary. There might be only one operation in a branch or several operations; there might even be no operations in one of the branches. With this structure, flow proceeds through one or the other of the branches and continues on into a later section of the flowchart. Therefore during a single pass through the structure, one of the branches is not used.

Another useful combination of symbols denotes the looping structure. Here a step or series of steps is performed repeatedly until some condition is satisfied, at which time the next step after the loop is executed. Two common looping structures are presented,

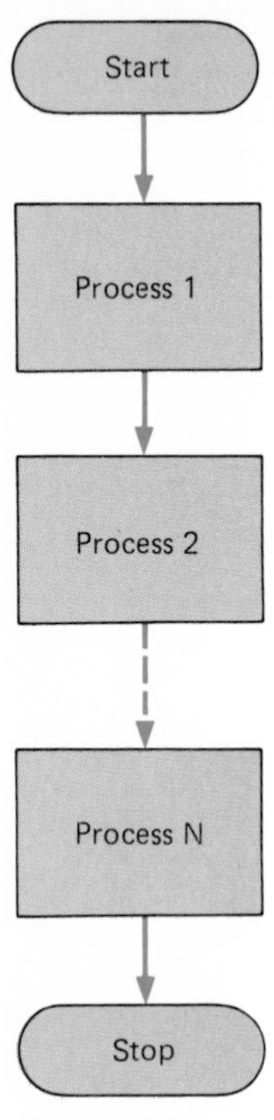

Figure 8.3
Simple sequence.

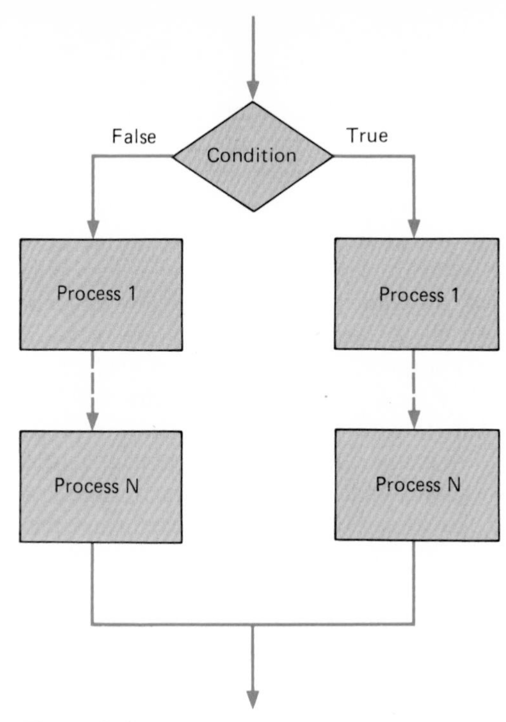

Figure 8.4
Decision structure.

one where the conditional test is performed as the last step of the structure and one where the test is the first step of the structure.

The looping structure where the conditional test is the last step is illustrated in Fig. 8.5. One or more processes are placed on the forward section of the structure. The conditional test could be reversed; that is, the true and false flow lines could be interchanged depending on the nature of the condition to be tested. There may be no need for the process block in the reverse loop, depending on the action in the process blocks of the forward section. Include the reverse process block or not as your logic dictates. Frequently the reverse-loop block performs the action of a counter. Operations such as $X = X + 1$ or $Z = Z + 5$ might appear there. These are not algebraic equations since they are clearly not mathematically correct. They are instructions to replace the current value of X by a new value 1 greater or to replace Z by $Z + 5$. Therefore they can count the number of times through the loop, as does X, or can increment a variable by a constant, as in the case of Z. Negative increments are also possible so you can count backwards, or decrement a variable.

Figure 8.6 shows the looping structure where the conditional test is performed first. Again the true and false branches can be reversed

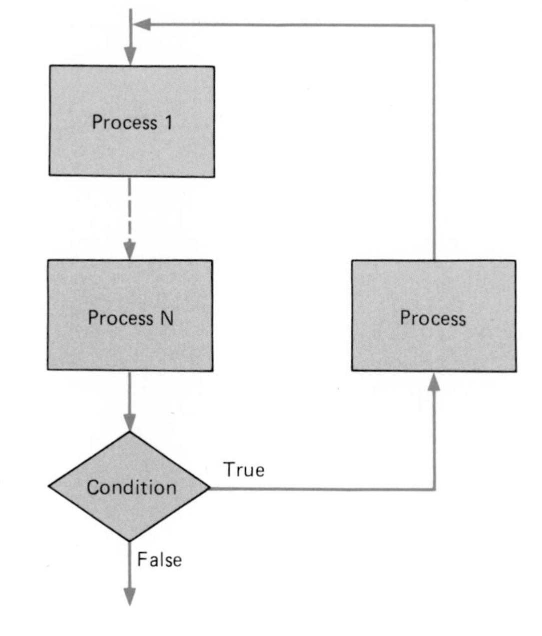

Figure 8.5
Looping structure with conditional test last.

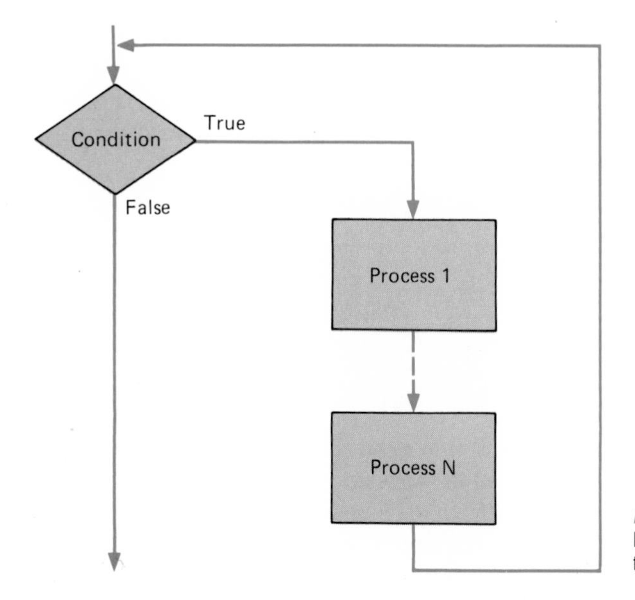

Figure 8.6
Looping structure with conditional test first.

to match the chosen conditional test. As many process blocks as desired may be used. One of them could be a counter or incrementing block.

Several decision structures or looping structures can be combined by a method called *nesting*. In this way one or more loops can be contained within a loop. Similarly, a decision structure can be placed within another decision structure. Figure 8.7 shows an example of

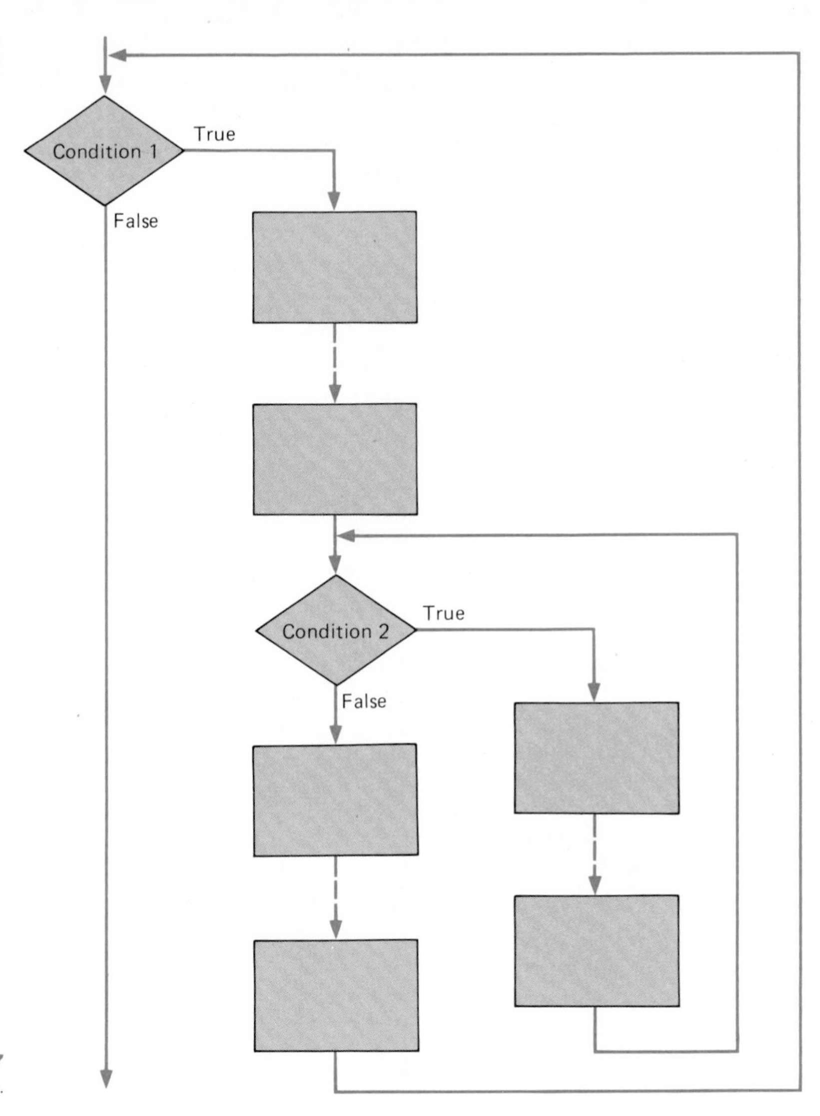

Figure 8.7
Nested loop.

how a nested loop might appear. The inner loop is performed until condition 2 is satisfied; then flow returns to the decision block for condition 1. In each pass through the outer loop, the inner loop will be completed until condition 2 is satisfied. Eventually condition 1 will be satisfied and flow will pass to the next part of the flowchart.

Examples of nested decision structures and other combinations of the various structures just discussed can be seen in the examples that follow.

Example problem 8.1 Construct a flowchart for calculating the sum of the squares of the even integers from N1 to N2.

Procedure For purposes of this example, N1 and N2 will be restricted to even integers only, and N2 > N1. We can use the sequence structure shown in Fig. 8.8. This flowchart will result in a variable called SUM as the desired value; SUM is then output. Because of the repetitive nature of the steps, it is far more convenient to use a looping structure. This form of flowchart for our problem is found in Fig. 8.9.

Study Fig. 8.9 carefully; several important flowcharting concepts are introduced there. First, the flowchart is useful for any pair N1, N2 as long as each is an even integer and N2 > N1. (The flowchart is also useful for N1 and N2 as odd integers, although that is not the specification in this problem.) Second, there are two variables, SUM and Y, that will take on numerous values. They must be initialized outside of the looping structure. If the process block con-

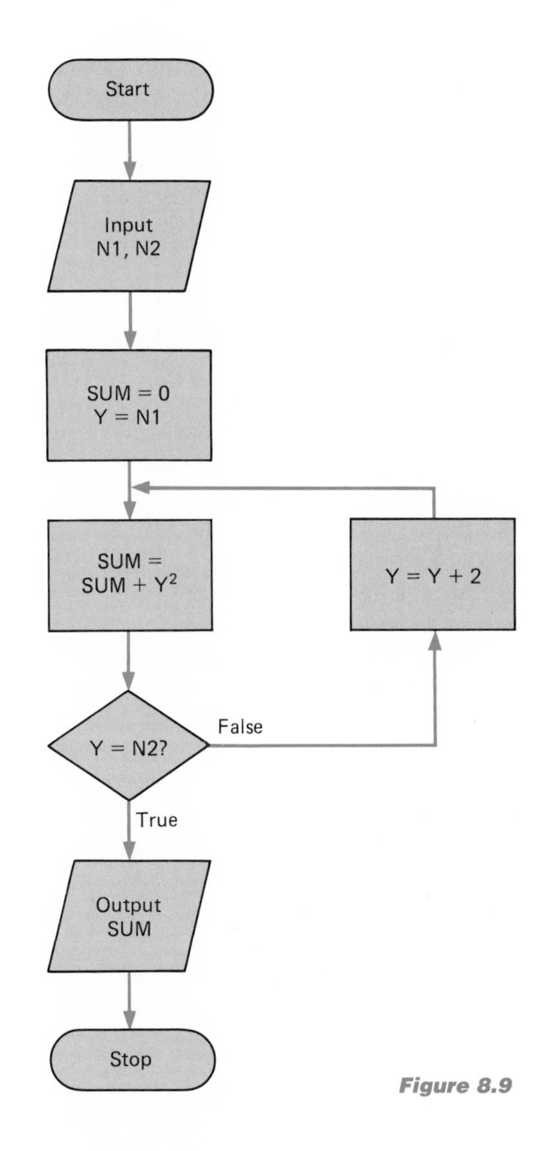

Figure 8.9

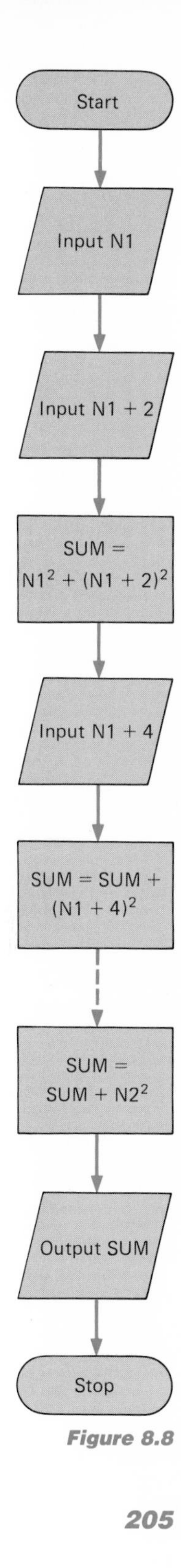

Figure 8.8

taining SUM = 0 and Y = N1 were inside the loop, SUM and Y would be reset to their initial values each time through the loop and the decision block (Y = N2?) could never be satisfied, thereby creating an infinite loop.

The output block is also outside the loop, which provides a single output value. It would be sensible to include N1 and N2 in the output block also so the user would have a more complete statement of the problem. The "Output SUM" block could be placed just prior to the decision block, in which case a running sum would be output, with the last value being the desired total.

The difference between initializing a variable and inputting a variable is an important one. As a general rule, variables that must have initial values but whose values will not change from one use of the flowchart to another (one run of the resulting program to another) should be placed in a process block. Variables that one wishes to change from one run to another should be placed in an input block. This provides the user with the necessary flexibility of using the flowchart (or program) without having to input the variables that do not change from run to run.

Example problem 8.2 Draw a flowchart that will calculate the future sum of a principal (an amount of money) for a given interest rate and number of interest periods. Allow the user to decide if simple or compound interest is to be used and to compute as many future sums as desired.

Procedure One solution is given in Fig. 8.10. The user is asked to input T = 0 for simple interest and T = 1 for compound interest. A decision block checks on P before performing any further calculation. If P < 0 (a value not expected to be used), the process terminates. Thus, using a unique value for one of the input variables is one method of terminating processing. A logical alternative would be to construct a counter and check to see if a specified number of variable sets has been reached.

The connector symbol has been used to avoid drawing a long flow line in this example. A letter or number (the letter A was used in this case) is placed in the symbols to uniquely define the pair of symbols that are meant to be connected. Connector symbols are also used when a flowchart occupies more than one page and flow lines cannot physically connect portions of the flowchart.

Example problem 8.3 The sine of an angle can be approximately calculated from the following series expansion.

$$\sin x \cong \sum_{i=1}^{N} (-1)^{i+1}\left[\frac{x^{2i-1}}{(2i-1)!}\right]$$

$$= x - \frac{x^3}{3!} + \frac{x^5}{5!} - \frac{x^7}{7!} + \cdots + (-1)^{N+1}\left[\frac{x^{2N-1}}{(2N-1)!}\right]$$

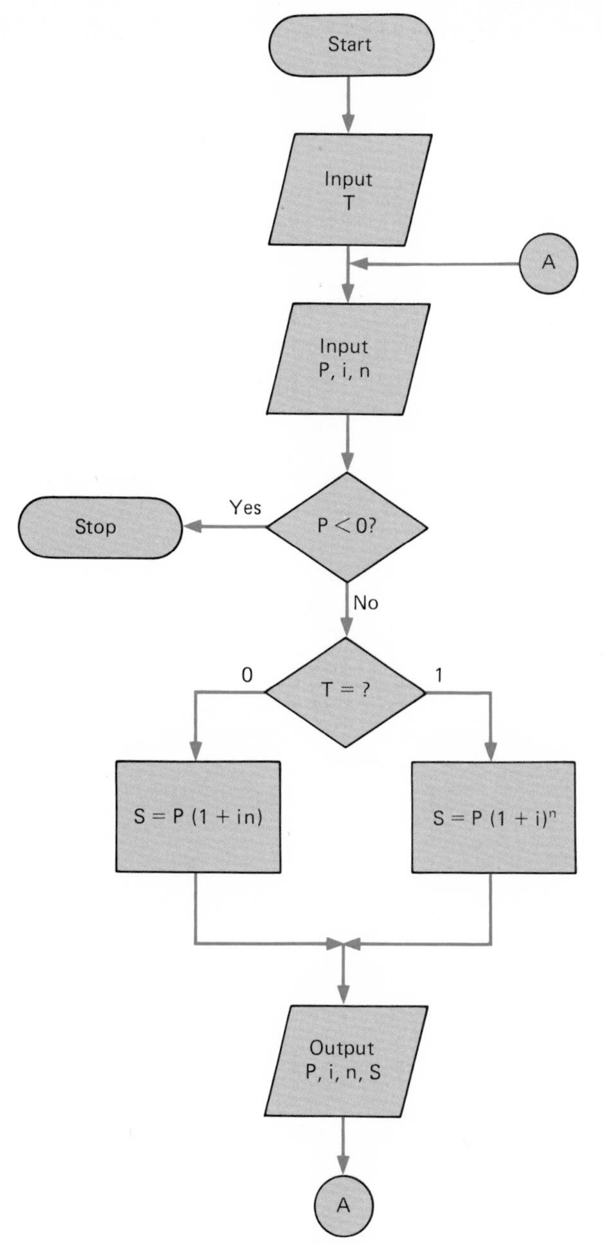

Figure 8.10

where x is the angle in radians. The degree of accuracy is determined by the number of terms in the series that are summed for a given value of x. Prepare a flowchart to calculate the sine of an angle of P degrees and cease the summation when the last term in the series calculated has a magnitude less than 10^{-7}. Of course, an exact answer for the sine of the angle would require the summation of an infinite number of terms. Assume that the computer used does not have the capability of computing a factorial directly, so the flowchart must indicate that procedure.

Procedure One possible flowchart is shown in Fig. 8.11. We will discuss several features of this flowchart, after which you should evaluate on paper the first three or four terms of the series expansion to make sure that you understand the problem. Note that the general term has been used in the loop and that specific values of each variable are calculated in order to produce the required term each time through the loop.

1. The magnitude of the quantity controlling the number of terms summed is called ERR. It is input so that a magnitude of other than 10^{-7} can be used in a future run.

2. i denotes the summation variable and is initialized as 1.

3. M represents $(2i - 1)!$ and is initialized as 1 (its value in the first term of the series).

4. SUM is the accumulated value of the series as each term is added to the previous total. It is initialized as 0.

5. The angle is input in degrees and then immediately converted to radians by multiplying by $\pi/180$.

6. TERM is the value of each term beginning with $(+1)\{x^{2(1)-1}/[2(1)-1]!\}$, or simply x. The factor $(-1)^{i+1}$ causes TERM to alternate signs.

7. SUM is equal to $0 + x$ the first time through the procedure.

8. The absolute value of TERM is now checked against the control value ERR to see if computations should cease. Note that the absolute value must be used because of the alternating signs. If a "no" answer is received to this conditional test, the appropriate incrementing of the variable must be undertaken.

9. i is increased by 1. This time through, i becomes 2, since it was initialized as 1.

10. M becomes $(1)(2 \times 2 - 2)(2 \times 2 - 1) = 1 \times 2 \times 3 = 3!$

11. Following the flowchart directions, we now return to the evaluation of TERM with the new value of i.

$$\text{TERM} = \frac{(-1)^{2-1}x^{2(2)-1}}{3!} = -\frac{x^3}{3!}$$

12. SUM then becomes $x - x^3/3!$. TERM is again tested against ERR.

13. The process repeats until the magnitude of the last term is less than ERR (10^{-7} in this case) and the machine is instructed to report the value of x, $\sin x$, and the value of the last term to make certain the standard of accuracy has been attained.

You should check several terms for a given value of x and then repeat the process for different angles. You will note that as the size of the angle varies, the number of terms required to achieve the standard of accuracy also varies. For example, when using a value of $P = 5°$, the third term of the series is about $4(10^{-8})$, much less than the 10^{-7} requirement. But for $P = 80°$, the third term is approximately $4.4(10^{-2})$. When $P = 80°$, seven terms of the series are required before the magnitude of the last term becomes less than 10^{-7}.

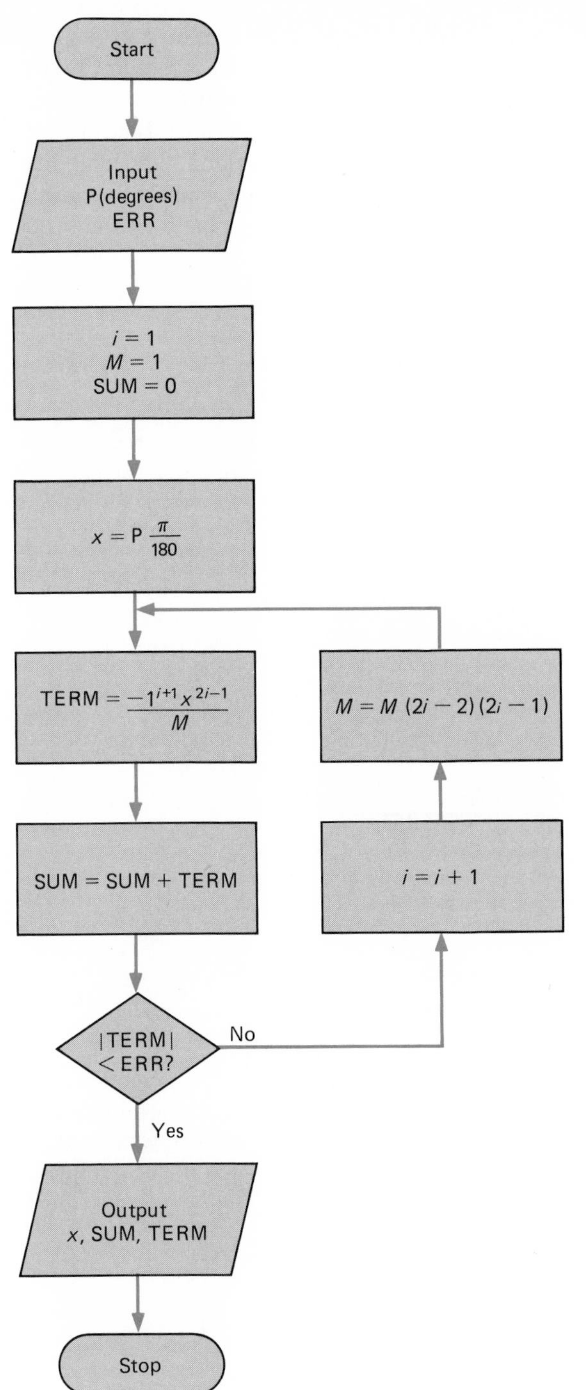
Start
Input
P(degrees)
ERR
$i = 1$
$M = 1$
SUM = 0
$x = P \frac{\pi}{180}$
$\text{TERM} = \frac{-1^{i+1} x^{2i-1}}{M}$
SUM = SUM + TERM
|TERM| < ERR?
No
Yes
$M = M\,(2i - 2)(2i - 1)$
$i = i + 1$
Output
x, SUM, TERM
Stop

Figure 8.11

The preceding examples should give you some insight into the construction of a flowchart as a prelude to writing a computer program for your particular computer.

The mechanisms for calculating, testing, incrementing, looping, etc., vary with the computational device and the programming language. A flowchart, however, should be valid for all computer systems, because it graphically portrays the steps that must be completed to solve the problem.

Problems

8.1 You have available a list of the heights of the 8 000 first-year students at your university. Draw a flowchart that will count the number of students whose height h is in each of the following categories. Heights must be input one at a time, and the height ranges and counts should be output.

$h < 5'$
$5' \leqq h < 5'6''$
$5'6'' \leqq h < 6'$
$6' \leqq h < 6'6''$
$h \geqq 6'6''$

8.2 Assume that the sales tax structure for your state is (S = sale value)

Price range	Tax
$0 < S \leqq \$0.25$	1¢
$\$0.25 < S \leqq \$0.50$	2¢
$\$0.50 < S \leqq \$0.75$	3¢
$\$0.75 < S \leqq \$1.00$	4¢
$\$1.00 < S \leqq \$1.25$	5¢

and so forth.

Produce a flowchart that will compute the appropriate sales tax and total price for any sale less than or equal to $1 000.

8.3 Draw a flowchart to calculate the areas of N circles whose smallest radius R is x cm and whose radii increase in increments of z cm. A looping structure is required. Provide the user with values for both the radius and the area of each circle. Allow the user to specify the number of circles, smallest radius, and radius increment.

8.4 The buckling load for a long column is given by Euler's equation:

$$F_B = \frac{n\pi^2 EA}{(L/r)^2}$$

where F_B = buckling load
E = modulus of elasticity
A = cross-sectional area
L = length of column
r = least radius of gyration

The factor n depends on the end conditions of column as follows: both ends hinged, $n = 1$; both ends fixed, $n = 4$; one end fixed, the other hinged, $n = 2$.

Draw a flowchart to compute the buckling load based on input values of E, A, L, r, and end conditions (H-H, F-F, or F-H).

8.5 Draw a flowchart to compute the approximate value of

$$\cos x \cong 1 - \frac{x^2}{2!} + \frac{x^4}{4!} + \cdots + (-1)^{N+1} \frac{x^{2N-2}}{(2N-2)!}$$

where x is the angle in radians. Design the flowchart so that the user specifies an angle in degrees and the number of terms desired and receives as output the angle in degrees and radians, the approximate value of cos x, and the magnitude of the first term in the series not included in the approximation. Explicitly compute the factorials.

8.6 A regular polygon of n sides can be inscribed in a circle of radius R. The area and perimeter of the polygon are given by

$$\text{Area} = \tfrac{1}{2} nR^2 \sin \frac{360°}{n}$$

$$\text{Perimeter} = 2\pi R \sin \frac{180°}{n}$$

Draw a flowchart that will compute the area and perimeter of a series of polygons beginning with $n = 3$ ($\Delta n = 1$) for a specified R. Processing should terminate either when the polygon area is within Z percent of the area of the associated circle or when the polygon perimeter is within Y percent of the perimeter of the circle. Y and Z must be input variables. There should be an indication of which condition resulted in the process termination.

8.7 The series for the hyperbolic sine of a number x is given by

$$\sinh x = x + \frac{x^3}{3!} + \frac{x^5}{5!} + \frac{x^7}{7!} + \cdots \qquad (-\infty < x < \infty)$$

By using a looping structure based on the general term of the series and by explicitly computing factorials, draw a flowchart to calculate the value of sinh x such that the last term included in the series has a magnitude less than a user-defined number. Provide output giving x, sinh x, and an indication that the last term meets the user's specifications.

8.8 For $|z| < 1$, the arcsine of z is given by

$$\sin^{-1} z = z + \left(\frac{1}{2}\right)\left(\frac{z^3}{3}\right) + \left(\frac{1 \times 3}{2 \times 4}\right)\left(\frac{z^5}{5}\right) + \left(\frac{1 \times 3 \times 5}{2 \times 4 \times 6}\right)\left(\frac{z^7}{7}\right) + \cdots$$

Draw a flowchart to calculate $\sin^{-1} z$ for a user-specified value of z and the maximum magnitude of the first term to be dropped. Output z, $\sin^{-1} z$, the value of the first term dropped, and the number of terms included in the approximation.

8.9 Draw a flowchart that will produce a table of temperature in kelvins vs. temperature in degrees Fahrenheit for temperatures from −50 to 2 000°F in increments of T°F as specified by the user. Assume that T is an even divisor of 50 and 2 000.

8.10 The sum of a sinking fund, S_n, that results from an annual investment of A dollars at an interest rate of i (decimal) for n years is given by

$$S_n = A\left[\frac{(1 + i)^n - 1}{i}\right]$$

Draw a flowchart that will produce a series of tables of S_n vs. n ($0 \leqq n \leqq 10$) for $A = A_1$, A_2, and A_3 and $i = i_1$ and i_2. Be sure each table is identified by the appropriate annual investment and interest rate.

8.11 Draw a flowchart that will input the exam scores (one score at a time) of 100 students taking an engineering-problems course; count the number of grades (based on 100 percent) in the ranges 90 to 100, 80 to 89, 70 to 79, 60 to 69, and <60; and output the results as a table of grade ranges vs. count within the range. Also compute and output the average score on the exam.

8.12 Draw a flowchart that will compute a student's grade point average for the term given the number of courses, number of credits for each course, and the letter grade earned in each course. Assume A, B, C, D, and F grades in a 4-point system with A = 4, B = 3, C = 2, D = 1, and F = 0.

8.13 The current i flowing in a series circuit consisting of a resistor (resistance = R), inductor (inductance = L), and capacitor (capacitance = C) but having no voltage source is given by the following equations. Assume that the capacitor has an initial charge of Q at time $t = 0$.

$$i = \frac{2\pi f^2 Q e^{-Rt/2L}}{f_d} \sin 2\pi f_d t$$

$$\text{where } f = \frac{1}{2\pi}\sqrt{\frac{1}{LC}}$$

$$f_d = \frac{1}{2\pi}\sqrt{\frac{1}{LC} - \frac{R^2}{4L}}$$

Draw a flowchart to compute the current flow i for $0 \leqq t \leqq 1$ s with $\Delta t = 0.05$ s. Allow the user to specify Q, R, L, and C. Produce a table of current vs. time.

8.14 In the example problem shown in Fig. 2.4, the cost C of the tank as a function of radius R was found to be

$$C = \frac{300\ 000}{R} + 400\pi R^2$$

Using an interactive technique, draw a flowchart to find and output the radius that is most economical and the minimum cost for the project.

8.15 The method of interval halving can be used to find the root of a function written as $f(x) = 0$ (see Fig. 8.12). The single real root must be between $x = a$ and $x = b$; that is, $f(a)$ and $f(b)$ must have opposite signs. If the midpoint of a to b is found as $x_m = (a + b)/2$, then $f(x_m)$ will be of the same sign as either $f(a)$ or $f(b)$. If $f(x_m)$ is of the same sign as $f(a)$, the root lies in the interval $(a + b)/2$ to b. If it has the same sign as $f(b)$, the root is in the interval a to $(a + b)/2$. One then discards the half interval where the root

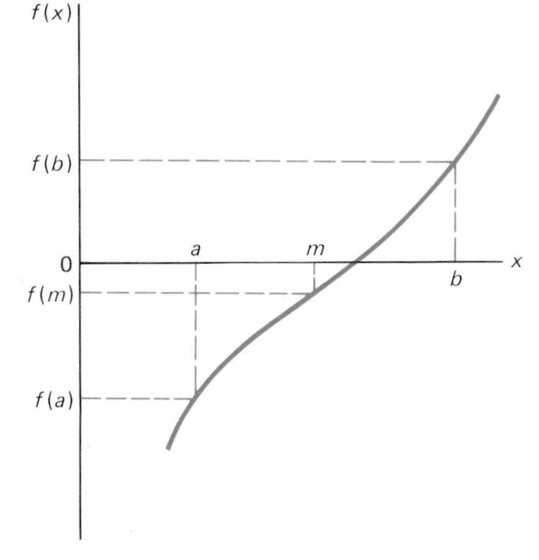

Figure 8.12

cannot be found and repeats the process until $|b - a|$ is less than some desired value.

Draw a flowchart to determine the root of $x^2 = 5x - 2$ between 2 and 6. Assure that the value is accurate to three significant figures. Output the root and a count of the number of trials that were needed.

8.16 Draw a flowchart to solve the quadratic equation $Ax^2 + Bx + C = 0$. The user may specify any (positive, negative, or zero) values of constants A, B, and C. Output the equation coefficients as well as the roots calculated.

8.17 The tangent of an angle $|x| < \pi/2$ can be approximately computed from the series

$$\tan x \cong x + \frac{x^3}{3} + \frac{2x^5}{15} + \frac{17x^7}{315} + \cdots + \frac{2^{2n}(2^{2n} - 1)B_n x^{2n-1}}{(2n)!}$$

where B_n are the Bernoulli numbers ($B_1 = 1/6$, $B_2 = 1/30$, $B_3 = 1/42$, $B_4 = 1/30$, $B_5 = 5/66$, $B_6 = 691/2730$, $B_7 = 7/6$, $B_8 = 3617/510, \ldots$).

Draw a flowchart to calculate an approximation for tan x for a given angle in degrees and number of terms less than or equal to 7. Give the user an indication of the accuracy of the approximation by providing the value of the first term not included in the approximation.

8.18 An investment of P dollars produces an annual income of A dollars for n years. The rate of return i (equivalent interest rate per year as a decimal) can be found from the equation

$$P = A\left[\frac{(1 + i)^n - 1}{i(1 + i)^n}\right]$$

Using an iterative scheme—interval halving, for example—draw a flowchart to compute the rate of return for given values of P, A, and n. Provide output of P, A, n, and i as an annual percent.

8.19 The method of least squares gives the slope and intercept for a linear fit $y = mx + b$ to a set of data as

$$m = \frac{n(\Sigma x_i y_i) + (\Sigma x_i)(\Sigma y_i)}{n(\Sigma x_i^2) - (\Sigma x_i)^2}$$

$$b = \frac{\Sigma y_i - m(\Sigma x_i)}{n}$$

where n is the number of data pairs (x_i, y_i) and the summations are over all data pairs.

Draw a flowchart to read in n (user-defined) data pairs, one pair at a time, and to compute and output m and b.

8.20 Draw a flowchart that will read in a set of n positive integers, one at a time; reorder them largest to smallest; and output the ordered list, one at a time.

Table 9.1

$\overline{F}$ = 700.00 N at 30°	Vector quantity showing both magnitude and direction
$\overline{F}$ = 700.00 N (slope 1:3) F_x = 606.2 N	Scalar component in the x direction
F_y = 350.0 N	Scalar component in the y direction
$\overline{F}_x$ = 606.2 N→	Vector component in the x direction
$\overline{F}_y$ = 350.0 N ↑	Vector component in the y direction
$\overline{F}$ = 606.2$\hat{i}$ + 350.0$\hat{j}$	Vector in terms of unit vectors

$$|\overline{R}| = (R_x^2 + R_y^2)^{0.5} \qquad 9.7$$

$$\theta = \tan^{-1}\frac{R_y}{R_x}$$

where $\overline{R}$ represents the resultant force. The arrows with the summation signs in Eqs. (9.6) indicate the conventional positive directions for the scalar components.

Example problem 9.2 Given the two-dimensional, concurrent, coplanar vector system illustrated in Fig. 9.8, determine the resultant vector $\overline{R}$.

Solution

1. Construct a table listing each vector and its components, assuming conventional positive x and y directions.

Vector	x component (→)	y component (↑)
$\overline{F_1}$	$F_1 \cos \theta_1$ = 95.0(cos 30) = +82.27 N	$F_1 \sin \theta_1$ = 95.0(sin 30) = +47.50 N
$\overline{F_2}$	$F_2 \cos \theta_2$ = 70.0(cos 90) = 0	$F_2 \sin \theta_2$ = 70.0(sin 90) = +70.00 N
$\overline{F_3}$	$F_3 \cos \theta_3$ = 62.0(cos 175) = −61.76 N	$F_3 \sin \theta_3$ = 62.0(sin 175) = +5.404 N
$\overline{F_4}$	$F_4 \cos \theta_4$ = 50.0(cos 240) = −25.00 N	$F_4 \sin \theta_4$ = 50.0(sin 240) = −43.30 N
$\overline{F_5}$	$F_5 \cos \theta_5$ = 115(cos 315) = +81.32 N	$F_5 \sin \theta_5$ = 115(sin 315) = −81.32 N

2. Use Eqs. (9.6) to determine the scalar components of $\overline{R}$:

$$R_x = \rightarrow \sum F_x = +82.27 - 61.76 - 25.00 + 81.32$$
$$= 76.83 \text{ N}$$

$$R_y = \uparrow \sum F_y = +47.50 + 70.00 + 5.40 - 43.30 - 81.32$$
$$= -1.72 \text{ N}$$

3. Determine the vector $\overline{R}$ using Eqs. (9.7). See Fig. 9.9.

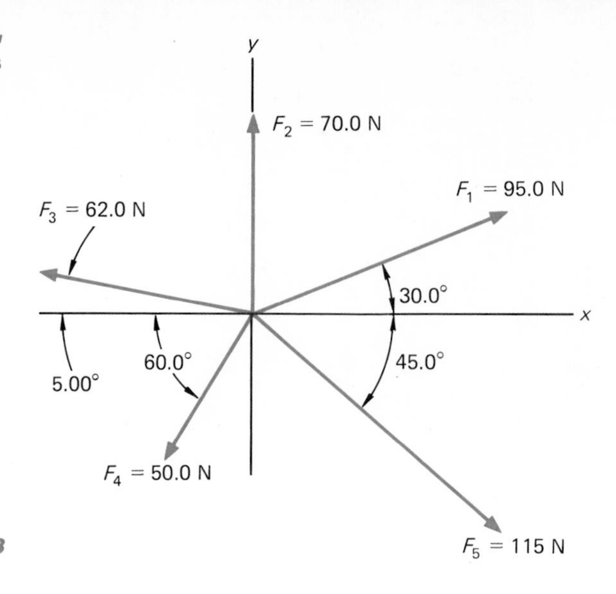

Figure 9.8

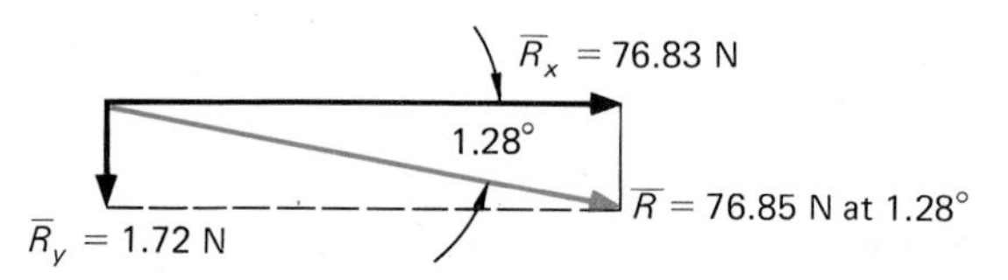

Figure 9.9

$$R = |\overline{R}| = [(76.83)^2 + (-1.72)^2]^{0.5}$$
$$= 76.849 \text{ N}$$
$$\theta_R = \tan^{-1}\frac{R_y}{R_x}$$
$$= \tan^{-1}\frac{-1.72}{76.83}$$
$$= -1.28°$$
$$\overline{R} = 76.8 \text{ N at } \searrow 1.28°$$

9.7 Moments and Couples

In order to solve problems involving complete force systems and their applications it is necessary to understand moments and couples. When was the last time you approached an exit in a public building and pushed on the panic bar of the door only to find that you had chosen the wrong side of the door, the side next to the hinges? No big problem. You simply move your hands to the side opposite the hinges and easily push open the door. You are thereby demonstrating the principle of the turning moment. By definition, the tendency of a force to cause rotation about a point is called the *moment* of the force relative to that point. The magnitude of the moment is the product of the magnitude of the force and the per-

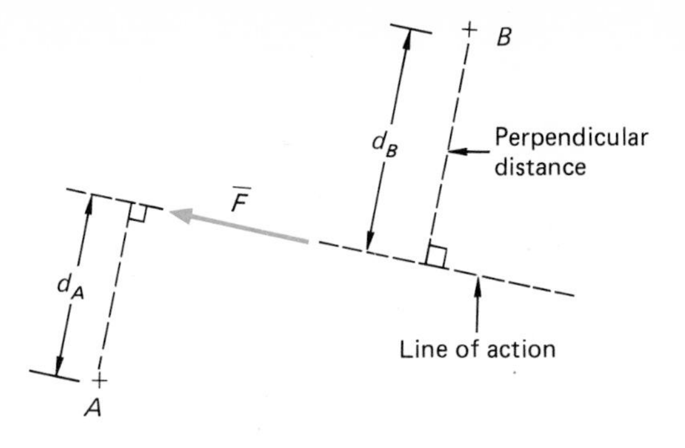

Figure 9.10
The force $\overline{F}$ and points A and B are in the same plane.

pendicular distance from the line of action of the force to the point. With respect to the door just mentioned, the same force may have been exerted in both attempts to open the door. In the second case, however, the moment was greater owing to the fact that you increased the distance from the force to the hinges, the point about which the door turns.

Figure 9.10 illustrates how a moment is evaluated for a specific problem. The magnitude of the moment of the force $\overline{F}$ about a point B is Fd_B. The distance d_B is the perpendicular distance, called the *moment arm*, from the point of application to the line of action of $\overline{F}$. The magnitude of the moment created by $\overline{F}$ about point A is Fd_A. The force $\overline{F}$ will tend to create a clockwise rotation about point B and a counterclockwise rotation about point A. The most common convention is to assign a positive sign to counterclockwise moments and a negative sign to clockwise moments. This convention is entirely arbitrary, demanding only consistency.

Figure 9.11 illustrates the concept of moments. If you neglect the mass of the beam and consider the force $\overline{F}$ = 135 N ↓ applied 5 m to the right of point O, then there will be a tendency for the beam to rotate clockwise about point O with the magnitude of the moment equal to (135 N)(5.00 m). To balance this situation and keep the beam level with a single force, you would need to apply a force $\overline{A}$ of 135 N ↓, a force $\overline{B}$ of 67.5 N ↓, or a force $\overline{C}$ of 45.0 N ↓. The less the force, the further the point of application needs to be from O in order to maintain balance.

A couple is similar to a moment. Figure 9.12 shows a hand wheel used to close and open a large valve. The two forces shown form a *couple* because they are parallel, equal in magnitude, and opposite in direction. The perpendicular distance between the two forces is

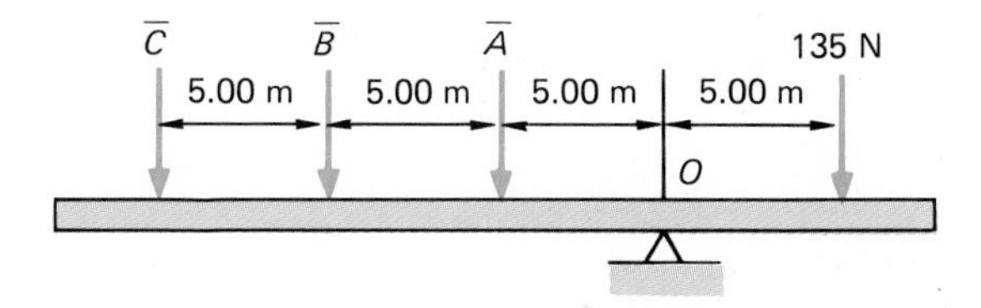

Figure 9.11
The forces are in the same vertical plane.

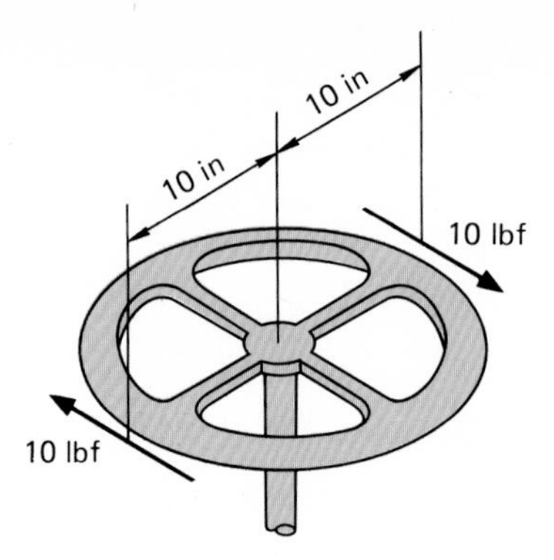

Figure 9.12
Illustration of a couple.

called the *arm* of the couple. The equivalent moment of a couple is equal to the product of one of the forces and the distance between the forces. In Fig. 9.12, the couple has a magnitude of (10 lbf) (20 in), or 200 in·lbf.

9.8 Free-Body Diagrams

The first step in solving a problem in statics is to draw a sketch of the body, or a portion of the body, and all the forces acting on that body. Such a sketch is called a *free-body diagram* (*FBD*). As the name implies, the body is cut free from all others; only forces that act upon it are considered. In drawing the free-body diagram, we remove the body from supports and connectors, so we must have an understanding of the types of reactions that may occur at these supports.

Examples of a number of frequently used free-body notations are illustrated in Fig. 9.13. It is important that you become familiar with these so that each FBD you construct will be complete and correct.

9.9 Equilibrium

Newton's first law of motion states that if the resultant force acting on a particle is zero, then the particle will remain at rest or move with a constant velocity. This concept is essential to statics.

Combining Newton's first law and the idea of equilibrium, we can state that a body will be in equilibrium when the sum of all external forces, moments, and couples acting on the body is zero. This requires the body to be at rest or moving with a constant velocity.

In this chapter, we will consider only bodies at rest. To study a body and the forces acting upon it, one must first determine what the forces are. Some may be unknown in magnitude and/or direction and quite often these unknown magnitudes and directions are information that is being sought. The conditions of equilibrium can be stated in equation form as follows:

$$\rightarrow \Sigma F_x = 0$$

$$\uparrow \Sigma F_y = 0 \qquad 9.8$$

$$\circlearrowleft \Sigma M = 0$$

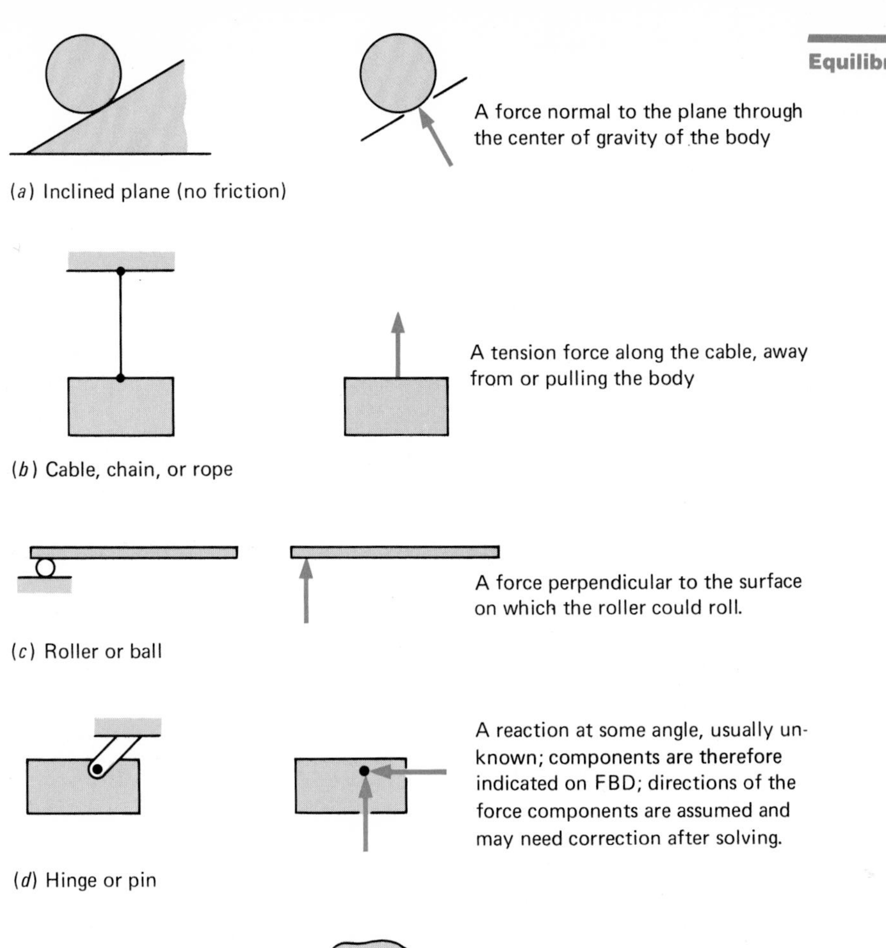

A force normal to the plane through the center of gravity of the body
(a) Inclined plane (no friction)
A tension force along the cable, away from or pulling the body
(b) Cable, chain, or rope
A force perpendicular to the surface on which the roller could roll.
(c) Roller or ball
A reaction at some angle, usually unknown; components are therefore indicated on FBD; directions of the force components are assumed and may need correction after solving.
(d) Hinge or pin

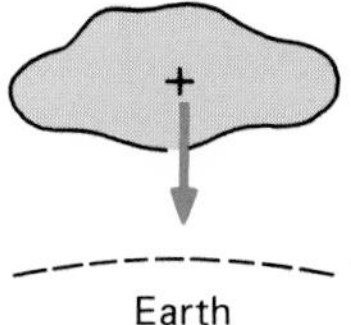

(e) Mass/earth

Mass multiplied by the earth's gravitational constant produces a force directed toward the center of the earth.

Figure 9.13
Free-body notations.

In order to assist in the tabulation and summation of force components and moments, a convention is necessary. In this text, F_x is positive to the right (→) and F_y is positive upward (↑) and counterclockwise moments are positive (↺). When evaluating moments in a particular problem, the point from which the moment arms are measured is arbitrary. A point is usually selected that simplifies the computations.

The following example problems will illustrate the concept of equilibrium.

Example problem 9.3 A uniform 12.00-m beam has a mass of 10.00 kg/m and is supported by two rollers, one at the left end and one 2.00 m left of the right end. What are the reactions at the two rollers?

Solution

1. When a diagram is not given with the problem, it is best to begin by diagramming the system. See Fig. 9.14.

2. Prepare a free-body diagram of the beam, replacing the roller bearings with unknown reactions R_A and R_B. See Fig. 9.15. Note that the mass for a uniform beam may be concentrated at the geometric center. For this problem the mass results in a force of

$$(10.00 \text{ kg/m})(9.806\,7 \text{ m/s}^2)(12.00 \text{ m}) = 1\,176.8 \text{ N}$$

Round off an intermediate calculation no further than one place more than required by the final answer.

In general, if an object has a mass and no additional information is available, the mass may be assumed to be concentrated at the geometric center; that is, the mass center (centroid) and geometric center are the same.

3. Apply the equations of equilibrium.
   (*a*) Sum of the horizontal forces must be zero: $\Sigma F_x = 0 \rightarrow$. In this case there are no horizontal forces, since rollers cannot sustain horizontal forces.
   (*b*) Sum of the vertical forces must be zero: $\Sigma F_y = 0 \uparrow$. Therefore,

$$R_A + R_B - 1\,176.8 = 0$$

$$R_A + R_B = 1.1768 \text{ kN}$$

   (*c*) The sum of the moments must be zero: $\Sigma M = 0$. That is, if moments are calculated about any point, the sum must be zero. Select a point, say the right end of the beam, and write the equation:

$$\circlearrowleft \Sigma M_c = 0$$

$$= -12.00R_A + 6.00(1\,176.8) - 2.00R_B = 0$$

$$-12.00R_A + 7\,060.8 - (1\,176.8 - R_A)(2.00) = 0$$

$$-12.00R_A + 7\,060.8 - 2\,353.6 + 2.00R_A = 0$$

$$-10.00R_A = -4\,707.2$$

$$R_A = 470.72 \text{ N}$$

$$= 471 \text{ N}$$

$$\overline{R}_A = 471 \text{ N} \uparrow$$

Then

$$R_B = 1\,176.8 - R_A$$

$$= 1\,176.8 - 470.72$$

$$= 706 \text{ N}$$

$$\overline{R}_B = 706 \text{ N} \uparrow$$

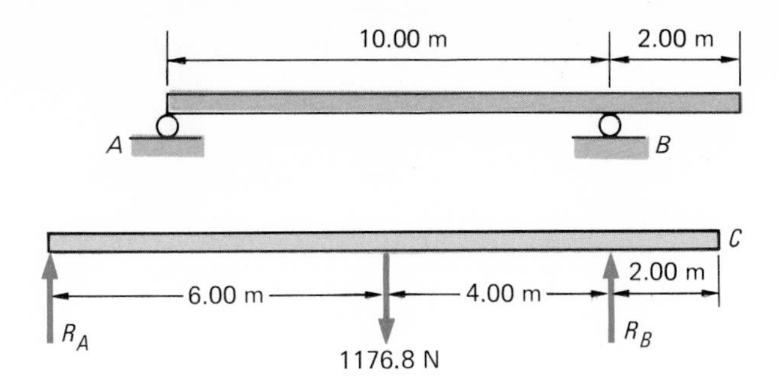

It would have simplified the solution if moments had been taken about the left end.

$$\Sigma M_A = 0$$

$$= (0)R_A - (6.00)(1\,176.8) + 10.00R_B = 0$$

$$10.00R_B = 7\,060.8$$

$$R_B = 706 \text{ N}$$

$$\overline{R}_B = 706 \text{ N} \uparrow$$

Then

$$R_A = 1\,176.8 - 706.08$$

$$= 470.72 \text{ N}$$

$$= 471 \text{ N}$$

$$\overline{R}_A = 471 \text{ N} \uparrow$$

Example problem 9.4 A 0.320-kg sphere is held by a cable attached to its surface. The sphere rests on an inclined plane that is 25° above the horizontal. The cable in turn is attached to a level plane, and the cable makes an angle of 40° with that plane. What is the tension in the cable, and what is the force on the inclined plane?

Solution

1. Construct a diagram of the system as described in the problem. See Fig. 9.16.
2. Observe:
   (*a*) The weight of the sphere acts vertically through the center of the sphere.
   (*b*) The tension in the cable acts upward along the cable's line and through the center of the sphere.
   (*c*) The force on the inclined plane is resisted by a force normal to the plane and through the center of the sphere.

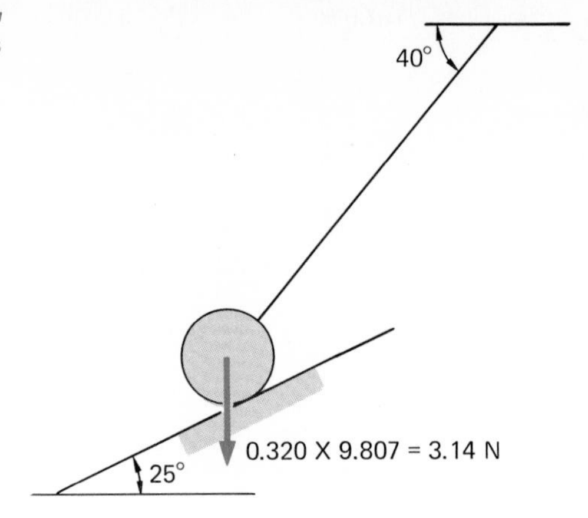

Figure 9.16

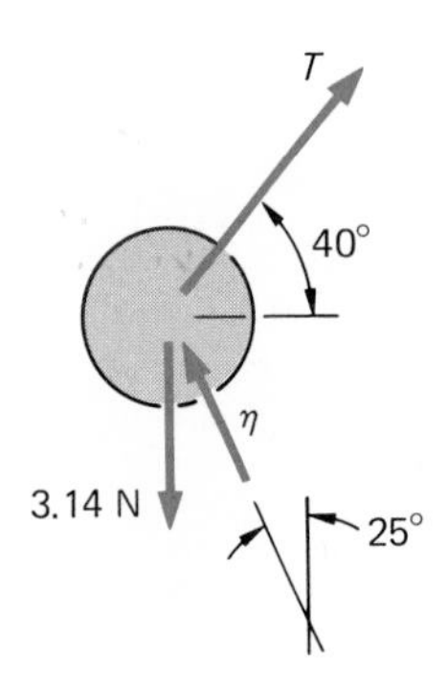

Figure 9.17

3. Make a free-body diagram (see Fig. 9.17).
4. Apply equations of equilibrium:

(a) $$\uparrow \Sigma F_y = 0$$

$$T_y + \eta_y - 3.14 = 0$$

(b) $$\rightarrow \Sigma F_x = 0$$

$$T_x - \eta_x = 0$$

(c) The moments equal zero because the forces are concurrent and if moments are taken about the center of the sphere, the moment arms are all zero. Thus the moment equation does not contribute useful information toward the solution.

(d) $T_y = T \sin 40° = 0.642\ 8T$

$T_x = T \cos 40° = 0.766\ 0T$

$\eta_y = \eta \cos 25° = 0.906\ 3\eta$

$\eta_x = \eta \sin 25° = 0.422\ 6\eta$

5. From these equations, we can write
(a) $\rightarrow \Sigma F_x = 0 \quad 0.776\ 0T - 0.422\ 6\eta = 0$
(b) $\uparrow \Sigma F_y = 0 \quad 0.642\ 8T + 0.906\ 3\eta = 3.14$

Multiplying 5*a* by 0.828 4, we get
(*c*) $0.642\,8T - 0.350\,1\eta = 0$

Subtracting 5*c* from 5*b*, we get

$1.256\,4\eta = 3.14$ N

Then

$\overline{\eta} = 2.50$ N 115°

From 5*a*,

$$T = \frac{0.422\,6}{0.766\,0}(2.50)$$

$$= 1.38 \text{ N}$$

Note that force on the inclined plane is $-\overline{\eta}$, so the force of the sphere on the plane = 2.50 N 65°

Example problem 9.5 Find the reaction at *G* and the cable tension. See Fig. 9.18.

Solution

1. Construct a free-body diagram. Neglect the mass of the beam. See Fig. 9.19.
2. Determine the geometry:

$$\theta = \measuredangle HGI = \tan^{-1}\frac{6}{8} = 36.87°$$

$$\alpha = 45 - \theta = 8.13°$$

3. Apply equations of equilibrium:

$$\circlearrowleft \Sigma M_G = 0$$

$$-20.0(9.807)(12 \cos 8.13°) + 8H = 0$$

$$H = 291.3 \text{ N}$$

$$\overline{H} = 291.3 \text{ N at } 45°$$

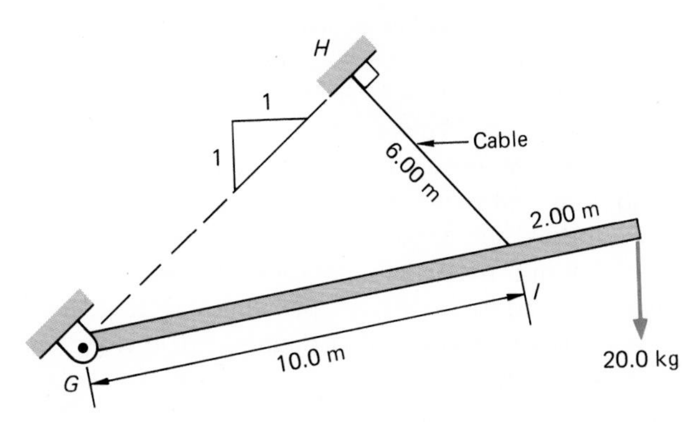

Figure 9.18

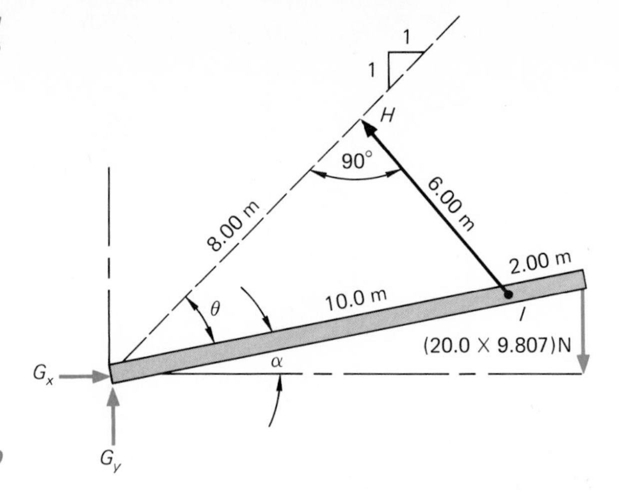

Figure 9.19

$\rightarrow \Sigma F_x = 0$

$G_x - H \cos 45 = 0$

$G_x = 205.9$ N

$\overline{G_x} = 205.9 \text{ N} \rightarrow$

$\uparrow \Sigma F_y = 0$

$G_y + H \sin 45 - 20.0(9.807) = 0$

$G_y = -9.840$ N

So $\overline{G_y} = -9.840 \text{ N} \uparrow$

or $\overline{G_y} = 9.840 \text{ N} \downarrow$

4. Combine vector components into resultant $\overline{G}$ using Eq. (9.7):

$\overline{G} = 206$ N at 2.74°

9.10 Stress

In statics, concepts were limited to rigid bodies. It is obvious that the assumption of perfect rigidity is not always valid. Forces will tend to deform or change the shape of any body, and an extremely large force may cause observable deformation. In most applications, slight deformations are experienced, but the body returns to its original form after the force is removed. One function of the engineer is to design a structure within limits that allow it to resist permanent change in size and shape so that it can carry or withstand the force (load) and still recover.

In statics, forces are represented as having a magnitude or an intensity in a particular direction. Structural members are usually characterized by their mass per unit length or their size in principal dimension, such as width or diameter. Consider the effect on a wire that is 5.00 mm in diameter and 2.50 m in length. It is suspended

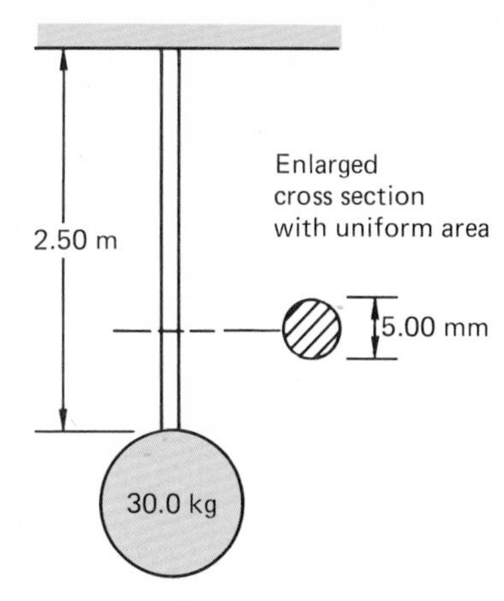

Figure 9.20

from a well-constructed support and has a ball with a mass of 30.0 kg attached to its lower end. See Fig. 9.20. The force exerted by the mass is (30.0)(9.807) = 294.2 N. The wire has a cross-sectional area of $\pi(5^2/4)$, or 19.63 mm^2. If it is assumed that every square millimeter equally shares the force, then each square millimeter supports 294.2 ÷ 19.63, or 15.0 N. This fact can be stated another way: The stress is 15.0 N/mm^2 (1.50×10^7 Pa); that is, the force in the wire is literally trying to separate the atoms of the material by overcoming the bonds that hold the material together.

The relationship above is normally expressed as

$$\sigma = \frac{F}{A} \qquad 9.9$$

where σ = stress, Pa
F = force, N
A = cross-sectional area, m^2

Since stress is force per unit area, it is obvious that the stress in the wire (ignoring its own weight) is unaffected by the length of the wire. Such a stress is called *tensile stress* (tending to pull the atoms apart). If the wire had been a rod of 5.00-mm diameter resting on a firm surface and if the mass had been applied at its top, the force would have produced the identical stress, but it would be termed *compressive*, for obvious reasons.

The simple or direct stress in either tension or compression as described above results from an applied force (load) that is in line with the axis of the member (axial loading). Also the cross-sectional area in both examples was constant lengthwise. If you have an axial load but the cross section varies, the stresses in separate cross sections are different because the areas are different.

A third type of stress is called *shear*. While tension and compression attempt to separate or push atoms together, shear tries to slide

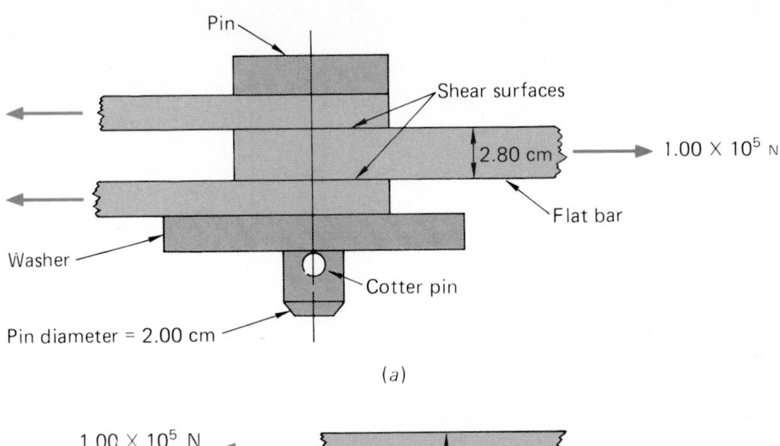

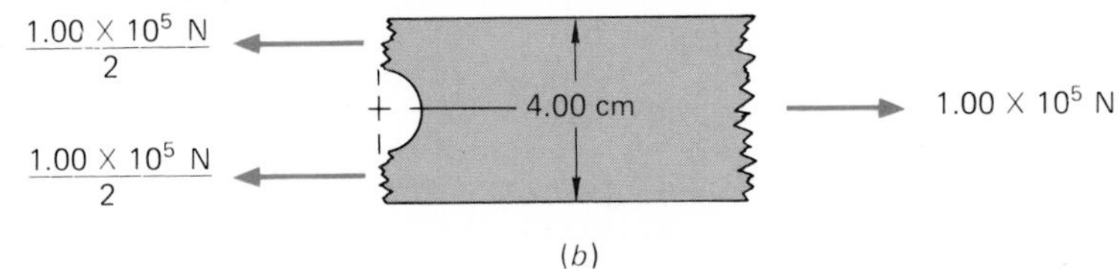

Figure 9.21

layers of atoms in the material across each other. (Imagine removing the top half of a stack of sheets of plywood without lifting.) Consider the pin in Fig. 9.21*a* as it resists the force of $1.00(10^5)$ N.

The shear in the pin is

$$\tau = \frac{F}{A} = \frac{1.00 \times 10^5}{(\pi)(2.00 \times 10^{-2})^2(2)/4} = 1.59 \times 10^8 \text{ N/m}^2$$

$$= 159 \text{ MPa}$$

Note: This is an example of double shear in that two cross sections of the pin resist the force; hence, the area is the cross-sectional area of two pins. The two pin shear surfaces are indicated in Fig. 9.21*a*.

To complete the computation, the tensile stress in the bar at the critical section through the pin hole, as shown in Fig. 9.21*b*, is

$$\sigma = \frac{F}{A} = \frac{1.00 \times 10^5}{(2.00 \times 10^{-2})(2.80 \times 10^{-2})}$$

$$= 179 \text{ MPa}$$

9.11 Strain

As an engineer you may be called upon to design both structures and mechanisms. In design work, it is important to consider not only the external forces but also the strength of each individual part or member. It is critical that each separate element be strong enough, yet not contain an excessive amount of material. Thus in the solution of many problems a knowledge of the properties of materials is essential.

One important test that provides designers with certain material

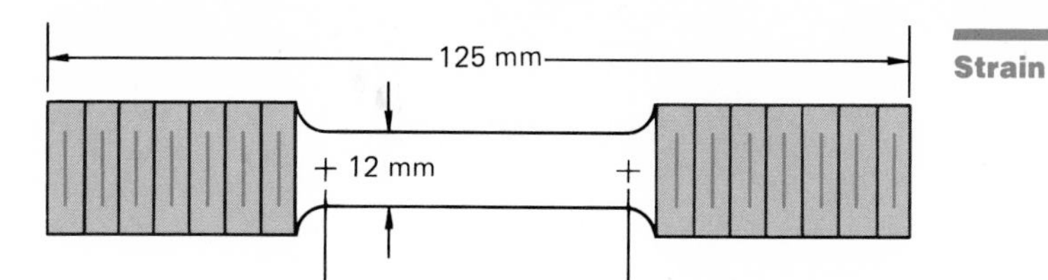

Figure 9.22

properties is called the *tensile test.* Figure 9.22 illustrates a schematic of a tensile-test specimen. When this specimen is loaded in an axial tensile-test machine, the force applied and the corresponding increase in material length can be measured. This increase in length is called the *elongation.* Next, in order to permit comparisons with standard values, the elongation is converted to a unit basis called *strain.*

Strain (ϵ) is defined as a dimensionless ratio of the change in length (elongation) to the original length:

$$\epsilon = \frac{\Delta l}{l} = \frac{\delta}{l} \qquad 9.10$$

where ϵ = strain, mm/mm
δ = deformation, mm
l = length, mm

A *stress-strain diagram* is a plot of the results of a tensile test (see Fig. 9.23). The shape of this diagram will vary somewhat for different materials, but in general there will first be a straight-line

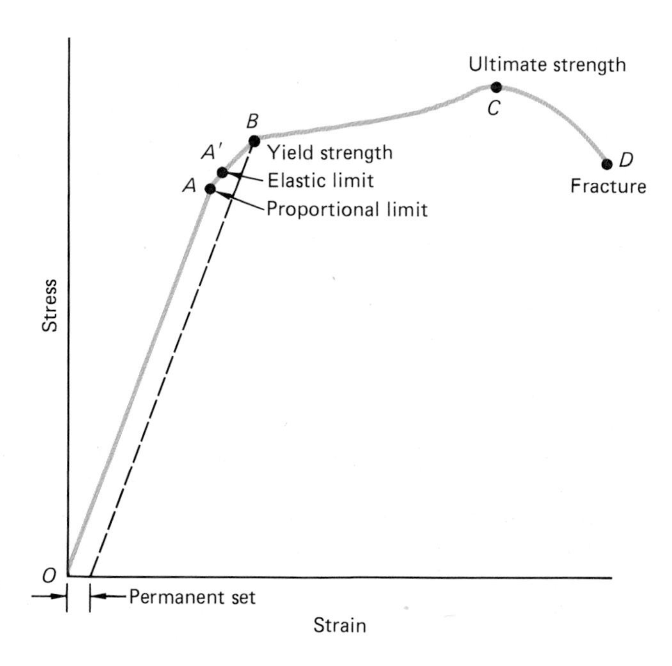

Figure 9.23
Stress-strain diagram.

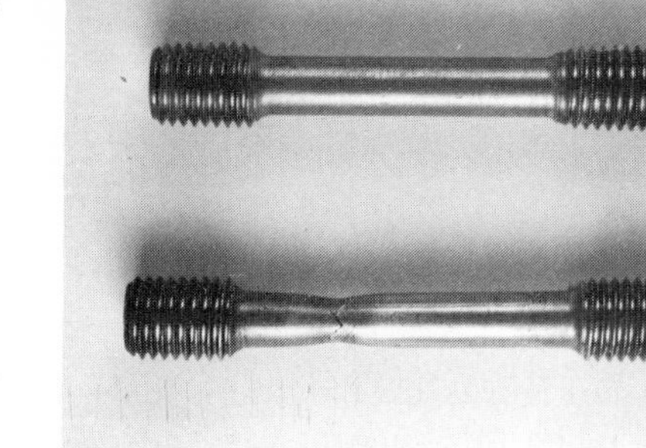

Figure 9.24
The lower photo shows the actual "necking-down" effect that occurs during a tensile test prior to actual failure of the steel member.

portion OA. Point A is the proportional limit—the maximum stress for which stress is proportional to strain.

At any stress up to point A', called the *elastic limit*, the material will return to its original size once the load has been removed. At stresses higher than A', permanent deformation (set) will occur. For most materials, points A and A' are very close together.

The stress at B in Fig. 9.23 that causes a permanent set of 0.05 to 0.3 percent (depending on material) is termed the *yield strength.* The corresponding strain is 0.0005 to 0.003.

Point C, called the *ultimate strength,* is the maximum stress that the material can withstand. Between points B and C, a small increase in stress causes a significant increase in strain. At approximately point C, the specimen will begin to neck down sharply; that is, the cross-sectional area will decrease rapidly, and fracture will occur at point D.

The Fig. 9.24 photo shows a typical specimen prior to test and the specimen after it was pulled apart.

9.12 Modulus of Elasticity

Approximately 300 years ago, Robert Hooke recognized the linear relationship between stress and strain. For stresses below the proportional limit, Hooke's law can be written

$$\epsilon = K\sigma \tag{9.11}$$

where K is a proportionality constant. The modulus of elasticity E (the reciprocal of K) rather than K is commonly used, yielding

$$\sigma = E\epsilon \tag{9.12}$$

Values of E for selected materials are given in Table 9.2.

9.13 Design Stress

Obviously, most products or structures that engineers design are not intended to fail or become permanently deformed. The task facing the engineer is to choose the proper type and size of material that will perform correctly under the conditions likely to be imposed. Since the safety of the user and the liability of the producer (including the engineer) are dependent on valid assumptions, the en-

Table 9.2 Modulus of elasticity for selected materials

	E, psi	E, MPA
Cold rolled steel	30×10^6	21×10^4
Cast iron	16×10^6	11×10^4
Copper	16×10^6	11×10^4
Aluminum	10×10^6	7×10^4
Stainless steel	27×10^6	19×10^4
Nickel	30×10^6	21×10^4

gineer typically selects a design stress that is less than the yield strength. The ratio of the yield strength to the design stress is called the *safety factor*. For example, if the yield strength is 210 MPa and the design stress is 70 MPa, the safety factor, based on yield strength, is 3. Care must be exercised in reporting and interpreting safety factors because they are expressed in terms of both yield strength and ultimate (tensile) strength. Table 9.3 lists typical values used in structural design. It should be noted that the United States still lists most of its standards in the English system. Conversions in this area will be necessary for some time to come.

Example problem A round bar is 40.0 cm long and must withstand a force of 20.0 kN. What diameter must it have if the stress is not to exceed 140.0 MPa?

Solution

$$\sigma = \frac{F}{A}$$

$$A = \frac{F}{\sigma} = \frac{20.0 \times 10^3 \text{ N}}{140.0 \times 10^6 \text{ N/m}^2} \left| \frac{10^6 \text{ mm}^2}{1 \text{ m}^2} \right. = 143 \text{ mm}^2$$

$$= \frac{\pi d^2}{4}$$

$$143 = \frac{3.14d^2}{4}$$

$$d = 13.5 \text{ mm}$$

Table 9.3 Ultimate and yield strength

	Ultimate strength		Yield strength	
	psi	MPa	psi	MPa
Cast iron	45×10^3	310	30×10^3	210
Wrought iron	50×10^3	345	30×10^3	210
Structural steel	60×10^3	415	35×10^3	240
Stainless steel	90×10^3	620	30×10^3	210
Aluminum	18×10^3	125	12×10^3	85
Copper, hard drawn	66×10^3	455	60×10^3	415

Example problem 9.7 Assume that in Example Prob. 9.6 the allowable elongation is 0.125 mm when the modulus of elasticity (E) is 2.00×10^2 GPa. Determine the diameter (in millimeters) of the bar.

Theory:

$$E = \frac{\sigma}{\epsilon} = \frac{F/A}{\Delta l/l}$$

so

$$\Delta l = \frac{Fl}{AE}$$

Δl is usually written δ.

$$\delta = \frac{Fl}{AE}$$

so

$$A = \frac{Fl}{\delta E}$$

Solution

$$A = \frac{20.0 \times 10^3 \text{ N} \mid 0.400 \text{ m} \mid 1 \text{ m}^2 \mid 10^6 \text{ mm}^2}{125 \times 10^{-6} \text{ m} \mid \quad \mid 2.00 \times 10^{11} \text{ N} \mid 1 \text{ m}^2}$$

$$= 320 \text{ mm}^2 = \frac{\pi d^2}{4}$$

$$d = 20.2 \text{ mm}$$

Example problem 9.8 Given the configuration in Fig. 9.25*a*, calculate the load that can be supported under the following design conditions.

(*a*) The pin at point R, enlarged in Fig. 9.25*b*, is 10.0 mm in diameter. What load can be supported by the pin if the ultimate shear strength of the pin is 195 MPa and a safety factor of 2.0 is required?

(*b*) Using the load condition from (*a*) with the same factor of safety, size cable ST if it is manufactured from structural steel.

Solution

(*a*) 1. Construct a free-body diagram, neglecting the mass of the beam (see Fig. 9.26).

2. Determine the geometry:

$$\tan 15° = \frac{H}{6.00}$$

$$H = 1.608 \text{ m}$$

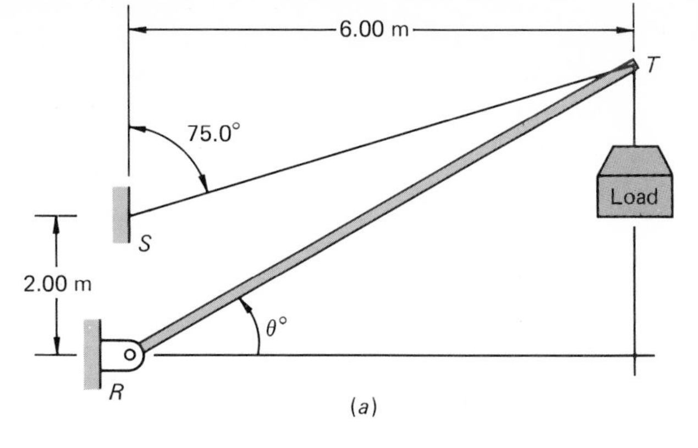

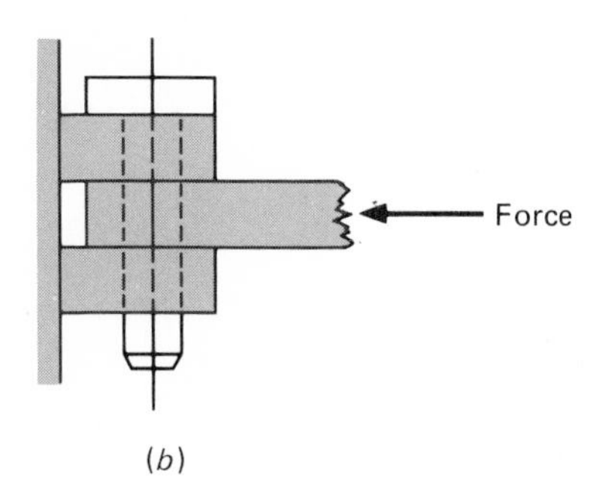

Figure 9.25

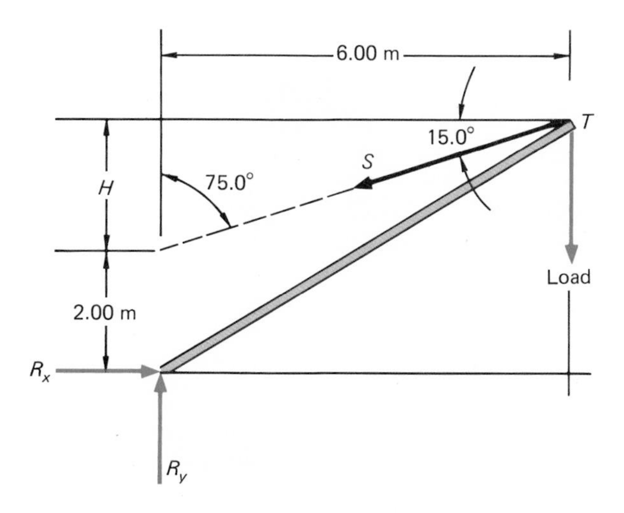

Figure 9.26

3. Calculate shearing stress τ on the pin in double shear:

$$\text{Design stress} = \frac{\text{ultimate strength}}{\text{safety factor}}$$

$$= \frac{195 \text{ MPa}}{2.0} = 97.5 \text{ MPa}$$

$$\tau = \frac{F}{A}$$

$$F(\text{load}) = \frac{(97.5 \times 10^6)(\pi)(0.010)^2(2)}{4}$$

$$= 15\,315 \text{ N}$$

$$\cong 15.3 \text{ kN}$$

Note: 15 300 N is the total force that can be supported by the pin with its given size, ultimate shear strength, and factor of safety. In this example, the only forces applied to the beam are at point *T*. Therefore the force at point *R* is along the beam. If the mass of the beam were considered, the force at *R* would no longer be along the beam.

4. Apply equation of equilibrium:

$$\circlearrowleft \Sigma M_R = 0 = -6.00(\text{load}) - 6.00(S) \sin 15° + 3.608(S \cos 15°)$$

$$1.932\,1S = 6.00(\text{load})$$

$$S = 3.105\,4(\text{load})$$

$$\rightarrow \Sigma F_x = 0$$

$$R_x - S \cos 15° = 0$$

$$R_x = 2.999\,6(\text{load})$$

$$\uparrow \Sigma F_y = 0$$

$$R_y - S \sin 15° - \text{load} = 0$$

$$R_y = 1.803\,7(\text{load})$$

$$\overline{R} = 3.500(\text{load}) \text{ at } 31.02°$$

5. Determine the maximum load:

$$R = 15.315 \text{ kN} = 3.500(\text{load})$$

$$\text{Max load} = 4.375\,7 \times 10^3 \text{ N}$$

(*b*) **1.** Determine force in the cable for maximum load from (*a*):

$$S = 3.105\,4(\text{load})$$

$$= 13\,588 \text{ N}$$

2. From Table 9.3, the ultimate strength of structural steel is 415 MPa.

$$\text{Design stress} = \frac{\text{ultimate strength}}{\text{safety factor}}$$

$$= \frac{415}{2.0}$$

$$= 207.5 \text{ MPa}$$

$$= 207.5 \times 10^6 \text{ N/m}^2$$

3. Calculate cable size:

$$\sigma = \frac{F}{A}$$

$$A = \frac{F}{\sigma} = \frac{13\,588\ \text{N}}{} \bigg| \frac{1\ \text{m}^2}{207.5 \times 10^6\ \text{N}}$$

$$\frac{\pi D^2}{4} = 6.548\,4 \times 10^{-5}\ \text{m}^2$$

$$D = 0.009\,131\ \text{m}$$

$$\cong 9.13\ \text{mm}$$

Problems

9.1 Find the resultant of two concurrent forces. One has a magnitude of 4.937 kN and acts 35.3° counterclockwise from vertical, and the other has a magnitude of 2.468 kN and acts 57.9° clockwise from vertical.

9.2 Find the resultant of two concurrent forces. One has a magnitude of 1.135 kN and acts 21.6° clockwise from vertically downward, and the other has a magnitude of 2.357 kN and acts 42.8° counterclockwise from vertical.

9.3 The resultant of two concurrent forces has a magnitude of 8.642 kN and acts 15.7° clockwise from vertical. One of the forces has a magnitude of 6.369 kN and acts 12.6° counterclockwise from the positive horizontal. Find the magnitude and direction of the other force.

9.4 A cable has a length of 14.0 m. It is attached at points R and S, which have the same elevation at the top of an upright cylindrical tank. The tank has a volume of 1 005 m³ and a depth of 20.0 m. Points R and S are on the circumference a maximum distance apart. A mass of 876 kg is attached to the cable 5.50 m along the cable from point R. The cable hangs into the tank. What is the tension in each of the two segments of the cable?

9.5 Given the data in Fig. 9.27, find the tension in PR and in RS.

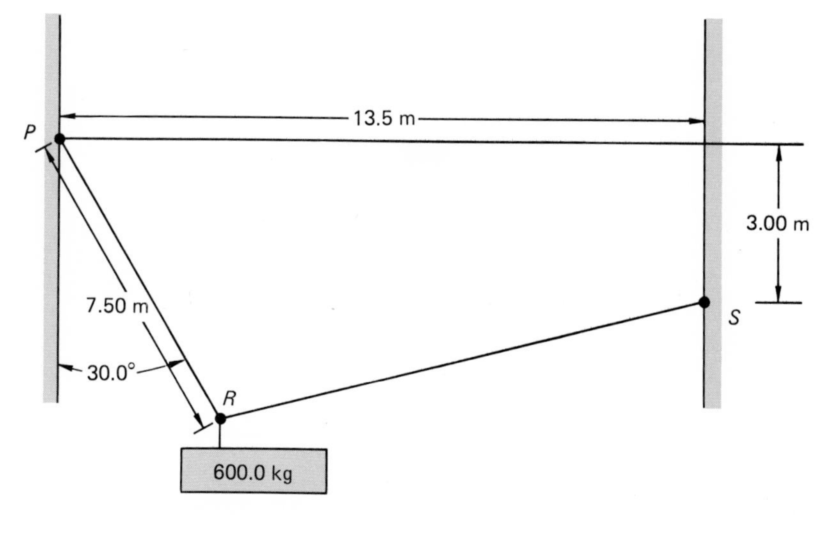

Figure 9.27

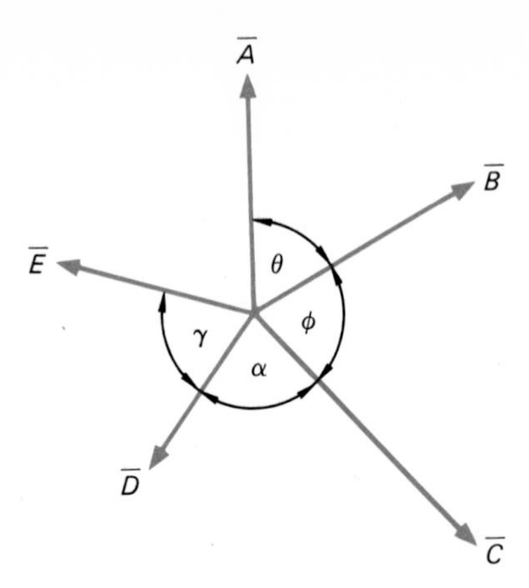

Figure 9.28

In Probs. 9.6 through 9.12, find the resultant of the concurrent, coplanar force systems. Express each answer as the magnitude and direction of the resultant. Refer to Fig. 9.28 for the definition of the variables. Vector $\bar{A}$ is vertical.

	$\bar{A}$	$\bar{B}$	$\bar{C}$	$\bar{D}$	$\bar{E}$	θ	ϕ	α	γ
9.6	1.23 kN	2.39 kN	816 N	3.15 kN	4.16 kN	55.2°	81.2°	75.0°	62.1°
9.7	2.46 kN	1.726 kN	3.33 kN	6.53 kN	777 N	49.7°	84.1°	81.2°	66.6°
9.8	3.15 kN	973 N	1.79 kN	2.46 kN	1.53 kN	58.9°	79.3°	72.3°	59.5°
9.9	4.68 kN	6.52 kN	3.18 kN	3.57 kN	0.678 kN	62.1°	78.1°	68.6°	60.0°
9.10	1.35 kN	715 N	4.22 kN	683 N	3.17 kN	65.3°	72.7°	62.3°	63.4°
9.11	2.72 kN	1.02 kN	1.06 kN	2.98 kN	2.37 kN	56.7°	80.0°	70.0°	68.3°
9.12	815 N	4.17 kN	979 N	3.03 kN	2.78 kN	48.3°	83.7°	73.6°	71.3°

In Probs. 9.13 through 9.20, force $\bar{F}$ goes through point O and makes an angle of θ° with the horizontal, as shown in Fig. 9.29. Calculate the moment of $\bar{F}$ about points A and B, assigning positive values to counterclockwise moments.

Figure 9.29

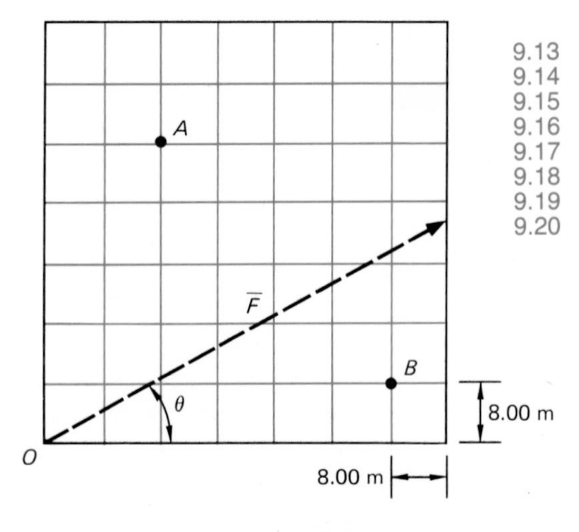

	F	θ
9.13	658 N	5.26°
9.14	982 N	15.7°
9.15	1.17 kN	22.8°
9.16	6.27 kN	34.6°
9.17	8.34 kN	56.2°
9.18	1.12 MN	63.8°
9.19	3.14 MN	78.3°
9.20	4.82 MN	84.8°

9.21 Given the force system in Fig. 9.30, find the reactions at A and B. Neglect the beam mass.

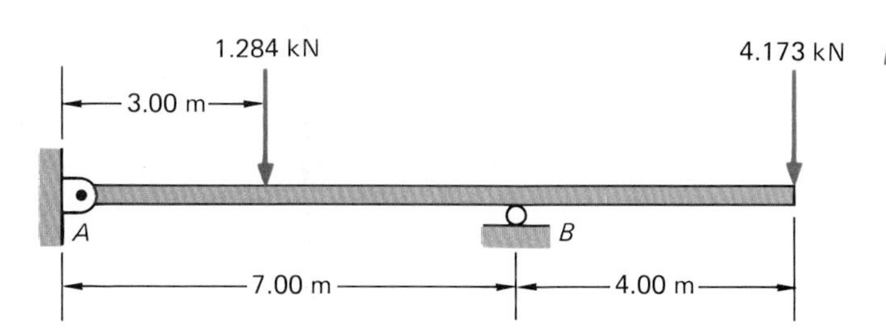

Figure 9.30

9.22 Given the force system in Fig. 9.31, find the reactions at C and D. The uniform beam has a mass of 4.37 kg/m.

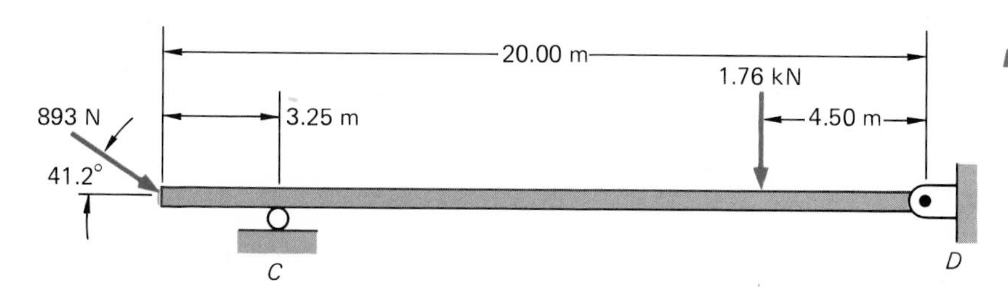

Figure 9.31

9.23 Given the force system in Fig. 9.32, find the reaction at E and F. The uniform beam has a mass of 8.67 kg/m.

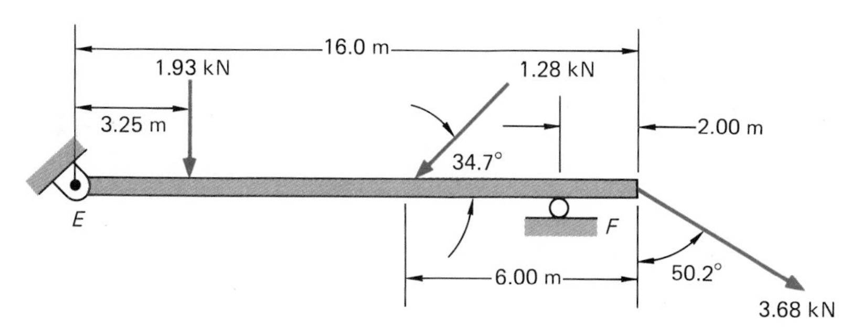

Figure 9.32

9.24 Given the force system in Fig. 9.33, find the reactions at G and H. The uniform beam has a mass of 17.2 kg/m.

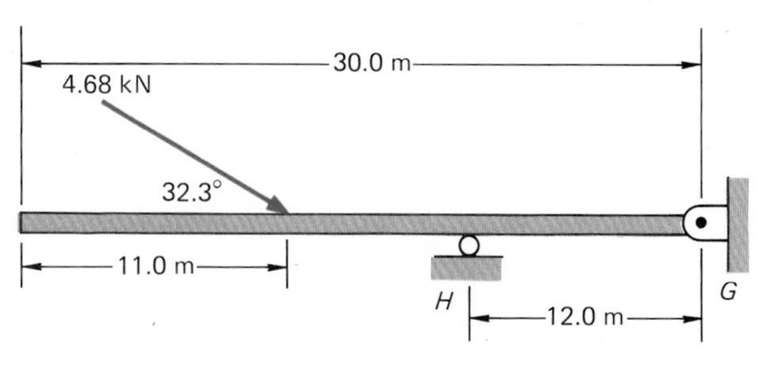

Figure 9.33

9.25 Debbie's mass is 50.5 kg and Bill's is 88.0 kg. They seat themselves on a teeter-totter as shown in Fig. 9.34. The teeter-totter beam has a mass of 2.73 kg/m. To balance the beam in a horizontal and motionless position, two friends pull on the ropes that are attached to the teeter-totter.

(a) What force must be applied by Bill's friend and by Debbie's friend?
(b) What is the force exerted on the roller?

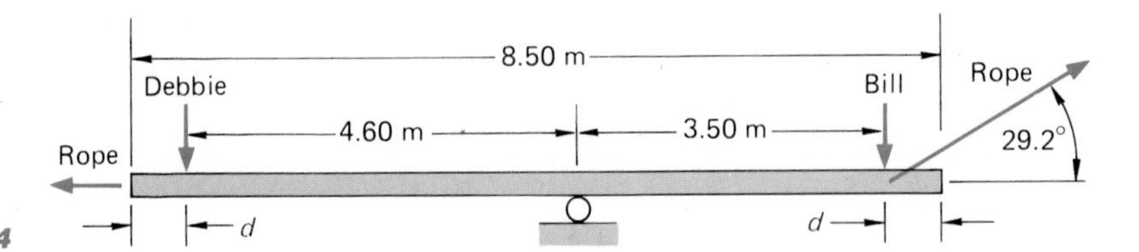

Figure 9.34

9.26 What is the magnitude, direction, and location of the force that must be added to the system shown in Fig. 9.35 to put the system into equilibrium?

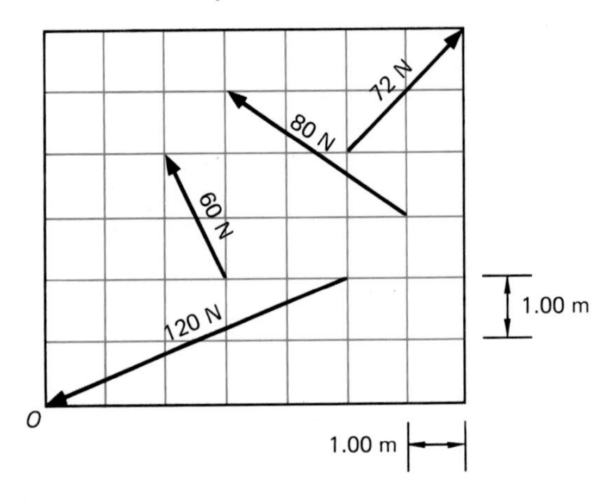

Figure 9.35

9.27 Each of the members in the truss shown in Fig. 9.36 has a mass of 12.6 kg/m. Find the reactions at L and M.

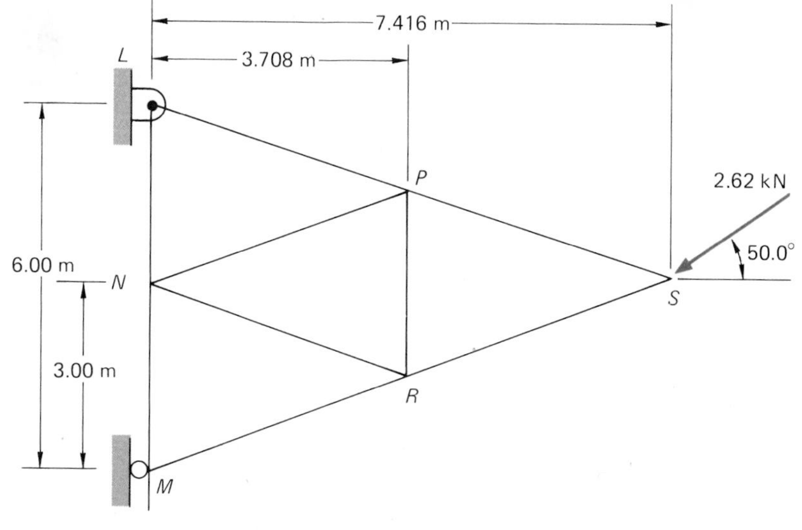

Figure 9.36

9.28 Two hooks are located on opposite walls of a hallway. One hook is 30.0 cm higher than the other. The width of the hallway is 2.00 m. A string that has a total length of 2.50 m is attached to the two hooks. A sample of this string was tested by attaching masses to it when it was hanging vertically. It broke when a total of exactly 17 kg had been added. If a mass is attached to the midpoint of the string when it is connected to the hooks as described above, what minimum mass will cause the string to break?

9.29 A hot-air balloon has a lifting force of 2.873 kN. A single rope is used to hold it in a gentle breeze, causing the rope to make an angle of 17.3° with the vertical.

(*a*) What is the tension in the rope?
(*b*) If the rope is 1.00 cm in diameter, what is the stress?

9.30 The beam *RS* shown in Fig. 9.37 has a mass of 2.11 kg/m.

(*a*) Find the tension in the cable *ST*.
(*b*) What is the magnitude and direction of the reaction at *R*?
(*c*) Allowing for a safety factor of 4.0 and an ultimate strength of 5.0×10^2 MPa, what diameter cable should be provided?

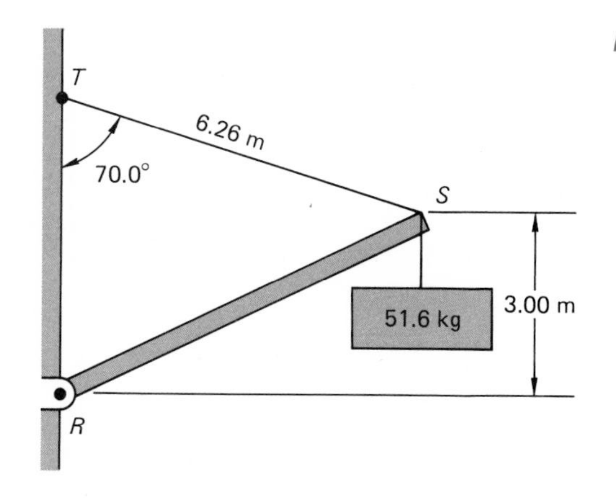

Figure 9.37

9.31 Beam *SM* in Fig. 9.38 has a mass of 3.12 kg/m.

(*a*) Find the tension in cable *KJ*.
(*b*) What is the reaction at *S*?
(*c*) If cable *KJ* has a diameter of 4.27 mm, what is the stress in the cable?
(*d*) If the ultimate strength for the cable material is 1 000 MPa, what is the safety factor?

Figure 9.38

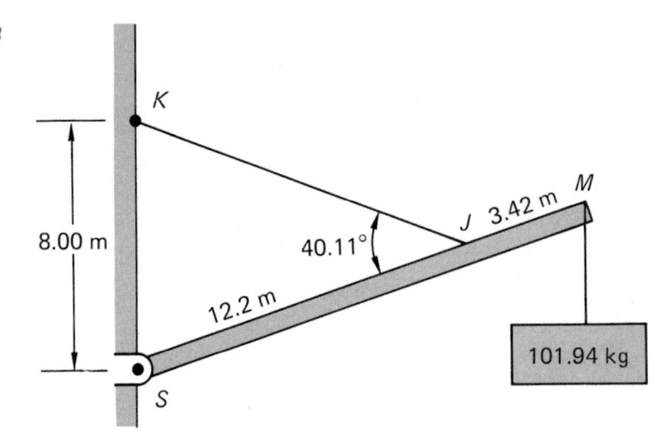

9.32 Each member in the frame shown in Fig. 9.39 has a mass of 5.27 kg/m.

(*a*) Find the tension in the guy wire.
(*b*) Find the reaction at *B* (magnitude and direction).

Figure 9.39

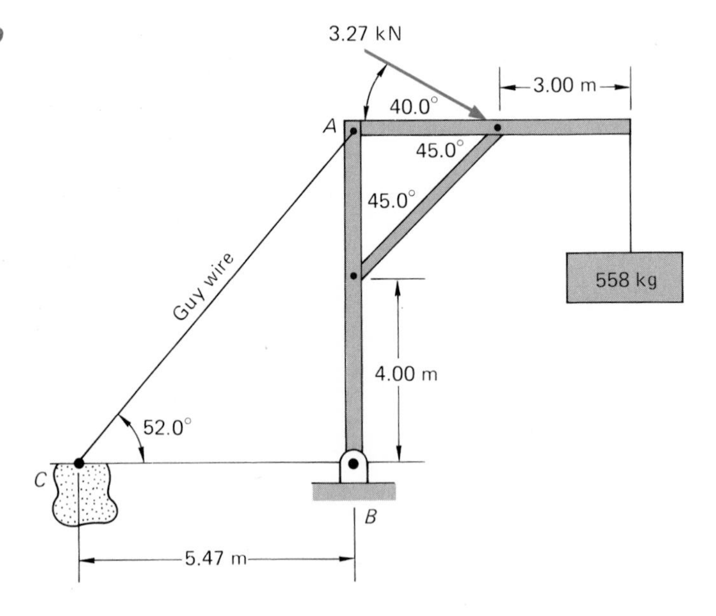

9.33 The beam in Fig. 9.40 has a mass of 7.89 kg/m.
(a) Find the tension in the cable.
(b) What is the reaction at the pinned connection?
(c) If the allowable design stress is 25 MPa, what diameter cable is needed?

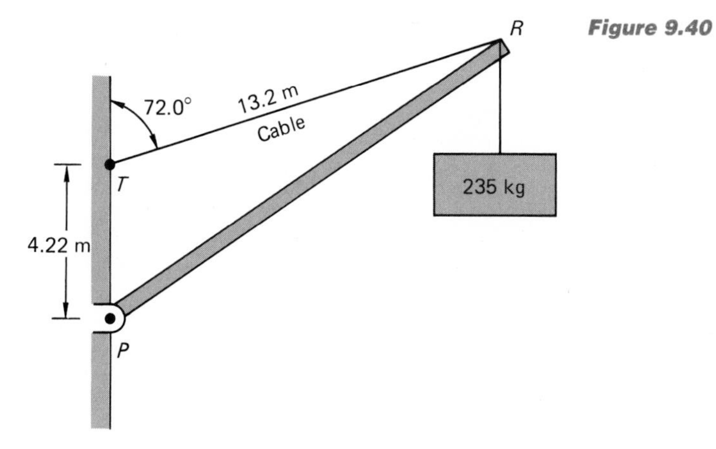

Figure 9.40

9.34 Assume the frame in Fig. 9.41 to have negligible mass. Find the reactions at the two connections, J and K.

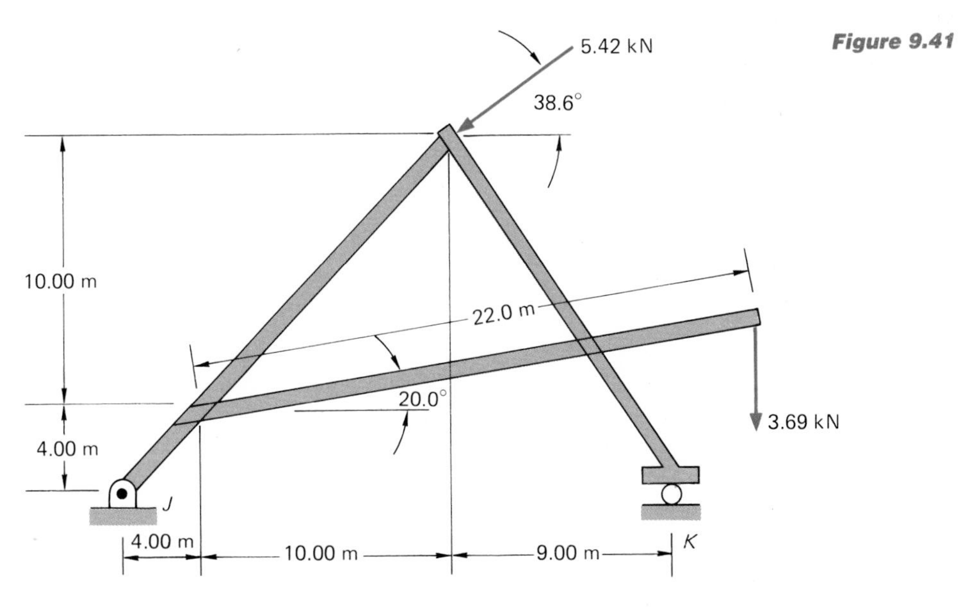

Figure 9.41

9.35 Assume the frame in Fig. 9.42 to have negligible mass. Find the reactions at the two connectors, *P* and *R*.

Figure 9.42

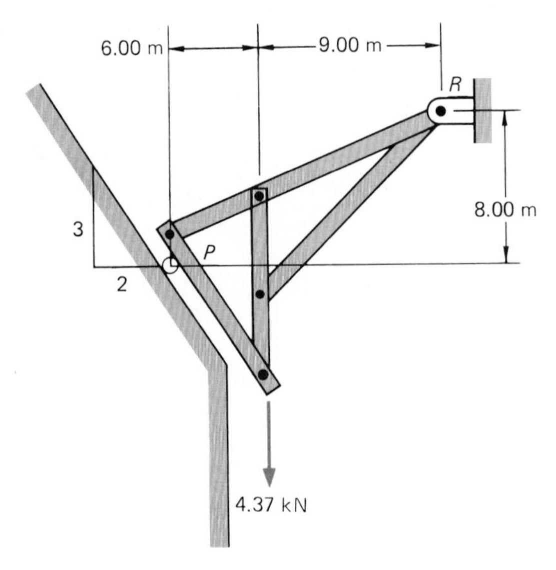

9.36 A business wants to hang a sign from the end of a horizontal pole. The pole will be attached to a wall by a pin and supported by a cable, as shown in Fig. 9.43. The pole and cable are each 8.0 m long. The sign has a mass of 100.0 kg. Ultimate strength is 1.0×10^9 Pa. A safety factor of 5 should be employed when designing the cable. At what point should the cable be attached to the pole in order to minimize the required cable diameter *D*? You should:

(*a*) Show the derivation of *D* as a function of *X*.
(*b*) Plot *D* vs. *X*.
(*c*) Find the value of *X* which minimizes *D* and state the minimum *D*.

As an alternate approach, write a computer program that permits the user to input the mass of the sign, the lengths of the cable and pole, the ultimate strength, and the safety factor. The program should print a table of values of *D* for corresponding values of *X*. In addition, the program should calculate the value of *X* and *D* to some specified accuracy where *D* is a minimum.

Figure 9.43

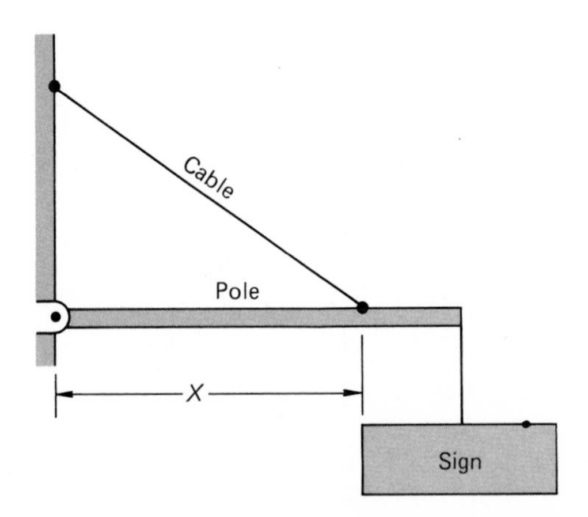

10

Chemistry—Concepts and Calculations

10.1 Introduction

Chemistry is one of the foundation sciences upon which engineering is built. Applications of chemistry are obvious in such fields as chemical engineering, ceramic engineering, and petroleum engineering. The use of and need for chemistry in a field such as electrical engineering is somewhat less apparent. Even so, in this field the transistor and integrated circuit technologies rely heavily on chemistry. Nearly everywhere one looks, some portion of each product or process depends on the application of chemistry.

This chapter discusses some general chemistry concepts and calculation procedures. The coverage of topics is limited to material

Figure 10.1
Student engineers, as well as graduate engineers, work in chemical industries. A process engineer directs the work of an engineering cooperative-education student. (*Sun Company, Inc.*)

that is commonly used in engineering courses. The particular concepts selected for discussion are the following:

Absolute zero
Temperature scales
Atomic and molecular weights
Avogadro's number
Density
Boyle's law
Charles' law
Equations of state
Balance of chemical equations
Composition calculations
Determination of empirical formulas from composition data

10.2 Chemical Concepts

10.2.1 Temperature, Temperature Scales, and Absolute Zero

The concept of temperature must be defined by examining the flow of heat from one body to another. Temperature is a physical property that determines which way heat will flow if two bodies are placed in thermal contact with one another. When bodies are in thermal contact, heat will flow from the body with the higher temperature to the one with the lower temperature. It follows then that if the objects have the same temperature, no heat will flow. Temperature can be measured on an empirical scale such as Celsius or Fahrenheit or on an absolute scale such as Kelvin or Rankine. More will be said about these scales later. It is important to point out that temperature is a fundamental, or basic, dimension and therefore cannot be expressed in terms of other fundamental dimensions such as mass, length, or time.

To state a temperature quantitatively, a reproducible scale must be developed. Two such scales, the Celsius and the Fahrenheit, are in common use both by technical workers and by the general public. Each of these scales is derived by dividing the difference in temperature between the freezing point and boiling point of water at standard pressure (1 atm or 0.101 MPa) into a specified number of equally spaced divisions.

The Celsius scale is obtained by dividing this temperature span into 100 equal parts. In addition, the freezing point of water is defined to be zero degrees on the scale. The result is that the boiling point of water is 100°C. The scale is linear and can be extended above 100°C for higher temperatures. It can also be extended to temperatures below 0°C by using negative values.

The Fahrenheit scale divides the temperature difference between the freezing point and boiling point of water into 180 parts. Here the freezing point of water is defined to be 32°F. Calculation then gives the boiling point of water as 32 + 180, or 212°F.

It is apparent then that the number of Fahrenheit units compared with Celsius units is 180 to 100. This makes the Celsius degree larger than the Fahrenheit degree by a factor of 9/5. The conversion from one temperature scale to the other is complicated by the fact that 0°C and 32°F represent the same physical temperature—that of freezing water or melting ice.

To convert from the Celsius scale to the Fahrenheit scale, you must use the following formula:

$$X°\text{F} = \tfrac{9}{5}Y°\text{C} + 32 \qquad 10.1$$

Conversely,

$$Y°\text{C} = \tfrac{5}{9}(X°\text{F} - 32) \qquad 10.2$$

Jacques A. C. Charles, a French physicist, discovered in the late 1780s that the volume of a fixed mass of gas held at constant pressure decreases by about $\frac{1}{273}$ of its volume at 0°C for each degree Celsius that its temperature is decreased. This suggests that at a temperature of −273°C, the volume of the gas will theoretically be reduced to zero. In the actual case, the gas will first liquefy and then solidify with decreasing temperature so that the zero-volume prediction does not actually occur.

This "zero-volume" temperature point is also the temperature where all molecular activity would theoretically cease as the temperature is lowered. (Higher temperature implies greater molecular activity, and vice versa.)

Absolute zero is thus defined in terms of an ideal gas that would follow the same laws near the zero molecular activity point as real gases follow near room temperature. The accepted value of absolute zero is −273.15°C. The corresponding point on the Fahrenheit scale is

$$\tfrac{9}{5}(-273.15°\text{C}) + 32 = -459.67°\text{F}$$

A new set of temperature scales, *absolute temperature scales*, uses the absolute zero temperature as the zero point of the scales. Two such scales are commonly used. They are the Kelvin scale based on the Celsius degree and the Rankine scale based on the Fahrenheit degree. Thus the unit size on the Kelvin scale is identical with the Celsius degree, and the unit size on the Rankine scale is identical with the Fahrenheit degree.

1 Kelvin unit = 1 Celsius degree

1 Rankine unit = 1 Fahrenheit degree

The conversion from the Celsius scale to the Kelvin scale is given by

$$Y\ \text{K} = X°\text{C} + 273.15 \qquad 10.3$$

and that from the Fahrenheit scale to the Rankine scale is

$$B°\text{R} = A°\text{F} + 459.67 \qquad 10.4$$

Because the Kelvin and Rankine scales use the same zero temperature point, the conversion between them is

$$\text{Kelvin temperature} = \tfrac{5}{9}\ \text{Rankine temperature} \qquad 10.5$$

A comparison of the four temperature scales discussed here is shown graphically in Fig. 10.2.

Note that in SI units, the kelvin is the base unit for temperature. It is not capitalized nor is it preceded by the symbol for degrees (°). If the unit is abbreviated, it is capital K—again without the degree symbol. For example,

$$100\ \text{kelvins} = 100\ \text{K}$$

Conventional practice is to use the degree symbol with the other temperature units, such as 20°C, 72°F, or 320°R.

Example problem 10.1 Determine normal body temperature, 98.6°F, on the Celsius, Kelvin, and Rankine scales.

Solution

$$\tfrac{5}{9}(98.6°\text{F} - 32) = 37°\text{C}$$

$$37.0°\text{C} + 273.15 = 3.10 \times 10^2\ \text{K}$$

$$98.6°\text{F} + 459.67 = 5.58 \times 10^{2}°\text{R}$$

10.2.2 Atomic Weight and Moles

In 1805, John Dalton proposed an atomic theory that stated that all matter is made up of particles called atoms, with atoms of the same element being all alike but differing from atoms of another element. Later developments have shown that even atoms of the same element may differ. In fact, over three-fourths of all naturally occurring chemical elements are mixtures of between two and ten different kinds of atoms. The atoms of different weight belonging to a given chemical element are called *isotopes* of that element. For example, hydrogen is made up of three isotopes: protium (common hydrogen), deuterium (heavy hydrogen), and tritium.

To better understand isotopes requires an understanding of the basic composition of atoms. *Atoms* are composed of positively charged particles (protons), negatively charged particles (electrons), and electrically neutral particles (neutrons). Electrons, protons, and

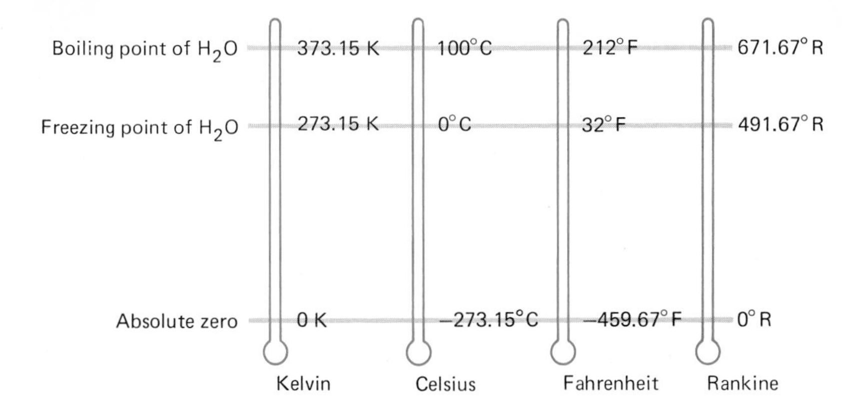

Figure 10.2

neutrons are always the same, no matter what element they are a part of; but the number of each of these particles varies from one element to another. The electrons form a revolving cloud around a nucleus of protons and neutrons. The mass of each electron is extremely small compared with the mass of either a proton or a neutron, so that the mass of the nucleus is nearly the same as the mass of the entire atom.

Each isotope of an element contains the same number of electrons and protons, but the number of neutrons differs. This is the factor that causes differences in the weights of the isotopes of the same element.

The number of protons in the nucleus of an atom is called the *atomic number*. The total number of protons and neutrons in the nucleus is the *mass number*. Obviously, the difference between the mass number and the atomic number is the number of neutrons present. From these definitions, isotopes of an element differ in mass number but have the same atomic number. In order to distinguish between the different isotopes of an element, the mass number is added to the symbol of an element as a superscript on the right or left. The symbol ^{12}N or N^{12} represents the nitrogen atom of mass number 12. N^{13}, N^{14}, and N^{15} refer to other isotopes with larger mass numbers (more neutrons in the nucleus).

Tabulated values of the masses of isotopes or elements are not given in absolute mass units such as grams because numerical factors of 10^{-22}, 10^{-23}, and so forth, would have to be used because of the extremely small size of an atom. Masses are instead given in relative units, called *unified atomic mass units*. One unified atomic mass unit is defined to be $\frac{1}{12}$ of the mass of a carbon 12 atom. Therefore, the mass of C^{12} is given as 12. Using this scale, the mass of the nitrogen 14 atom has been measured to be 14.003 07. A tabulation listing the isotope masses of the elements is referred to as a *table of nuclidic masses*.

Chemical calculations that require knowledge of the masses of the atoms involved normally do not utilize the masses of isotopes. Usually chemical reactions don't depend on which isotope is present because isotopes have very similar chemical properties. Therefore, since most elements occur in nature (and likewise in chemical reactions) as a mixture of isotopes, the mass used for chemical calculations is a weighted average of the masses of the isotopes involved. The average mass of the atoms of an element is also given in unified mass units. These tabulations of masses are called *relative atomic weights, chemical atomic weights*, or *atomic weights*. It is important to distinguish between nuclidic masses and atomic weights. Atomic weights will be used for calculations in this text.

The extremely small mass of an individual atom makes it inconvenient to make calculations for each atom. Moreover, chemical experiments usually involve many, many atoms. This would suggest that a mass unit be defined that is of the order of magnitude of the mass that might be present in a practical experiment.

The amount of material contained in 12 g of C^{12} is taken as a basis. It is called a *mole*. A mole of any other substance has exactly the same number of atoms as are present in 12 g of C^{12}. That number of atoms, *Avogadro's number*, has been experimentally determined to be 6.023×10^{23}. Therefore, a mole of 0^{16} or a mole of H^1 each contains 6.023×10^{23} atoms, although it is obvious that the mole of 0^{16} has a larger mass than the mole of H^1 because each oxygen atom is more massive than each hydrogen atom.

The definition of 12 g as the mass of one mole of C^{12} makes it possible to calculate the size of the unified mass unit. Since C^{12} has a mass of 12 unified mass units and a mole of C^{12} has a mass of 12 g that contains 6.023×10^{23} atoms, one unified mass unit must be

$$\frac{1}{6.023 \times 10^{23}} = 1.660 \times 10^{-24} \text{ g}$$

It is simple to calculate the mass of a mole of any element because the mass of a mole of the element in grams is numerically equal to its atomic weight in unified mass units. Therefore, a mole of H has a mass of 1.007 97 g because 1.007 97 is the atomic weight of hydrogen. A mole of 0 is then 15.999 4 g of oxygen. For this reason, atomic weights are sometimes shown in units of grams per gram-atom or grams per mole.

10.2.3
Chemical Formulas and Equations

Each element is identified by an elementary symbol. When elements react to form new substances, combinations of these symbols (formulas) are used to describe the new substances. An *empirical formula* shows the simplest combination of elements that form the substance. Ordinary water has the empirical formula H_2O; that is, two atoms of the element hydrogen react with one atom of the element oxygen to become water, which has properties quite different from either of the parent elements.

The empirical formula for hydrogen peroxide is HO, meaning that for each atom of hydrogen there is also one atom of oxygen. This, however, is not the formula normally associated with hydrogen peroxide. Rather, the *molecular formula*, H_2O_2, is used, because it gives additional information about the substance. It specifies that the molecule of hydrogen peroxide is actually made up of two atoms each of hydrogen and oxygen. The molecular formula is always a whole number multiple (1, 2, 3, . . .) of the empirical formula.

The weight of the atoms shown in the molecular formula is called the *molecular weight*. It is determined by adding up the atomic weights of each of the atoms involved. In the case of hydrogen peroxide, the molecular weight is

$$2(1.007\ 97) + 2(15.999\ 4) = 34.014\ 7$$

A mole of H_2O_2 contains 6.023×10^{23} molecules and therefore has a mass of 34.014 7 g (the molecular weight expressed in grams).

In general, a mole of any substance is a collection of 6.023×10^{23} entities, whether they be atoms, molecules, ions, and so on. When the mass of a mole is expressed in grams, it is referred to as a *gram-mole*.

Chemical reactions take place when many types of substances are brought together under appropriate conditions. The process of combining substances (called *reactants*) into new substances (called *products*) can be shown by writing a chemical equation. Convention dictates that the reactants be placed on the left side of the equation and the products on the right side. Arrows are used to separate the two sides and show the direction in which the reaction is occurring. Arrows directed both ways mean that some of the products are themselves reacting to form the reactants anew. Chemical formulas represent all substances involved in the reaction; and the coefficients ahead of the formulas specify the relative number of molecules of each substance required.

An example is the exposure of iron to oxygen to form ferric oxide, commonly known as rust:

$$4Fe + 3O_2 \rightarrow 2Fe_2O_3$$

This chemical formula illustrates that both iron (Fe) and oxygen (O_2) must be present as the reactants to form, in this case, the single product ferric oxide (Fe_2O_3). In addition, the arrow to the right shows that the reaction takes place only in that direction (and is not reversible). Also in the equation are coefficients explaining that four atoms of Fe and three molecules of O_2 combine to form two molecules of Fe_2O_3.

The equation does not mean that every iron atom will combine with an appropriate number of oxygen atoms to form ferric oxide, but rather that if the reaction does take place, the reaction shown will occur. Everyone is aware that iron parts will rust on the surface, but iron atoms in the interior may not react. Moreover, the equation

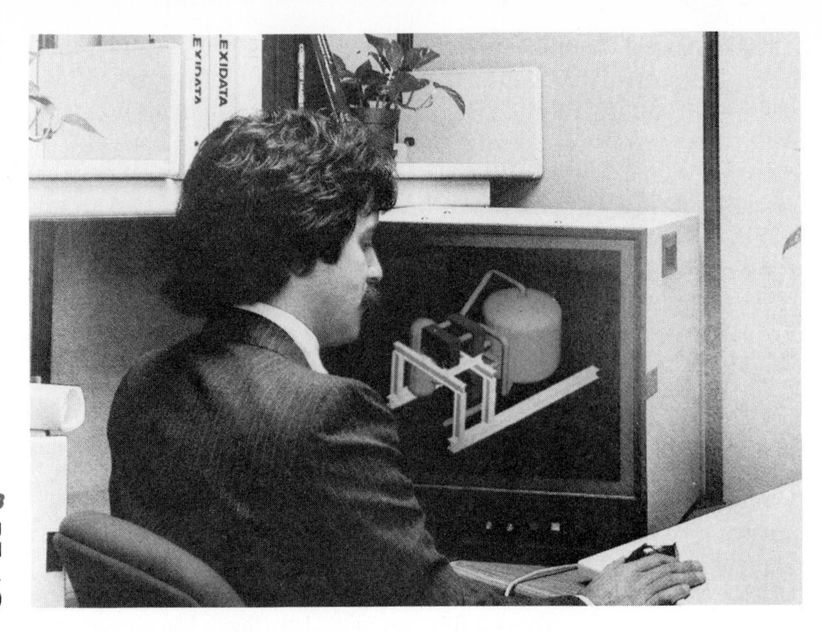

Figure 10.3
Computer graphics modeling techniques assist a chemical engineer with a plant design. (*Lexidata.*)

does not give any information about how rapidly the reaction occurs. Some reactions take years to complete, whereas others occur so rapidly that they are explosive. What the equation above does say is that if large quantities of Fe and O_2 are brought together, some of the iron and oxygen will react to form Fe_2O_3.

10.2.4 Density

The density of a substance is a measure of the quantity of material contained in a given volume. The definition is

$$\text{Density} = \frac{\text{mass of substance}}{\text{volume of substance}} = \frac{\text{mass}}{\text{unit volume}} \qquad 10.6$$

Typical units for density would be kilograms per cubic meter (kg/m^3) in SI units or slugs per cubic foot ($slug/ft^3$) in the British system.

Example problem 10.2 What is the density of cast iron if 3.00 m^3 has a mass of 21.6 Mg?

Solution

$$\text{Density} = \frac{\text{mass of cast iron}}{\text{volume of cast iron}} = \frac{21.6\ \text{Mg}}{3.00\ \text{m}^3} = 7.20\ \text{Mg/m}^3$$

Example problem 10.3 If the density of water is 1.000 Mg/m^3, what is the mass of 4 cm^3 of water?

Solution

$$\text{Mass}_{H_2O} = (\text{density}_{H_2O})(\text{volume}_{H_2O})$$

$$= \frac{1.000 \times 10^6 \text{ g}}{1 \text{ m}^3} \Bigg| \frac{4 \text{ cm}^3}{} \Bigg| \frac{1 \text{ m}^3}{(1.0 \times 10^2)^3 \text{ cm}^3}$$

$$= 4 \text{ g}$$

Another term used in the measure of density is *specific gravity*, which is the ratio of the mass of a substance to the mass of an equal volume of a standard substance. Water at 4°C is typically taken as the standard for comparison of liquids and solids. The density of water at this temperature is about 1.000 Mg/m^3. Thus, if the specific gravity of a solid or liquid is known, its density can be found by simply multiplying the specific gravity by 1.000 Mg/m^3.

Specific gravity has one advantage in that its value is independent of the system of units being used. Density can be found in any set of units if the density of water is known in those units.

Example problem 10.4 The specific gravity of cast iron is 7.20. What is its density in kilograms per cubic meter?

Solution

$$\text{Density of cast iron} = (\text{specific gravity})(\text{density of water})$$

$$= (7.20)(1.000 \text{ Mg/m}^3)$$

$$= 7.20 \text{ Mg/m}^3$$

$$= 7.20 \times 10^3 \text{ kg/m}^3$$

10.2.5 Perfect-Gas Relationships

The volume and temperature of a gas are much more sensitive to the pressure exerted on it than are those of a liquid. For many calculations with liquids, the change in pressure can be ignored as long as the pressure is near atmospheric pressure. For gases, however, even small pressure changes can result in significant volume or temperature changes.

The definition of pressure is

$$\text{Pressure} = \frac{\text{force acting on a surface}}{\text{area of surface}} \qquad 10.7$$

Pressure as a force per unit area is typically given in newtons per square meter (N/m^2) or pounds force per square inch (lbf/in^2). Normal sea-level atmospheric pressure is 0.101 MN/m^2 or 14.7 lbf/in^2—the pressure exerted by the column of air above a 1-m^2 or 1-in^2 area of the earth at sea level, respectively. One newton per square meter is called a pascal (Pa).

A perfect gas is an idealized behavioral model of a real gas. All gases follow the ideal model at sufficiently low pressures and high temperatures. Although the pressure and temperature range where ideal gas behavior is expected will not be defined numerically here (it depends on the gas and accuracy requirements, among other things), it can be said that the ideal range is approached at conditions removed from the point where the gas starts to condense to a liquid (high pressure, low temperature, or a combination of these). The ideal range is also restricted on the high-temperature side below the temperature where dissociation (breakup of molecules) or ionization (removal of electrons from atoms) begins to occur. Fortunately, the pressures and temperatures where gases are used in many engineering applications are in the ideal behavioral range for the gases.

Three basic laws are used to describe the behavior of a perfect, or ideal, gas: Boyle's law, Charles' law, and Gay-Lussac's law.

Boyle's law states that pressure and volume are inversely proportional when the temperature of a given mass of gas is held constant. In equation form,

$$P_1V_1 = P_2V_2 \tag{10.8}$$

where P_1 and V_1 are the pressure and volume of the gas at state 1 and P_2 and V_2 are the pressure and volume at state 2.

Charles' law relates the absolute temperature and volume of a fixed mass of a perfect gas undergoing a constant-pressure process:

$$\frac{V_1}{T_1} = \frac{V_2}{T_2} \tag{10.9}$$

For a constant-pressure process, an increase in temperature T results in a proportional increase in volume V.

Gay-Lussac's law establishes the relation between pressure and absolute temperature for a constant-volume process:

$$\frac{P_1}{T_1} = \frac{P_2}{T_2} \tag{10.10}$$

Note carefully that the temperature in the ideal-gas laws *must* be expressed as absolute temperature (Kelvin or Rankine scales). Any temperature given in degrees Celsius or degrees Fahrenheit must first be converted to the appropriate absolute scale before being substituted into the gas-law equations.

The gas laws stated above can be combined and a *general ideal-gas law* can be found which allows for variations in pressure, temperature, and volume:

$$\frac{P_1V_1}{T_1} = \frac{P_2V_2}{T_2} \tag{10.11}$$

or more generally, for a given mass of gas,

$$\frac{PV}{T} = \text{constant} \tag{10.12}$$

10.3 Chemical Calculations

Several calculation techniques will be described which use the concepts introduced in Sec. 10.2 to achieve engineering results for practical problems.

10.3.1 Method of Balancing Chemical Equations

Balancing a chemical equation is simply applying the *law of conservation of mass*. Stated in terms relating to chemical equations, it is as follows: The total number of atoms of each element represented among the reactants must equal the total number of atoms of the same element found in the products.

The balancing process can begin only after you have a skeleton equation. That is, you must know which products are produced from a given set of reactants. Then unknown numbers can be assigned as the coefficients of each reactant and each product. The problem reduces to the determination of this set of unknown numbers such that the number of atoms of each element is conserved and the coefficients are the set of smallest integers that can be found.

The process of producing elemental iron in a blast furnace is given by the following skeleton equation:

$$Fe_2O_3 + CO \rightarrow Fe + CO_2$$

In this form, the proper reactants and products are shown but the relative number of molecules of each needed is not correct, because the law of conservation of mass has been violated. The equation shows two Fe atoms on the left side and only one on the right. Likewise, there are four oxygen atoms on the left and two on the right.

Balancing this or any other equation is a bit of an art, often requiring a trial-and-error process. First, assign an unknown set of numbers as the coefficients of all reactants and products:

$$a\ Fe_2O_3 + b\ CO \rightarrow c\ Fe + d\ CO_2$$

As a general rule, it is best to begin by looking at the most complex molecule. In this case, it is the Fe_2O_3 molecule. Note that there are a minimum of two atoms of Fe among the reactants because a cannot be smaller than 1. Then as a first trial, choose $a = 1$, which means that $c = 2$, to provide a balance in Fe atoms. It is apparent that carbon atoms appear in CO on the reactant side and CO_2 on the product side only. Therefore, it can be concluded that b and d must be equal.

Selection of $b = 1$ results in four O's on the left side and two on the right. With $b = 2$, there are five O's on the left side and four on the right; but with $b = d = 3$, there are six oxygen atoms on both the left and right sides. The balanced equation is then

$$(1)Fe_2O_3 + (3)CO \rightarrow (2)Fe + (3)CO_2$$

A somewhat more systematic approach to the same problem is to write the set of algebraic equations in the coefficients found by the application of the law of conservation of mass. The equations resulting from a balance of each element are

$$\begin{aligned} \text{Fe:}\quad & 2a = c \\ \text{C:}\quad & b = d \\ \text{O:}\quad & 3a + b = 2d \end{aligned}$$

There are three equations in four unknowns, so a choice must be made for one of the unknowns. (The extra unknown is merely a result of the fact that one may multiply through a chemical equation by any number without altering the validity of the equation.) As before, select $a = 1$, then $c = 2a = 2$. Combining the C and O equations results in

$$\begin{aligned} 3a + d &= 2d \\ d &= 3a = 3 \\ b &= d = 3 \end{aligned}$$

Therefore, the set is $a = 1$, $b = 3$, $c = 2$, and $d = 3$, as previously determined.

The number set will not always come out as integers on the first try as it did this time. Just remember, however, that if each coefficient is multiplied by the same number, the relative results are not changed. So, it is always possible to find a multiplier that will give a final set of integers. Finding the multiplier is really the trial-and-error part of the solution.

Figure 10.4
A pilot plant may be necessary to demonstrate the feasibility of a new process before full-scale equipment is built. (*Phillips Petroleum Company.*)

The simple reaction of hydrogen and oxygen producing water is given by

$H_2 + O_2 \rightarrow H_2O$ (unbalanced)

Assigning coefficients gives

$r\ H_2 + s\ O_2 \rightarrow t\ H_2O$

The balance equations are

$H{:}\quad 2r = 2t$

$O{:}\quad 2s = t$

If t is selected as 1, then the equations above give

$s = \frac{1}{2}$

$r = t = 1$

The equation is then

$(1)H_2 + (\frac{1}{2})O_2 \rightarrow (1)H_2O$

The coefficients are not all integers, but multiplication of each coefficient by 2 provides the final balanced chemical equation:

$(2)H_2 + (1)O_2 \rightarrow (2)H_2O$

10.3.2 Determination of the Mass of a Constituent from the Molecular Formula

The molecular formula for a compound specifies the relative numbers of each element contained in the molecule. The atomic weights of the elements are known, so it is possible to calculate the mass of a specified element in a given amount of the compound. Likewise, it is possible to determine the percentage composition by weight (mass) from the formula for the compound and the atomic weights of the elements involved.

Example problem 10.5 Calculate the mass of potassium in a metric ton (megagram) of potassium permanganate.

Solution The formula of the compound is $KMnO_4$. The molecular weight is given by

$$\overset{K}{39.102} + \overset{Mn}{54.938\,0} + \overset{O_4}{4(15.999\,4)} = 158.04$$

The fraction of $KMnO_4$ that is due to the element potassium is

$$\frac{39.102}{158.04} = 0.247\,42$$

The amount of potassium in 1.000 Mg (metric ton) of $KMnO_4$ is

$$(1.000 \times 10^3 \text{ kg})(0.247\,42) = 247.4 \text{ kg}$$

Example problem 10.6 Glycerin is a common substance made of carbon, hydrogen, and oxygen atoms. Its formula is $C_3H_8O_3$. Determine the percentage by weight (mass) of each constituent.

Solution

$$\text{Molecular weight} = \overset{C_3}{3(12.011\,15)} + \overset{H_8}{8(1.007\,97)} + \overset{O_3}{3(15.999\,4)} = 92.095\,4$$

$$\text{Percentage of carbon} = \frac{\text{weight of carbon}}{\text{molecular weight}}\,100\% = \frac{3(12.011\,15)}{92.095\,4}\,100\% = 39.126\%$$

$$\text{Percentage of hydrogen} = \frac{\text{weight of hydrogen}}{\text{molecular weight}}\,100\% = \frac{8(1.007\,97)}{92.095\,4}\,100\% = 8.756\%$$

$$\text{Percentage of oxygen} = \frac{\text{weight of oxygen}}{\text{molecular weight}}\,100\% = \frac{3(15.999\,4)}{92.095\,4}\,100\% = 52.118\%$$

The sum of each constituent percentage = 39.126 + 8.756 + 52.118 = 100.00 percent. This provides a check on the percentage calculations, since the sum must be 100 percent.

This problem could have asked for the fraction of each constituent element, giving 0.391 26, 0.087 56, and 0.521 18 for carbon, hydrogen, and oxygen, respectively.

10.3.3 Determination of the Empirical Formula of a Compound from Its Percentage Composition

Various techniques are available to determine which elements are present in an unknown compound and to find the weights (masses) of each of them in a given amount of the compound. This data can then be used to calculate the empirical formula for the compound. The weight (mass) data may be given in either percent of total or in number of grams of each element.

A procedure for finding the empirical formula is as follows:

1. Choose an amount of compound to use as a basis for the calculations if the data are given in percentage terms. (An excellent choice is 100 g.) Use the actual amount if amounts of constituents are given in grams, pounds mass, etc.
2. List the amount of each element per 100 g or per actual amount if known.
3. Determine and list the atomic weight of each element in grams per gram-atom.
4. Calculate the ratio of the mass of the element to the atomic weight of the element for each element present. (The result is the relative number of gram-atoms of each element.)
5. Choose the smallest ratio found in step 4 and divide all ratios by this number.
6. Multiply all quantities calculated in step 5 by a constant, if each is not already an integer, in order to obtain all integer values. Note that if there are inaccuracies in the quantitative analysis or the calculations, some rounding of figures may be necessary at this point. Be certain to carry out all computations to as many significant figures as possible and reserve the rounding process for the last step of the procedure.
7. Write the empirical formula from the relative gram-atoms of the elements.

Example problem 10.7 Determine the empirical formula for the compound made up of 88.82% oxygen and 11.18% hydrogen.

Solution Select 100 g of the compound as a basis:

Element	Amount (g/100 g)	Atomic weight	Gram-atoms	Relative gram-atoms (g-atoms/5.5515)
O	88.82 g	15.999 4	5.551 5	1.000
H	11.18 g	1.007 97	11.091 6	1.998

Round off the number of relative gram-atoms to 1 and 2. The empirical formula is H_2O.

Example problem 10.8 A compound has been found to contain the following percentages of its elements: sulfur = 24.52%, oxygen = 48.96%, and chromium = 26.52%. What is its empirical formula?

Solution Again select 100 g of the compound as a basis with which to perform the calculations:

Element	Amount (g/100 g)	Atomic weight	Gram-atoms	Relative gram-atoms (g-atoms/0.5100)
S	24.52 g	32.064	0.764 7	1.499 4
O	48.96 g	15.999 4	3.060 1	6.000 2
Cr	26.52 g	51.996	0.510 0	1.000 0

In this case, the relative gram-atoms are not integer values, but if each is multiplied by 2, the S:O:Cr relative values after rounding become 3:12:2. The empirical formula is then written $Cr_2S_3O_{12}$, or $Cr_2(SO_4)_3$.

Example problem 10.9 A compound has been experimentally found to contain 68.97 g sodium, 3.024 g hydrogen, 96.19 g sulfur, and 192.0 g oxygen. What is its empirical formula?

Solution In this case, it is unnecessary to assume a basis on which to work because the actual quantities of each element are already available.

Element	Amount, g	Atomic weight	Gram-atoms	Relative gram-atoms (g-atoms/3.000)
Na	68.97	22.989 8	3.000	1.000
H	3.024	1.007 97	3.000	1.000
S	96.19	32.064	3.000	1.000
O	192.0	15.999 4	12.000	4.000

The empirical formula for the compound is $NaHSO_4$.

Figure 10.5
This desaltation plant in the West Indies, shown during construction and after completion, demonstrates the kind of impact that chemical engineers can have on people throughout the world. (*Stanley Consultants.*)

Problems

10.1 What are the equivalents of 72°F on the Celsius and Kelvin scales?

10.2 Convert 600 K to Fahrenheit and Celsius degrees.

10.3 Express the following temperatures in degrees Celsius:

(*a*) 200 K	(*b*) 127°F	(*c*) 460°R	(*d*) 3 000°F
(*e*) 29 K	(*f*) 29°F	(*g*) 29°R	(*h*) 100°F

10.4 Express the following temperatures in kelvins:
(*a*) 200°C (*b*) 460°R (*c*) −4°F (*d*) 4 000°R (*e*) 72°F

10.5 Express the following temperatures in degrees Rankine:
(*a*) 72°F (*b*) −20°C (*c*) 300 K (*d*) 0°F

10.6 Express the following temperatures in degrees Fahrenheit:
(*a*) −20°C (*b*) 200°C (*c*) 440°R (*d*) 300 K

10.7 How many atoms are there in 1 mg of NaCl?

10.8 Determine the formula weight of each of the following:
(*a*) H_2O (*b*) C_6H_6 (*c*) K_2CO_3 (*d*) $Ba(OH)_2$

10.9 What is the molecular weight of each of the following compounds?
(*a*) CH_2O (*b*) $C_6H_{12}O_6$ (*c*) $(NH_4)_3PO_4$ (*d*) $Al_2(SO_4)_3$

10.10 What is the mass of 1 mol of each of the following?
(*a*) H_2 (*b*) H_2SO_4 (*c*) $C_{12}H_{22}O_{11}$ (*d*) $NaHCO_3$

10.11 How many moles of each of the following are contained in 100 g of the pure substance?

(*a*) H_2O	(*b*) CH_4	(*c*) C_8H_{18}
(*d*) $CaCO_3$	(*e*) $MgSO_4$	(*f*) $H_3C(CH_2)_{14}CH_2OH$

10.12 In the English system, the density of water is 62.4 lbm/ft_3. What is the density of water in grams per cubic centimeter?

10.13 What is the density of aluminum if 60.0 cm^3 has a mass of 162.0 g?

10.14 Determine the weight (mass) of 1.0 gal of gasoline if gasoline has a density of 0.71 g/cm^3.

10.15 Calculate the mass of 1.00 m^3 of a substance whose specific gravity is
(*a*) 1.98 (*b*) 2.70 (*c*) 0.178 (*d*) 0.5

10.16 A part weighs 0.50 N in air and 0.30 N when immersed in water. What is the specific gravity of the part?

10.17 A bottle that weighs 120 N empty weighs 260 N when filled with water and 243 N when filled with ammonia. What is the specific gravity of ammonia? What is the capacity of the bottle?

10.18 What will be the final volume of 2.00 L of a gas at 2.00-atm pressure if the pressure is increased to 6.00 atm? The temperature does not change.

10.19 Calculate the temperature of a fixed volume of gas after its pressure has been increased from 2.57 to 2.96 MPa. Its initial temperature was 20°C.

10.20 Oxygen is heated in a constant-pressure process from 25 to 170°C. The initial volume was 4.66 m^3. What is the final volume?

10.21 Hydrogen gas can be cooled from 72°F to −20°C by a constant-volume process. The original pressure of 14.7 lbf/in^2 has been changed to what pressure in pascals?

10.22 Ammonia occupies 8.9 m^3 at 2.05 MPa. What is its volume at standard pressure? The process is one of constant temperature.

10.23 A constant-pressure process that results in a volume increase of 2.0 L is performed on 6.0 L of neon. What is the final gas temperature if it was initially 25°C?

10.24 Nitrogen is contained in an 11-L tank at 4.2-atm pressure and 100°F. It is allowed to expand into a chamber where the pressure is 1.0 atm and the temperature is 72°F. What volume does the nitrogen occupy now?

10.25 Air, initially at 25°C and 1.0-atm pressure, is compressed to one-half of its original volume. If the new pressure is 1.75 atm, what is the new temperature?

10.26 A gas sample has a volume of 0.54 L measured at 97°C and 0.53 atm. What is its volume at 0°F and 14.7 lbf/in^2?

10.27 If a football is inflated to an absolute pressure of 21 lbf/in^2 in a room that has a temperature of 25°C, what will the pressure in the ball be during the game at 5°C?

10.28 Balance the following equations:

(*a*) $BCl_3 + P_4 + H_2 \rightarrow BP + HCl$
(*b*) $C_7H_6O_2 + O_2 \rightarrow CO_2 + H_2O$
(*c*) $NH_3 + O_2 \rightarrow N_2 + H_2O$

10.29 Lead carbonate has the formula $PbCO_3$. How much lead is contained in 1 kg of this compound?

10.30 Ferric chloride ($FeCl_3$) is produced when iron (Fe) is heated in an atmosphere of gaseous chlorine (Cl_2). Determine

(*a*) The balanced equation for this reaction
(*b*) The mass of chlorine needed to produce 10 g of ferric chloride

10.31 One component of gasoline, octane, reacts with oxygen according to the following unbalanced equation:

$$C_8H_{18} + O_2 \rightarrow CO_2 + H_2O$$

(*a*) Balance the equation.
(*b*) Calculate the mass of water produced when 4 kg of octane is completely burned.

10.32 Hydrogen gas can be used to reduce iron ore, Fe_2O_3, to its metal as shown in the following unbalanced equation:

$$Fe_2O_3 + H_2 \rightarrow Fe + H_2O$$

(*a*) Balance the equation.
(*b*) Determine the amount of iron ore needed to produce 100 kg of pure iron.
(*c*) Indicate how much hydrogen would be consumed in the process of producing 100 kg of iron.

10.33 A lead storage battery has positive electrodes filled with lead dioxide, PbO_2, and negative electrodes filled with spongy lead, both dipped in a concentrated sulfuric acid solution. Upon discharge, the following reaction takes place:

$$PbO_2 + Pb + H_2SO_4 \rightarrow PbSO_4 + H_2O \qquad \text{(unbalanced)}$$

(*a*) Balance this equation.
(*b*) If 1 g of sulfuric acid is used up, state how much water is produced.

10.34 Liquid hydrogen and liquid oxygen have been a popular fuel combination for large rocket engines. The reaction is simply $2H_2 + O_2 \rightarrow 2H_2O$. If 20 000 kg/s of water must be formed to produce the necessary thrust, how much hydrogen and how much oxygen must be carried aboard to allow a 60-s burn?

10.35 Sodium lauryl sulfate, $C_{12}H_{25}OSO_3Na$, is a detergent produced in a two-step process from lauryl alcohol, $C_{12}H_{25}OH$, sulfuric acid, H_2SO_4, and sodium hydroxide, NaOH:

$$C_{12}H_{25}OH + H_2SO_4 \rightarrow C_{12}H_{25}OSO_3H + H_2O$$

$$C_{12}H_{25}OSO_3H + NaOH \rightarrow C_{12}H_{25}OSO_3Na + H_2O$$

If you as an engineer are operating a process to produce 25 t (metric tons) of detergent per day, what are the minimum amounts of lauryl alcohol, sulfuric acid, and sodium hydroxide that you would need each day?

10.36 Table sugar, $C_{12}H_{22}O_{11}$, can be decomposed into carbon and water.
(*a*) Write the equation for this reaction and balance it.
(*b*) State how many molecules of water are produced by the decomposition of 10 molecules of sugar.
(*c*) Indicate how much carbon and water the decomposition of 1 g of sugar will produce.

10.37 Hydrogen can be produced in the laboratory by reacting CaH_2 with water to produce $Ca(OH)_2$ and H_2.
(*a*) Write a balanced equation for this reaction.
(*b*) State how many grams of H_2 will result from 100 g of CaH_2.

10.38 Acetic acid has the formula CH_3COOH. Determine the percentage by weight of each constituent.

10.39 NH_4NO_3 is the formula for ammonium nitrate. What is the mass percentage of each element it contains?

10.40 Calculate the percentage composition of anhydrous sodium phosphate, Na_2HPO_4.

10.41 What is the percentage of water in hydrated sodium phosphate, $Na_2HPO_4 \cdot 5H_2O$?

10.42 A salt has been found to contain 60.6% chlorine and 39.4% sodium by weight. What is its empirical formula?

10.43 Determine the empirical formula for teflon, which is 70.37% fluorine and 29.63% carbon.

10.44 A compound known as isopropanol has a molecular weight of 60. It is composed of 60% carbon, 13.3% hydrogen, and the remainder oxygen by mass. Write:

(*a*) Its empirical formula
(*b*) Its molecular formula

10.45 Determine the empirical formula for a compound containing 21.6% boron, 22.8% sodium, and 55.9% oxygen by weight.

10.46 Calculate the empirical formula for a compound found to contain 65.3% oxygen, 32.7% sulfur, and 2.04% hydrogen.

10.47 In 30 g of a pure compound, there are 8.73 g of sodium, 9.12 g of oxygen, and 12.15 g of sulfur. What is the simplest formula for this compound?

10.48 Ten grams of a flammable gas was analyzed and found to be composed of 7.49 g of carbon and 2.51 g of hydrogen. What is its formula?

10.49 A gaseous compound of hydrogen and carbon is burned to provide 1.798 g of water and 6.60 g of carbon dioxide. Find the empirical formula for the compound.

10.50 Derive the empirical formula for a compound containing 56.33% oxygen and 43.67% phosphorus by weight.

11

Material Balance

11.1 Introduction

We depend a great deal on industries that produce food, household cleaning products, energy for heating and cooling homes, fertilizers, and many other products and services. These process industries, as they are called, are continually involved with the distribution, routing, blending, mixing, sorting, and separation of materials. (See Fig. 11.1 for one example of a processing system.)

A typical process problem that an engineer might be called on to solve is exemplified by the drying process, shown schematically in Fig. 11.2. A process engineer designing a system to dry grain would most likely know the percent moisture (on a mass basis) of the wet grain, the desired moisture content for the dried grain, and the amount of grain to be dried in a specific amount of time. The engineer would then have to calculate the flow rate of dry heated air required to be forced through the grain. Knowing the air flow rate, the engineer could then design the mechanical system of heaters, motors, blowers, and ducting.

To perform computations involving material flow in a process, you must use an engineering analysis technique called *material balance*, which is based on the principle of conservation of mass.

Figure 11.1
A portable processing unit that separates quarry material according to aggregate specifications. The system can perform required crushing and screening operations. (*Iowa Manufacturing Company.*)

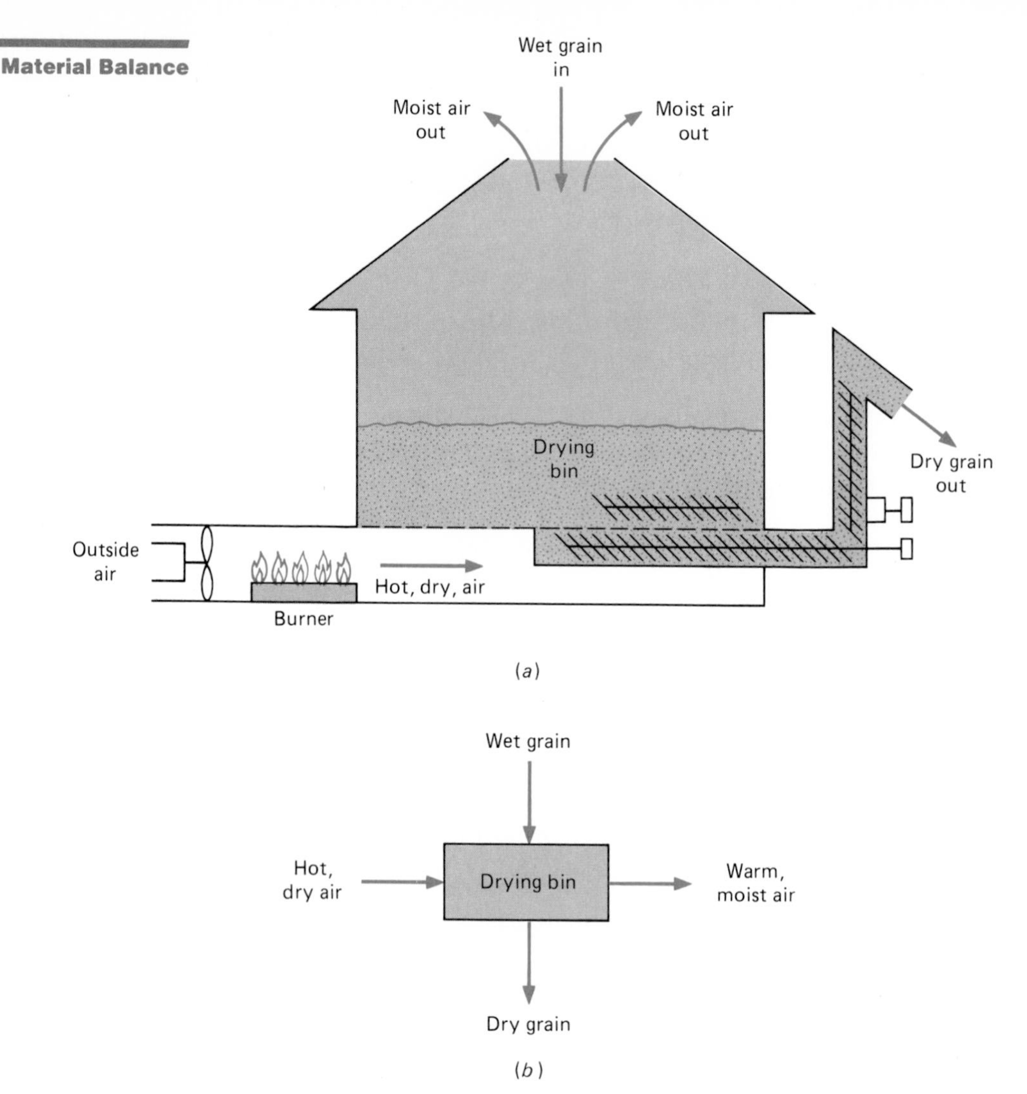

Figure 11.2
(*a*) Cross section of a typical grain-drying process; (*b*) working schematic used to depict the actual process so that the necessary calculations can be performed.

11.2 Conservation of Mass

The conservation-of-mass principle is a very useful concept in the field of engineering analysis. Simply stated, it says that, excluding nuclear reactions, mass is neither created nor destroyed. We know that mass is converted to energy in a nuclear reactor so the conservation-of-mass principle does not apply to the reaction itself.

Before we apply the conservation-of-mass principle to a material balance problem, additional concepts and terminology must be introduced. Figure 11.3 illustrates a number of these terms.

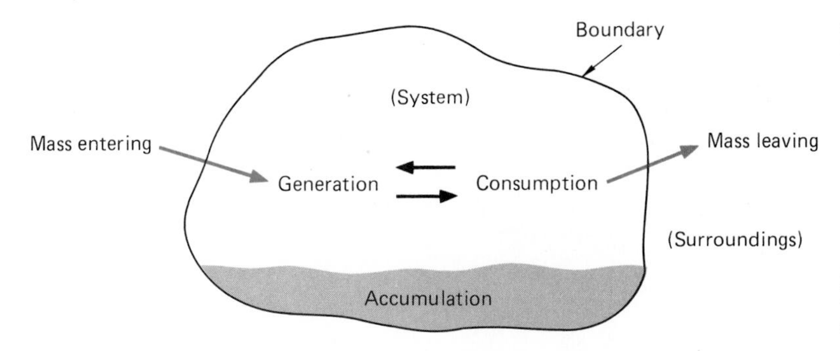

Figure 11.3
Illustration of a typical system with a defined boundary.

A *system* can be defined as any designated portion of the universe with a definable boundary. Whenever mass crosses the boundary either into or out of the system, it must be considered. In certain situations, the amount of mass entering the system is greater than the amount leaving. This results in an increase of mass within the system called *accumulation*. If the mass leaving the system is greater than that entering, the accumulation is negative.

When chemical reactions are included within the system boundaries, chemical reactants are consumed in the formation of reaction products. A simple example would be the dehydrogeneration of ethane in a reactor, i.e., $C_2H_6 \rightarrow C_2H_4 + H_2$. One constituent is consumed and two others are generated. Thus, if chemical reactions occur, it is necessary to account for the consumption of some elements or compounds and the generation of others. It is important to understand that, even considering chemical reactions, mass is conserved. In the example above ($C_2H_6 \rightarrow C_2H_4 + H_2$), the number of atoms of carbon and hydrogen remains constant.

Before using the conservation-of-mass principle to analyze a system, you must understand the terminology explained above and be able to account for all the constituents that enter, leave, or change within the system boundaries.

Consider this familiar example to more clearly visualize the definition of a system. Let the system boundary be the city limits of a large metropolitan area. When a person enters the city limits, we have an accumulation of plus one until someone leaves. The generation and consumption terms are harder to explain, but suppose that within the city there are three electronic-component manufacturers. Employees move from one company to the other two, thus maintaining the same number of people within the boundary but resulting in different constituents (employment per company) within the city (system).

When all of the above considerations are included, the conservation-of-mass principle applied to a system or to system constituents can be expressed as

$$\text{Input} + \text{generation} - \text{output} - \text{consumption} = \text{accumulation} \qquad 11.1$$

Figure 11.4
This arrangement of duct work directs the air flow in a large building. It illustrates the complicated network of an actual situation. (*Stanley Consultants.*)

11.3 Processes

Two types of processes typically analyzed are the batch process and the rate process. In a *batch process*, materials are put or placed into the system before the process begins and are removed after the process is complete. Cooking is a familiar example. Generally, you follow a recipe that calls for specific ingredients to be placed into a system that produces a processed food.

A *rate-flow process* involves the continuous time rate of flow of inputs and outputs. The process is performed continuously as mass flows through the system. An example of a rate-flow process is a pipe delivering water to a tank at the rate of 2.0 kg/s.

Rate processes may be classified as either uniform or nonuniform, steady or unsteady. A process is uniform if the input rate equals the output rate. It is steady if the rates do not vary with time. Solution of material balance problems involving nonuniform and/or unsteady flow may require the use of differential equations (that is, mathematical expressions that contain derivatives of functions as variables). However, many important processes can be classified as uniform and steady and thus be analyzed with more elementary mathematical skills.

Since this chapter is intended to be an introductory look at conservation of mass, we will now make a number of simplifying assumptions.

Many engineering problems involve chemical reactions, but if we assume no such reactions, then Eq. (11.1) can be reduced to

$$\text{Input} - \text{output} = \text{accumulation} \qquad 11.2$$

If we assume for a batch process that we take out at the end of the process all of the mass we placed into the system at the beginning, then the accumulation term is zero and Eq. (11.2) can be written

$$\text{Total input} = \text{total output} \qquad 11.3$$

Figure 11.5 is an example of a batch process for a concrete mixer.

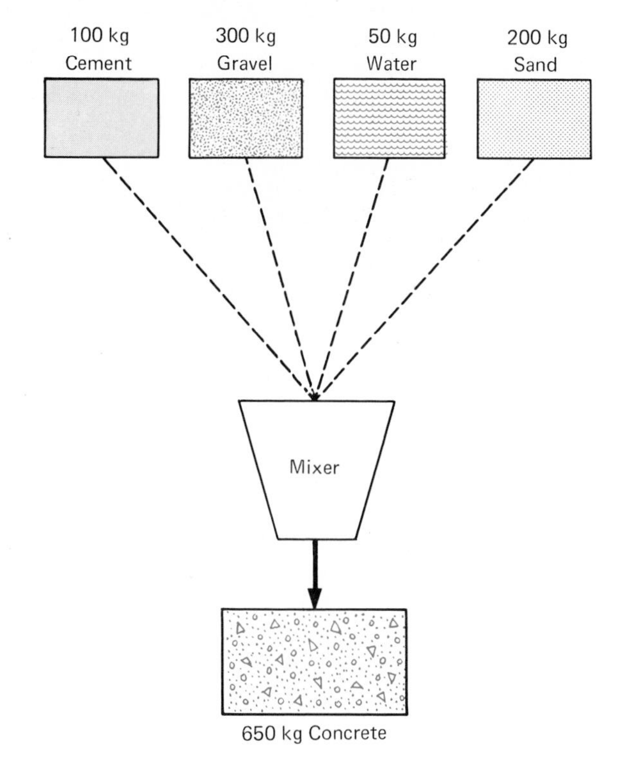

Figure 11.5
The constituents needed for a proper batch mix of concrete.

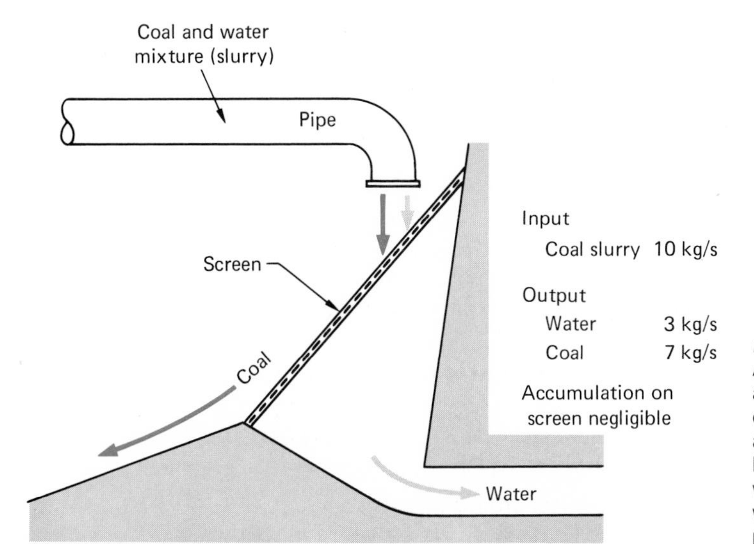

Figure 11.6
A coal dewatering system shown as a rate process. Sampling the downstream portion would provide a more detailed material balance because the coal would carry some water with it and the water side would contain some fine coal particles.

For continuous flow, if we assume a uniform, steady rate process, the accumulation term is also zero, so Eq. (11.2) reduces to

$$\text{Rate of input} = \text{rate of output} \qquad 11.4$$

Figure 11.6 illustrates a rate process for a coal dewatering process.

Although Eqs. (11.3) and (11.4) seem so overly simple as to be of little practical use, application to a given problem may be complicated by the need to account for several inputs and outputs as well as for many constituents in each input or output. The simplicity of the equations is in fact the advantage of a material balance approach, because order is brought to seemingly disordered data.

11.4 A Systematic Approach

Material balance computations require the manipulation of a substantial amount of information. Therefore, it is essential that a systematic procedure be developed and strictly followed. If a systematic approach is used, material balance equations can be written and solved correctly in a straightforward manner. The following list of steps is recommended as a procedure for solving material balance problems.

1. Identify the system(s) involved.
2. Determine whether the process is batch or rate and whether a chemical reaction is involved. If no reaction occurs, it is evident that the material balance involves compounds. If a reaction is to occur, elements must be involved and must be balanced. In a process involving chemical reactions, additional equations based on chemical composition may be required in order to solve for the unknown quantities.
3. Construct a diagram showing the feeds (inputs) and products (outputs) schematically.
4. Label known material quantities or rates of flow.
5. Identify each unknown input and output with an appropriate symbol.
6. Apply Eq. (11.3) or (11.4) for each constituent as well as for the overall process. Care must be taken to include only independent equations.
7. Solve the equations for the desired unknowns and express the results in suitable form.

To demonstrate each of these important steps, let's go through an example problem that specifically illustrates each step.

Example problem 11.1 Drinking water can be obtained from salt water by partially freezing the salt water to create salt-free ice and a brine solution. If salt water is 3.50% salt by mass and the brine solution is found to be an 8.00% concentrate by mass, determine how many kilograms of salt water must be processed to form 2.00 kg of ice.

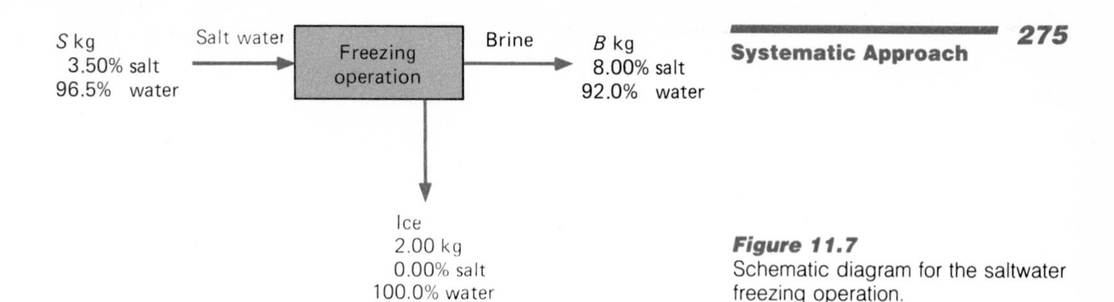

Figure 11.7
Schematic diagram for the saltwater freezing operation.

Solution

1. The system in this example problem involves a freezing operation.
2. The freezing operation is a batch process because a fixed amount of product (ice) is required. There are no chemical reactions.
3. A diagram of the process is shown in Fig. 11.7.
4. Salt water is the input to the system, with brine and ice taken out at the end (see Fig. 11.7).
5. Appropriate symbols are used to identify unknown quantities (see Fig. 11.7).
6. The material balance equation for each constituent as well as for the overall process is written. It is important to understand that the material balance equation [Eq. (11.3)] is applicable for each constituent as well as for the overall process. In this example, three equations are written, but only two are independent. That is, the overall balance equation is the sum of the salt and water balance equations. Thus we have a good method of checking the accuracy of the equations we have written.

Equation	$\boxed{\text{Input} = \text{output}}$
Overall balance	$S = B + 2.00$
Salt balance	$0.035S = 0.08B$
Water balance	$0.965S = 0.920B + 2.00$

7. The equations are solved by substitution:

$$0.035(B + 2.00) = 0.08B$$
$$0.045B = 0.070$$
$$B = 1.56 \text{ kg}$$

Since $S = B + 2.00$, then

$$S = 3.56 \text{ kg}$$

The following example problems will apply the seven steps discussed above but will not list each step separately.

Example problem 11.2 A process to remove water from solid material consists of a centrifuge and a dryer. If 35 t/h of a mixture containing 35% solids is centrifuged to form a sludge consisting of 65% solids and then the sludge is dried to 5% moisture in a dryer, how much total water is removed in a 24-h period?

Solution There are three possible systems involved in this problem: the centrifuge, the dryer, and the combination. (See Fig. 11.8.) The operation in this system is a continuous flow process. There are no chemical reactions.

Rate of input = rate of output

The following equations are written for the process illustrated in Fig. 11.8. The overall process is illustrated in Fig. 11.8*a*, with subsystem diagrams for the centrifuge and the dryer shown in Fig. 11.8*b* and *c*, respectively. The overall balance equation for a selected system is the sum of the constituent balance equations for that system. This means that the set of equations written for a selected system are not all independent when the overall balance equation and the constituent balance equations are included.

For the entire system (Fig. 11.8*a*):

1. Solid balance $0.35(35) = 0.45E$
2. Water balance $0.65(35) = B + D + 0.05E$
3. Overall balance $35 = B + D + E$

For the centrifuge (Fig. 11.8*b*):

Figure 11.8
Schematic diagrams illustrating flow process inputs and outputs depending on system boundaries.

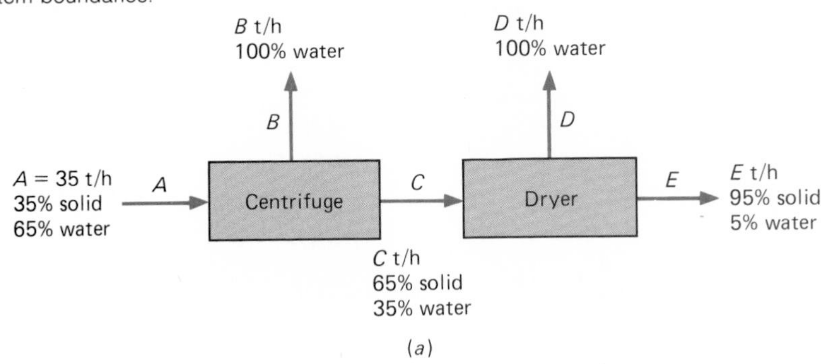

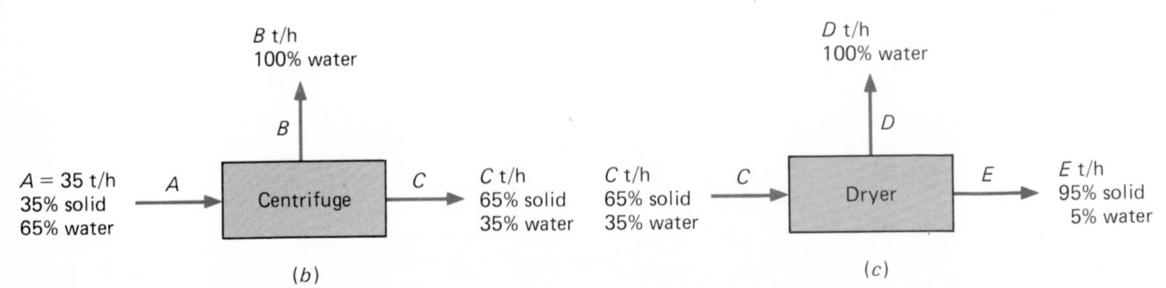

4. Solid balance $0.35(35) = 0.65C$
5. Water balance $0.65(35) = B + 0.35C$
6. Overall balance $35 = B + C$

For the dryer (Fig. 11.8*c*):

7. Solid balance $0.65C = 0.95E$
8. Water balance $0.35C = D + 0.05E$
9. Overall balance $C = D + E$

Solve for rate of mass out of centrifuge (C) from Eq. (4):

$$C = \frac{0\ 35(35)}{0.65}$$

$$= 18.85 \text{ t/h}$$

Solve for rate of water out of centrifuge (B) from Eq. (6):

$$B = 35 - C$$

$$= 35 - 18.85$$

$$= 16.15 \text{ t/h}$$

Solve for rate of mass out of dryer (E) from Eq. (7):

$$E = \frac{0.65C}{0.95}$$

$$= \frac{0.65(18.85)}{0.95}$$

$$= 12.90 \text{ t/h}$$

Solve for rate of water out of dryer (D) from Eq. (9):

$$D = C - E$$

$$= 18.85 - 12.90$$

$$= 5.95 \text{ t/h}$$

Calculate total water removed in 24 h:

$$\text{Total water} = (B + D)24$$

$$= (16.15 + 5.95)24$$

$$= (22.10)24$$

$$= 5.30 \times 10^2 \text{ t}$$

A general problem that would involve typical material balance consideration is a standard evaporation, crystallization, recycle process. Normally this type of system involves continuous flow of some solution through an evaporator. Water is removed, leaving the output stream more concentrated. This stream is fed into a crystallizer where it is cooled, causing crystals to form. These crystals are then filtered out, with the remaining solution recycled to join the feed stream back into the evaporator. This system will be more clearly illustrated in the following example problem.

Example problem 11.3 A solution of potassium chromate (K_2CrO_4) is to be used to produce K_2CrO_4 crystals. Twenty-five hundred kilograms per hour of 40% solution by mass is fed to an evaporator. The stream leaving the evaporator is 50% K_2CrO_4. This stream is then fed into a crystallizer and is passed through a filter. The resulting filter cake is 100% crystals. The remaining solution is 45% K_2CrO_4. The 45% solution that passed through the filter is recycled. Calculate the total input to the evaporator, the feed rate to the crystallizer, the water removed from the evaporator, and the amount of pure K_2CrO_4 produced each hour.

Solution See Fig. 11.9. There are different ways the system boundaries can be selected for this problem, i.e., around the entire system, around the evaporator, around the crystallizer-filter, etc. There are no chemical reactions that occur in the process.

Rate of input = rate of output

The following balance equations can be written, not all of which are independent.

For the entire system (Fig. 11.9*a*):

1. K_2CrO_4 balance $\quad 0.40(2\,500) = D$
2. H_2O balance $\quad 0.60(2\,500) = B$
3. Overall balance $\quad 2\,500 = B + D$

For the evaporator (Fig. 11.9*b*):

4. K_2CrO_4 balance $\quad 0.40(2\,500) = 0.50C$
5. H_2O balance $\quad 0.60(2\,500) = B + 0.50C$
6. Overall balance $\quad 2\,500 + E = B + C$

For the crystallizer-filter (Fig. 11.9*c*):

7. K_2CrO_4 balance $\quad 0.50C = D + 0.45E$
8. H_2O balance $\quad 0.50C = 0.55E$
9. Overall balance $\quad C = D + E$

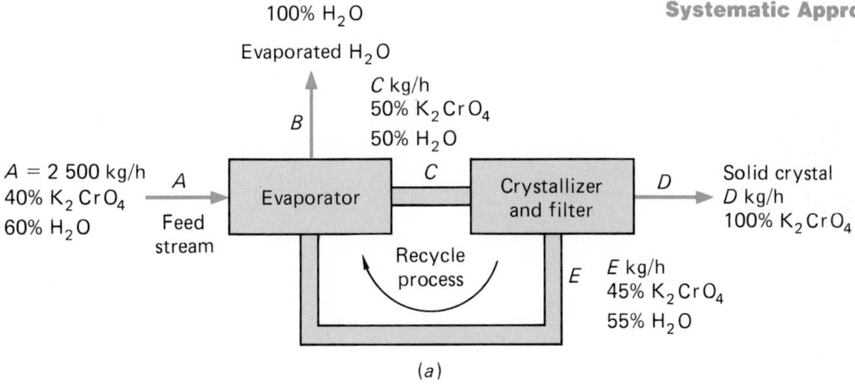

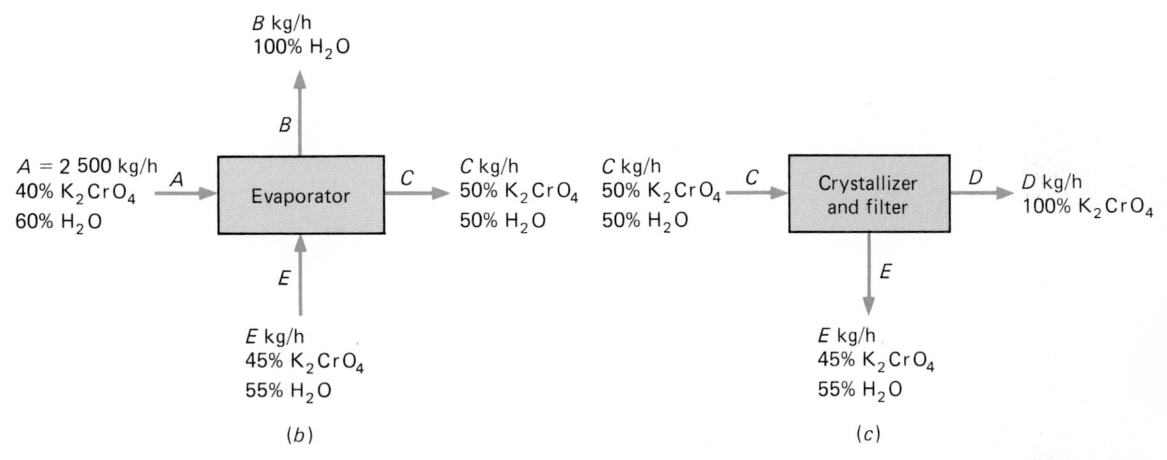

Figure 11.9
(*a*) Schematic of the overall system; (*b*) and (*c*) diagrams of systems where boundaries are selected around individual components.

Solve for K_2CrO_4 out of crystallizer-filter from Eq. (1):

$$D = 0.4(2\,500)$$
$$= 1.00 \times 10^3 \text{ kg/h}$$

Solve for H_2O out of evaporator from Eq. (2):

$$B = 0.6(2\,500)$$
$$= 1.50 \times 10^3 \text{ kg/h}$$

Solve for recycle rate from crystallizer-filter from Eqs. (9) and (7):

$$(9) \quad C = E + D$$
$$= E + 1\,000$$

$$(7) \quad 0.5C = D + 0.45E$$
$$0.5(E + 1\,000) = 1\,000 + 0.45E$$
$$E = 10\,000 \text{ kg/h}$$

Calculate total input for evaporator:

$$\text{Total input} = E + A = 10\,000 + 2\,500 = 12\,500 \text{ kg/h}$$

Calculate feed rate for crystallizer-filter from Eq. (6):

$$C = 2\,500 + E - B = 2\,500 + 10\,000 - 1\,500 = 1.10 \times 10^4 \text{ kg/h}$$

Problems

11.1 A quantity of pure dry salt is added to 82.7 kg of brine that is 11.3% salt. How much dry salt must be added to produce a 13.62% brine?

11.2 If 16.71 kg of water is evaporated from a vat containing 123.2 kg of a 12.3% brine and 2.37 kg of pure dry salt is added, what percentage of salt does the resulting brine contain?

11.3 A vat contains 1120 kg of brine that is 7.62% salt. If 233 kg of water is evaporated from the vat, how much pure dry salt must be added to produce a 9.82% brine?

11.4 Sap from a maple tree contains 5.12% sugar and 94.88% water. Your evaporator will boil away 1.263 kg of water per minute. How many hours must you operate the evaporator in order to produce 450.0 kg of maple sugar containing 20.5% moisture?

11.5 Syrups A, B, and C are mixed together. It is known that the quantity of A is equal to 40.0% of the total. Syrup A is 3.53% sugar, B is 4.27% sugar, and C is 5.76% sugar. To this mixture is added 17.34 kg of pure dry sugar, while 123 kg of water is boiled away. This results in 1 963 kg of syrup that is 5.50% sugar. How much of each syrup (A, B, and C) was added initially?

11.6 Cod-liver oil is produced by an extracting process in which ether dissolves the oil from the livers. In one process, the livers are fed into the extractor at the rate of 2 240 kg/h. These livers consist of 29.7% oil; the rest is inert material. The solvent is mostly ether (97.9%) with 2.10% oil. It is fed into the extractor at the rate of 3 950 kg/h. The extract leaves the extractor at the rate of 3 640 kg/h and consists of 17.3% oil and 82.7% ether. Determine the flow rate and composition of the product (processed livers).

11.7 Machine parts are often coated with a thick layer of grease while in storage. Before use, they must be degreased. Kerosene, with a specific gravity of 0.809, is added to a large vat containing 1.73 t of grease-coated parts. The average coating is 2.76 kg of grease for each 96.5 kg of parts (metal plus grease). The used kerosene, containing 8.76% grease, is withdrawn and sent to a separator. How many liters of kerosene were used? *Note:* 1 t = 1 000 kg.

11.8 Construction engineers choose different "mixes" of stone to produce desired strengths of concrete. One such "mix" consisted of 37.5% stones (by mass) between 16 and 25 mm in diameter and 26.8% between 10 and 15 mm; the remaining percentage of stones was less than 10 mm in diameter. The engineer decides to use this combination as a base supply to create a new mix by screening out all of the stones less than 10 mm in diameter and adding some stones between 16 and 25 mm. The resulting mixture has 74.6% of the largest size, and the total mix has a mass of 46.8 t. How much of the first mix was used?

11.9 A bottle of cherry wine contains 3.00 L and is 11.38% alcohol (by mass). The wine has a specific gravity of 0.978. This wine is mixed with 0.750 L of brandy that is 39.6% alcohol (by mass) and has a specific gravity of 0.920. Based on mass, what is the alcoholic content of the mixture?

11.10 If the mixture resulting in Prob. 11.9 has 0.637 kg of water removed and 212 mL of pure (100%) alcohol (specific gravity = 0.790) added, what is the alcoholic content of the resulting mixture?

11.11 A farmer makes alcohol from corn to blend with gasoline (gasohol). He is not satisfied with his production so he analyzes his operation. He feeds 8.00×10^2 kg/h into his still. This feed has been tested and it contains 11.3% alcohol, 83.9% water, and some inert material. From the still, a vapor is drawn off and passed through a condenser where it is cooled. The finished product is 12.6% of the feed and contains 73.1% alcohol, 26.2% water, and 0.7% inert materials. What is the quantity and composition of the bottoms, i.e., the waste from the bottom of the still?

11.12 Because of environmental concerns, your plant must install an acetone recovery system. Your task is to calculate the size of the various components of the system, which includes an absorption tank into which is fed 625 kg of water per hour and 3 500 kg/h of air containing 1.63% acetone. The water absorbs the acetone and the purified air is expelled. The water and acetone solution go to a distillation process where the solution is vaporized and then to a condenser. The resulting product is 98.9% acetone and 1.1% water. The bottoms (waste) of the distillation process contains 4.23% acetone and 95.77% water. To determine the volume of a holding tank, calculate how much product is generated in kilograms per hour.

11.13 Fish is used as animal feed by removing the fish oil and then drying the remainder into a cake that is mixed with other feed. One operation feeds 1 760 kg/h of fish that contains 5.27% oil, 73.82% water, and the balance dry fish cake. The fish cake produced has 0.123% oil and 12.7% water. How much water is evaporated during the process if the fish oil contains 1.25% water?

11.14 It is desired to produce a 60.0% solution of nitric acid (HNO_3) and water. The solution is produced by beginning with a dilute acid containing 22.3% HNO_3, and adding a quantity of 94.6% HNO_3. How many kilograms of the stronger acid must be added to 255 kg of the dilute acid?

11.15 An industry must clean up 3.2×10^3 kg of its by-product containing both toxic and inert materials. The toxic content is 11.2%; the remainder is inert material. Treatment with 3.60×10^4 kg of solvent results in dirty solvent containing 0.35% toxic material and a discard composed of 1.2% toxic material and all the inert materials. Determine the quantity of dirty solvent, the percentage of solvent in the discard, and the percentage of toxic substance removed in the process.

11.16 Benzene, toluene, and xylene can be separated by distillation. When 652 kg of a mixture containing 45% benzene, 32% toluene, and 23% xylene is separated into three streams, stream A contains 98.9% benzene and 1.1% toluene; stream B contains 95.8% toluene, 2.7% benzene, and 1.5% xylene; and stream C contains 93.6% xylene and 6.4% toluene. Find the mass of the three streams.

11.17 Water is often used to wash ore into a separator. In one such installation, water transports a mixture of dirt and pure iron ore into the separator (the mixed ore contains three times the percentage of iron ore as dirt) from which two streams emerge: one stream has 62.8% iron ore, 3.8% dirt, and 33.4% water; the waste stream has 3.6% iron ore, 33.7% dirt, and 62.7% water. The amount of iron ore contained in the first stream is 17.3 t/h. Determine the flow rate of all three streams.

11.18 A stream of fluid feeds 9.27 t/h of a mixture containing 41.3% ethane, 29.2% propane, and 29.5% butane into a still. Three streams are drawn off: Stream A is 93.7% ethane, 5.12% propane, and 1.18% butane; stream B is 92.1% propane, 0.82% ethane, and 7.08% butane; and stream C is 94.2% butane and 5.80% propane. What is the quantity of each of the streams?

11.19 Leftover acid from a nitrating process contains 24.0% nitric acid (HNO_3), 55.0% sulfuric acid (H_2SO_4), and 21.0% water (H_2O) (mass percents). The acid is to be concentrated (strengthened in acid content) by adding sulfuric acid with 92.0% H_2SO_4 and nitric acid containing 89.0% HNO_3. The final product is to contain 28.0% HNO_3 and 61.0% H_2SO_4. Compute the mass of the initial acid solution and the mass of the concentrated acids that must be combined to obtain 1.00×10^3 kg of the desired mixture.

11.20 A commercial dryer processes 4 500 kg/h of a wood pulp. If the pulp contains 36% water before it enters the dryer and 12 000 kg of water is removed in an 8-h day, what is the final moisture content of the pulp?

11.21 A drilling mud contains 60.0% water and 40.0% special clay. The driller wishes to increase the density of the mud, and a curve shows that 48% water will give the desired density. Calculate the mass of bone-dry clay that must be added per metric ton of original mud to give the desired composition.

11.22 A syrup contains 6.27% sugar. If some of the water is boiled away and 12.9 kg of dry sugar is added, leaving 873.6 kg of syrup that is 8.92% sugar, how much syrup was in the initial mixture and how much water was removed?

11.23 The water analysis in a flowing stream shows 180 ppm (parts per million) of sodium sulfate. If 10.0 lbm of sodium sulfate is added to the stream over a 1-h period, and the analysis downstream where mixing is complete indicates 3 300 ppm of sodium sulfate, how many gallons of water are flowing per hour?

11.24 A very sweet syrup is made by combining some beet syrup and some corn syrup. The beet syrup is 12.34% sugar (the remainder is water) and the corn syrup is 7.89% sugar. They are mixed, and 13.62 kg of pure dry sugar is added while 456.7 kg of water is boiled away. This leaves 891.2 kg of syrup that is 16.78% sugar. How much beet syrup and how much corn syrup did you have to start the process?

11.25 A fruit punch is to be made by mixing a grape wine that is 12.00% alcohol (the remainder is water), a cherry wine that is 19.00% alcohol, and some apricot brandy that is 42.20% alcohol. The initial mixture has twice as much apricot brandy as grape wine, and there is 50.00% more cherry wine than grape. To these ingredients are added 72.2 kg of pure water and 263.7 kg of pure grain alcohol. This produces a quantity of a punch that is 33.3% alcohol. How much grape wine, cherry wine, and apricot brandy were mixed together before adding the water and alcohol?

11.26 A fruit punch is made by mixing apple wine that is 14.4% alcohol (the remainder is water), a berry wine that is 19.0% alcohol, and some peach brandy that is 32.2% alcohol. The initial mixture is 22.2% apple wine, 33.3% berry wine, and 44.5% peach brandy. To these ingredients are added 272.2 kg of pure water and 363.7 kg of pure grain alcohol. This produces a quantity of punch that is 30.3% alcohol (the remainder is water). How much apple wine, berry wine, and peach brandy were mixed together before adding the water and grain alcohol?

11.27 Two brine solutions are mixed. Brine A is 68.7% of the total and brine B is 31.3%. Brine A is 42.1% salt (the remainder is water), and brine B is 15.8% salt (the remainder is water). Some of the water is then removed from the mixture, leaving 987.6 kg of a mixture (brine) that is 43.62% salt.

(*a*) How much of brine A and of brine B was mixed initially?

(*b*) How much water was removed?

12

Electrical Theory

12.1 Introduction

Electricity is one of our most powerful and useful forms of energy. It affects our lives through communication systems (TV, radio, etc.), computer systems (personal computers, video games, etc.), control systems (electronic fuel injection, auto ignition, etc.), and power systems (lighting, heating, etc.). Certainly electrical engineers, but also to some extent all engineers, must understand how to design, analyze, and operate such systems.

A common area that supports all of electrical engineering is circuit theory. Because of this, most engineering students are asked to take one or more courses in circuit theory. We have chosen to include an introduction to circuit theory because it is so fundamental to the study of engineering.

What does circuit theory do for us? It provides simple solutions to practical problems with sufficient accuracy to be useful. It allows us to reduce the analysis of large systems to a series of smaller problems that we can conveniently handle. It provides a means of

Figure 12.1 Integrated circuits and microelectronics are the basis for the rapid development of computer technology.

synthesizing (building up) complex systems from basic components. This, of course, is the design process.

In this chapter, we will review the concepts of electricity first learned in physics. We will also introduce and apply some of the fundamental circuit-analysis equations such as Ohm's law and Kirchhoff's laws. Applications, however, will be restricted to those involving direct current. Methods of solving alternating current problems can be found in many electrical engineering textbooks.

12.2 Structure of Electricity

First we must ask what electricity is. Most people today accept the idea that all matter consists of minute particles called molecules. *Molecules* are the smallest particles into which a substance can be divided and still retain all the characteristics of the original substance. Each of these particles will differ according to the type of matter to which it belongs. Thus, a molecule of iron will be different from a molecule of paper.

Looking more closely at the molecule, we find that it can be divided into still smaller parts called *atoms*. Each atom has a central core, or nucleus, that contains both *protons* and *neutrons*. In somewhat circular motion around the nucleus are particles of extremely small mass called *electrons*. In fact the entire mass of the atom is practically the same as that of its nucleus, since the proton is approximately 2.0×10^3 times more massive than the electron.

To understand how electricity works bear in mind that electrons possess a negative electric charge, and protons a positive electric charge. Their values are opposite in sign but numerically the same.

The neutron is considered neutral, being neither positive nor negative.

The atom in its entirety has no electric charge because the positive charge of the nucleus is exactly balanced by the negative charge of the surrounding electron cloud. That is, each atom contains as many electrons orbiting the nucleus as there are protons inside the nucleus.

The actual number of protons depends on the element of which the atom is a part. Thus, hydrogen (H) has the simplest structure, with one proton in its nucleus and one orbital electron. Helium (He) has two protons and two neutrons in the nucleus; and since the neutrons exhibit a neutral charge, there are two orbital electrons (see Fig. 12.2). More complex elements have many more protons, electrons, and neutrons. For example, gold (Au) has 79 protons and 118 neutrons, with 79 orbital electrons.

As the elements become more complex, the orbiting electrons arrange themselves into regions, or "shells," around the nucleus.

The maximum number of electrons in any one region is uniquely defined. The shell closest to the nucleus contains two electrons, the next eight, etc. There are a maximum of six shells, but the last two shells are never completely filled. Atoms can therefore combine by

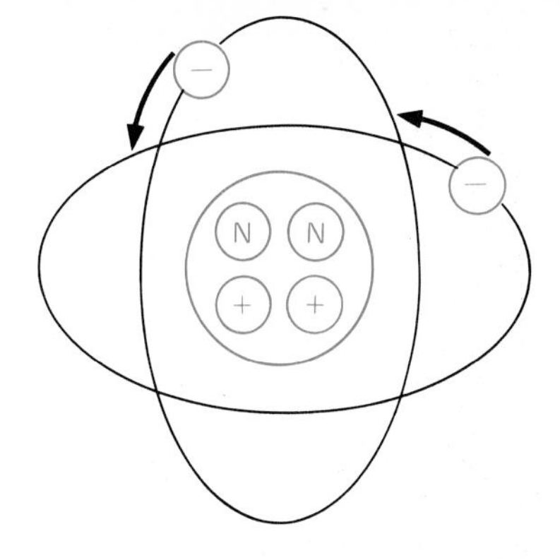

Figure 12.2
Schematic representation of a helium atom.

sharing their outer orbital electrons and thereby fill certain voids and establish unique patterns or molecules.

Atoms are extremely minute. In fact, it may be very difficult for one to imagine the size of an atom, since a grain of table salt is estimated to contain 10^{18} atoms. However, it is possible to understand the relation between the nucleus and the orbital electrons. Assume for purposes of visualization that the hydrogen nucleus is a 1-mm-diameter sphere; its orbiting electron appears to be the same size, but moves at an average distance of 25 m from the nucleus. Although the relative distance is significant, the electron is prevented from leaving the atom by an electric force of attraction that exists because the proton has a positive charge and the electron a negative charge.

How closely the millions upon millions of atoms and molecules are packed together will determine the state (solid, liquid, or gas) of a given substance. In solids, the atoms are packed closely together, generally in a very orderly manner. The atoms are held in a specific, lattice structure but vibrate around their nominal positions. Depending on the substance, some electrons may be free to move from one atom to another.

It is the nature of the molecular structure with the availability of free electrons that results in electricity.

12.3 Static Electricity

History indicates that the word "electricity" was first used by the Greeks. They discovered that after rubbing certain items together, the materials would exert a force on one another. It was concluded that during the rubbing process, the bodies were "charged" with some unknown element, which the Greeks called electricity. For example, it was believed that by rubbing silk over grass, electricity was added to each substance. We realize today that during the process of rubbing, electrons are removed from some of the surface atoms of the glass and added to the surface atoms of the silk.

The branch of science concerned with static, or stationary, charged bodies is called *electrostatics*.

The charge on an electron can be measured, but the amount is extremely small. In fact, it is inconvenient to use such a small quantity as a unit of electric charge. A larger and more practical unit, the coulomb, has thus been selected to denote electric charge. A *coulomb* is defined in terms of the force exerted between unit charges. A charge of one coulomb (C) will exert on an equal charge, placed 1 m away in air, a force of about 8.988×10^9 N.

12.4 Electric Current

Earlier we noted that electrons are prevented from leaving the atom by the attraction of the protons in the nucleus. It is entirely possible, however, for an electron to become temporarily separated from an atom. These free electrons drift around randomly in the space between atoms. During their random travel, many of them will collide with other atoms; when they do so with sufficient force, they dislodge electrons from those atoms. Since electrons are frequently colliding with other atoms, there is a continuous movement of free electrons in a solid. If the electrons drift in a particular direction instead of moving randomly, there is movement of electricity through the solid. This continuous movement of electrons in a direction is called an *electric current*. If the electron drift is in only one direction, it is called *direct current*. If the electrons periodically reverse direction of travel, then we have *alternating current*.

The ease with which electrons can be dislodged by collision as well as the number of free electrons available varies with the substance. Materials in which the drift of electrons can be easily produced are good conductors; those in which it is difficult to produce an electron drift are good insulators. For example, copper is a good conductor, whereas glass is a good insulator. Practically all metals are good conductors. Silver is perhaps the best, but it is expensive. Copper and aluminum are also good conductors and are commonly used in electric wire.

12.5 Electric Potential

From research and experimentation it is known that like charges repel and unlike charges attract one another. Consequently, to bring like charges together, an external force is necessary and therefore work must be done. The amount of work required to bring a positive charge near another positive charge from a large distance is used as a measure of the electric potential at that point. This amount of potential is measured in units of work per unit charge or joules per coulomb. By definition, one joule per coulomb is one volt, a unit of electric potential.

Certain devices, such as electric batteries or generators, are capable of producing a difference in electric potential between two points. Such equipment is rated in terms of the potential difference produced in volts. When these devices are connected to other components in a continuous circuit, electric current flows. You can also

Figure 12.3
A power-distribution system includes several substations like the one shown here.

think of batteries as devices that change chemical energy to electric energy while generators convert mechanical energy to electric energy.

12.6 Simple Electric Circuits

When electric charge and current were initially being explored, scientists thought that current flow was from positive to negative. They had no knowledge of electron drift. By the time it was discovered that the electron flow was from negative to positive, the idea that current flow was from positive to negative had become so well established that it was decided not to change the convention.

When the two terminals are connected to a conducting material, the battery or generator creates a potential difference across the load. It follows that the random movement of the negatively charged electrons will have a drift direction induced by this potential difference. The resulting effect will be the movement of electrons away from the negative terminal of the battery. Electrons will travel in a continuous cycle around the circuit, reentering the battery at its positive terminal. (See Fig. 12.4.)

The speed by which any individual electron moves is relatively slow, less than 1 mm/s. Once a potential difference is connected into a circuit, the "flow" of electrons starts almost instantaneously at all points. Individual electrons at all locations begin their erratic movement around the circuit, colliding frequently with other atoms in the conductor. Because electron activity starts at all points practically simultaneously, electric current appears to travel about 3×10^8 m/s.

Electric current is really nothing more than the rate at which electrons pass through the cross section of a conductor. The number of electrons involved is gigantic in magnitude in that approximately

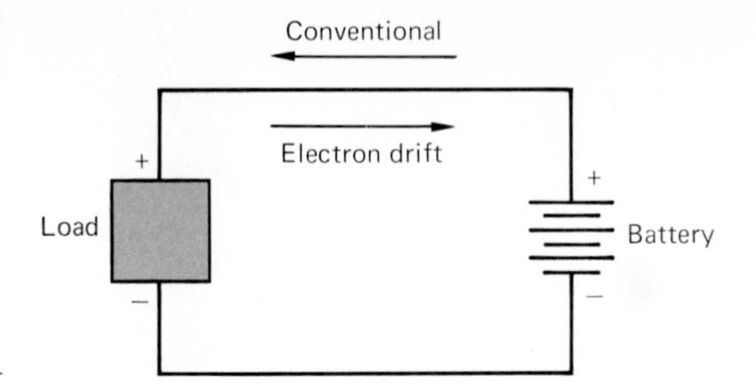

Figure 12.4
A simple electric circuit.

6.28×10^{18} electrons pass a point per second per ampere. Thus, it is not convenient to use the rate of electron flow as a unit of current measurement. Instead, current is measured in terms of the total electric charge (coulomb) that passes a certain point in a unit of time (second). This unit of current is called the *ampere*. That is, one ampere equals one coulomb per second.

12.7 Resistance

Ohm, a German scientist, investigated the relation between electric current and potential difference. He found that for a metal, the potential difference across the conductor was directly proportional to the current. This important relationship is concisely stated as Ohm's law:

At constant temperature, the current I in a conductor is directly proportional to the potential difference between its ends, E.

The ratio E/I is called *resistance* and is denoted by the symbol R:

$$R = \frac{E}{I} \qquad 12.1$$

or

$$\text{Resistance, or } R(\text{ohms, } \Omega) = \frac{\text{potential, or } E \text{ (volts, V)}}{\text{current, or } I \text{ (amperes, A)}}$$

This is one of the simplest but most important relations used in electric circuit theory. A conductor has a resistance R of one ohm when the current I through the conductor is one ampere and the potential difference E across it is one volt.

The reciprocal of resistance is called *conductance* (G):

$$G = \frac{1}{R} \qquad 12.2$$

Conductance is measured in siemens (S).

Example problem 12.1 The current in an electric aircraft instrument heater is 2.0 A when connected to a battery with a potential of 60.0 V. Calculate the resistance of the heater.

Solution

$$E = RI$$

$$R = \frac{E}{I}$$

$$= \frac{60.0}{2.0}$$

$$= 3.0 \times 10^1\ \Omega$$

12.8 Circuit Concepts

A considerable amount of information about electric circuits can be presented in a condensed form by means of circuit diagrams. Figure 12.5 illustrates some typical symbols that are frequently used in such diagrams.

Circuits may have resistors that are connected either end to end or in parallel to each other. When resistors are attached end to end, they are said to be connected in series, and the same current flows through each.

For the series circuit illustrated in Fig. 12.6, there will be a potential across each resistor, since a potential must exist between the ends of a conductor if current is to flow. This potential difference E is related to current and resistance by Ohm's law. For each unknown voltage (potential), we can write

$$E_1 = R_1 I \qquad E_2 = R_2 I \qquad E_3 = R_3 I$$

Since the total voltage drop E_T across the three resistors is the

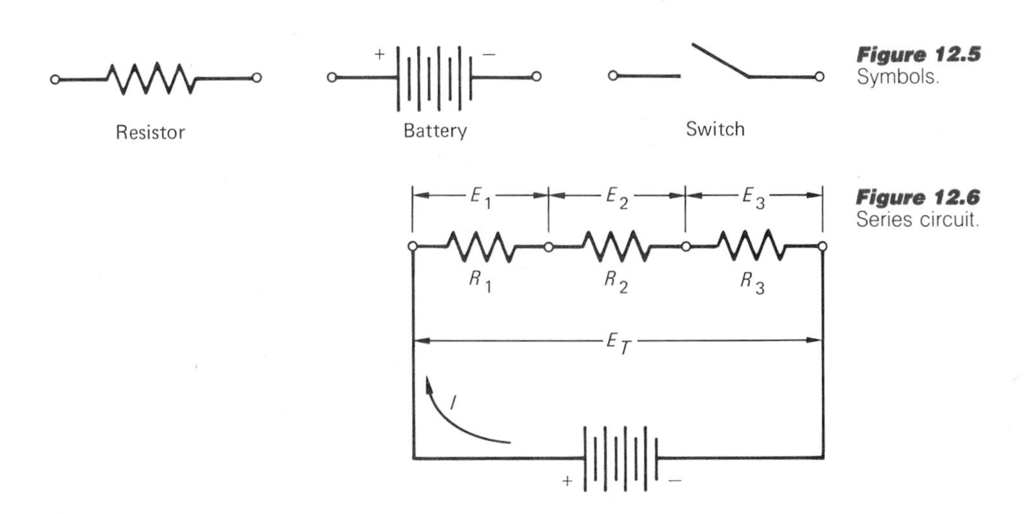

Figure 12.5 Symbols.

Figure 12.6 Series circuit.

sum of the individual drops, and since the current is the same through each resistor, then

$$E_T = E_1 + E_2 + E_3$$
$$= IR_1 + IR_2 + IR_3$$
$$= I(R_1 + R_2 + R_3)$$

or

$$R_1 + R_2 + R_3 = \frac{E_T}{I}$$

Since E_T is the total potential difference across the circuit and I is the circuit current, then the total resistance must be

$$R_T = \frac{E_T}{I}$$

Therefore,

$$R_T = R_1 + R_2 + R_3 \qquad 12.3$$

These steps demonstrate that when any number of resistors are connected in series, their combined resistance is the sum of their individual values.

Example problem 12.2 The circuit in Fig. 12.7 has three resistors connected in series with a 12.0-V source. Determine the line current and the voltage drop across each resistor.

Solution For resistors in series,

$$R_T = R_1 + R_2 + R_3$$
$$R_T = 4.0 + 8.0 + 12$$
$$= 24\ \Omega$$

Figure 12.7

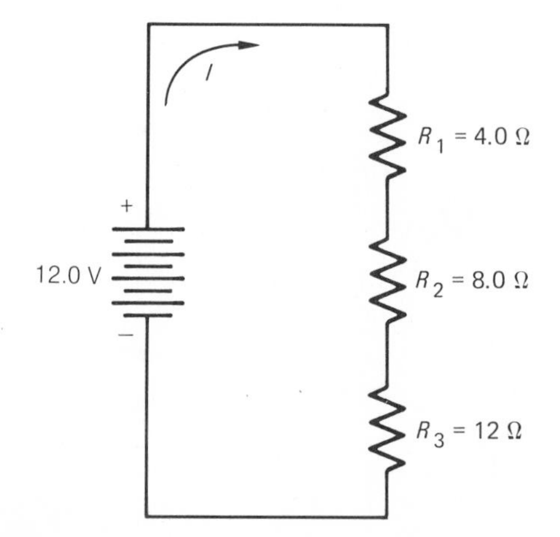

Ohm's law gives

$$E = RI$$

$$I = \frac{E_T}{R_T} = \frac{12.0}{24} = 0.50 \text{ A}$$

Then

$$E_1 = 0.50(4.0)$$
$$= 2.0 \text{ V}$$
$$E_2 = 0.50(8.0)$$
$$= 4.0 \text{ V}$$
$$E_3 = 0.50(12)$$
$$= 6.0 \text{ V}$$

Check: $E_T = 12.0 = 2.0 + 4.0 + 6.0$

When several resistors are connected between the same two points, they are in parallel. Figure 12.8 illustrates three resistors in parallel.

The current between points 1 and 2 divides among the various branches formed by the resistors. Since each resistor is connected between the same two points, the potential difference across each of the resistors is the same.

In analyzing the problem, let I_T be the total current passing through the points 1 and 2 with I_1, I_2, and I_3 representing the branch currents through R_1, R_2, and R_3, respectively.

Using Ohm's law we can write

$$I_1 = \frac{E_T}{R_1} \qquad I_2 = \frac{E_T}{R_2} \qquad I_3 = \frac{E_T}{R_3}$$

or

$$I_1 + I_2 + I_3 = E_T\left(\frac{1}{R_1} + \frac{1}{R_2} + \frac{1}{R_3}\right)$$

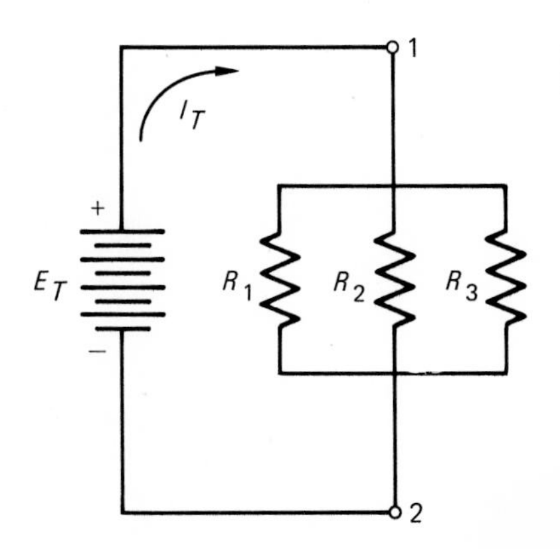

Figure 12.8
Parallel circuit.

But we know that

$$I_T = I_1 + I_2 + I_3$$

Applying Ohm's law to total circuit values reveals that

$$I_T = \frac{E_T}{R_T}$$

Therefore,

$$\frac{E_T}{R_T} = E_T\left(\frac{1}{R_1} + \frac{1}{R_2} + \frac{1}{R_3}\right)$$

or

$$\frac{1}{R_T} = \frac{1}{R_1} + \frac{1}{R_2} + \frac{1}{R_3} \qquad 12.4$$

This equation indicates that when a group of resistors are connected in parallel, the reciprocal of their combined resistance is equal to the sum of the reciprocals of their separate resistances.

Example problem 12.3 In Fig. 12.9 three resistors are connected in parallel across a 6.0×10^1 V battery. What is the equivalent resistance of the three resistors and the line current?

Solution For resistors in parallel,

$$\frac{1}{R_T} = \frac{1}{R_1} + \frac{1}{R_2} + \frac{1}{R_3}$$

$$= \frac{1}{5.0 \times 10^0} + \frac{1}{6.0 \times 10^0} + \frac{1}{1.0 \times 10^1}$$

$$= \frac{28}{6.0 \times 10^1}$$

Figure 12.9

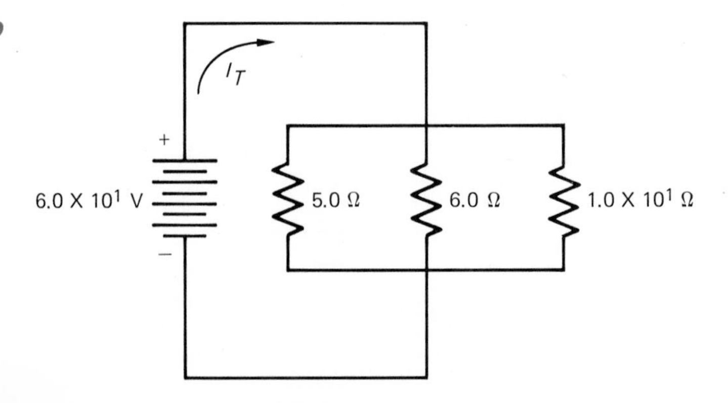

$$R_T = \frac{6.0 \times 10^1}{2.8 \times 10^1}$$

$$= 2.1\ \Omega$$

From Ohm's law

$$E = RI$$

$$E_T = R_T I_T$$

so,

$$I_T = \frac{E_T}{R_T}$$

$$= (6.0 \times 10^1)\frac{2.8 \times 10^1}{6.0 \times 10^1}$$

$$= 28\ \text{A}$$

Many electric circuits involve combinations of resistors in parallel and series. The next example problem demonstrates a solution of that nature.

Example problem 12.4 Determine the line current, the circuit equivalent resistance, and the voltage drop across R_4 for the circuit in Fig. 12.10. What resistance should be substituted for R_1 to reduce the line current by one-half?

Solution Ohm's law gives

$$E = RI$$

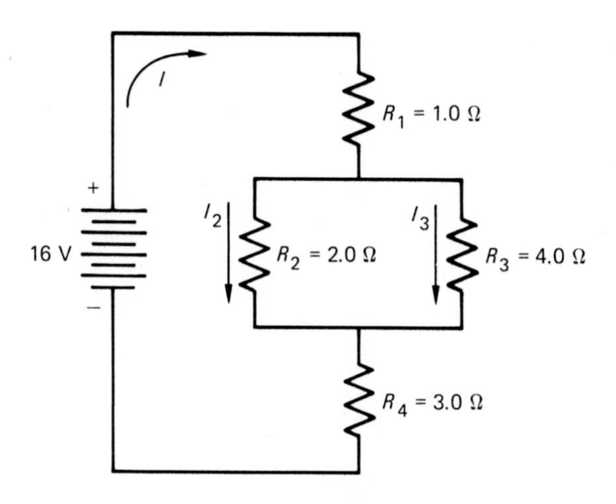

Figure 12.10

Also,

$$R_T = R_1 + R_2 + \cdots + R_N \text{ (series)}$$

$$\frac{1}{R_T} = \frac{1}{R_1} + \frac{1}{R_2} + \cdots + \frac{1}{R_N} \text{ (parallel)}$$

For parallel resistors,

$$\frac{1}{R_T} = \frac{1}{R_2} + \frac{1}{R_3}$$

$$= \frac{1}{2.0} + \frac{1}{4.0}$$

$$R_T = \frac{4.0}{3.0}\ \Omega$$

For series resistors,

$$R_E = R_1 + R_T + R_4$$

$$= 1.0 + \frac{4.0}{3.0} + 3.0$$

$$= \frac{16}{3.0}$$

$$= 5.3\ \Omega$$

Then the line current is

$$E = RI$$

$$I = E\left(\frac{1}{R_E}\right)$$

$$= 16\left(\frac{3.0}{16}\right)$$

$$= 3.0\ \text{A}$$

and the voltage drop is

$$E_4 = R_4 I$$

$$= 3.0(3.0)$$

$$= 9.0\ \text{V}$$

The new line current is $3.0/2 = 1.5$ A. The new equivalent resistance must then be

$$R_E = \frac{E}{I} = \frac{16}{1.5} = 10.667\ \Omega$$

But

$$R_E = R_1 + R_T + R_4$$

so the new R_1 is

$$R_1 = R_E - R_T - R_4 = 10.667 - \frac{4.0}{3.0} - 3.0 = 6.3\ \Omega$$

12.9 Electric Power

If a current flows for some time period as a result of a potential difference, then the amount of energy produced or used is

$$\text{Energy} = EIt \qquad 12.5$$

Power is the time rate at which energy is supplied or consumed. It is expressed in joules per second, but this unit has been given a special name after the scientist James Watt. Thus, one watt equals one joule per second.

Power can be expressed as

$$\text{Power} = \frac{\text{energy}}{\text{time}} = \frac{EIt}{t} = EI \qquad 12.6$$

By applying Ohm's law, we can express power in two other convenient forms:

$$P = \frac{E^2}{R} \qquad 12.7$$

and

$$P = I^2R \qquad 12.8$$

Figure 12.11
This satellite-signal receiving dish is a common sight in areas not able to obtain cable television.

Example problem 12.5 A five-cell flashlight contains a lamp with a measured resistance of 2.0 Ω. What is the power consumed by the lamp?

Solution Each battery in the flashlight produces a potential of approximately 1.5 V. They are in series so their potentials are added together, and their total potential is 5(1.5) V. Thus

$$P = \frac{E^2}{R}$$

$$= \frac{[5(1.5)]^2}{2.0}$$

$$= 28 \text{ W}$$

Example problem 12.6 A current of 4.0 A flows through a 50.0-Ω resistor. What is the power consumed in watts?

Solution

$$P = I^2R$$

$$= (4.0)^2(50.0)$$

$$= 8.0 \times 10^2 \text{ W}$$

12.10 Terminal Voltage

Figure 12.12 is a basic circuit in which there can be different sources of electric potential. The storage battery and the electric generator are two familiar examples.

The potential does work of amount E in joules per coulomb on charges passing through the voltage source from the negative to the positive terminal. This results in a difference of potential E across the resistor R, which causes current to flow in the circuit. The energy furnished by the voltage source reappears as heat in the resistor.

Current can travel in either direction through a voltage source. When the current moves from the negative to the positive terminal, some other form of energy (e.g., chemical energy) is converted into electric energy. If we were to impose a higher potential in the external circuit—for example, forcing current backward through the voltage source—the electric energy would be converted to some other form. When current is sent backward through a battery, electric energy is converted into chemical energy (which can, by the way, be recovered in certain types of batteries). When current is sent backward through a generator, the device becomes a motor.

A resistor, on the other hand, converts electric energy into heat no matter what the direction of current. Therefore, it is impossible to reverse the process and regain electric energy from the heat.

Figure 12.12
Basic circuit.

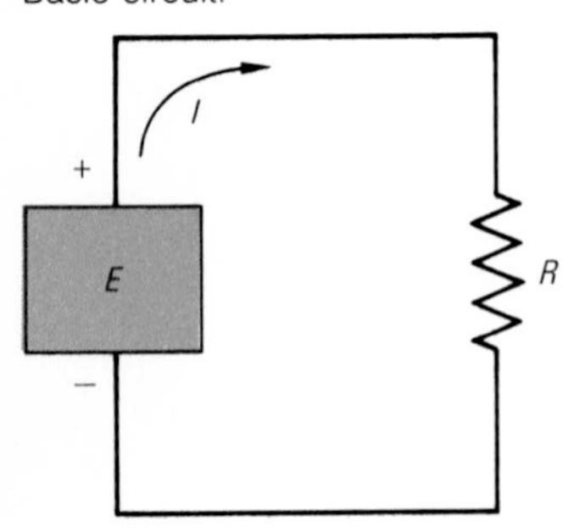

Figure 12.13
These electrical motors are part of a power laboratory for electrical engineering students.

Electric potential always drops by the amount IR as current travels through a resistor. This drop occurs in the direction of the current.

In a generator or battery with current flowing negative to positive, the positive terminal will be E above the negative terminal minus the voltage drop due to internal resistance between terminals. There will always be some energy converted to heat inside a battery or generator no matter which direction the current.

When a battery or a generator is driving the circuit, the internal current passes from the negative to the positive terminal. Each coulomb of charge gains energy E from chemical or mechanical energy but loses IR in heat dissipation. The net gain in joules per coulomb can be determined by $E - IR$.

For a motor or a battery being charged, the opposite is true, because the internal current passes from the positive to the negative terminal. Each coulomb loses energy E and IR. The combined loss can be determined by $E + IR$.

In the case of a motor, the quantity E is commonly called *back-emf* (the electromotive force, or potential), since it represents a voltage that is in a direction opposite the current flow.

Figure 12.14 shows a circuit wherein a battery is being charged by a generator. E_G, E_B, R_G, and R_B indicate the potentials and internal resistances, respectively, of the generator and battery. Each coulomb that flows around the circuit in the direction of the current I gains energy E_G from the generator and loses energy E_B in the form of chemical energy to the battery. Heat dissipation is realized as IR_G, IR_1, IR_B, and IR_2.

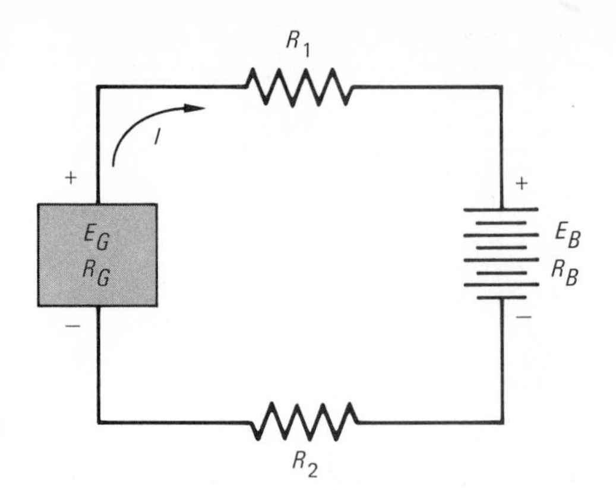

Figure 12.14
Generator charging a battery.

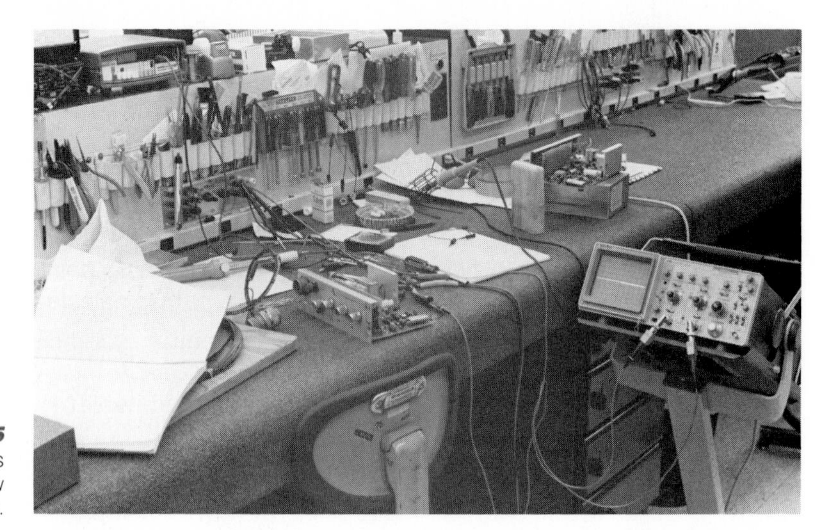

Figure 12.15
A test bench with oscilloscope is used to develop and prove new electrical components and devices.

12.11 Kirchhoff's Laws

Two fundamental principles that are frequently used in circuit analysis are known as Kirchhoff's laws:

1. The summation of the potential differences around a closed loop must be zero.
2. The summation of the currents at a junction must be zero.

To demonstrate these principles consider the following problem, illustrated in Fig. 12.16.

Example problem 12.7 A 220-V generator is driving a motor drawing 8.0 A and charging a 170-V battery. Determine the back-emf of the motor (E_m), the charging current of the battery (I_2), and the current through the generator (I_1).

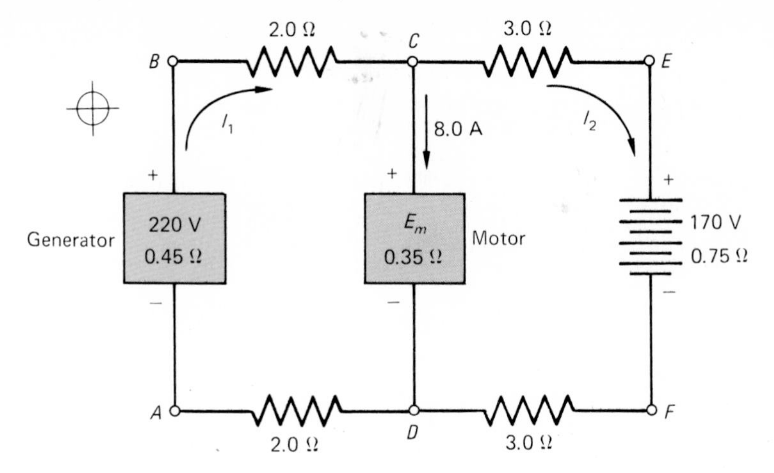

Figure 12.16

Solution Since the current through the motor is given as 8.0 A, we can see by applying Kirchhoff's second law to junction C that the current out of the junction must be $I_2 + 8.0$ if the current through the battery is I_2. That is, $I_1 = I_2 + 8.0$.

Kirchhoff's first law dictates that we select a beginning point and travel completely around a closed loop back to the starting point, thereby arriving at the same electric potential. As a path is selected and followed, note carefully all changes in potential. Once a loop has been completely traveled and all potential changes noted, the sum is set equal to zero.

The following voltage summation for the circuit $ABCDA$ in Fig. 12.16 results from Kirchhoff's first law:

$$(E_{A-B})_{\text{emf}} + (E_{A-B})_{\text{loss}} + E_{B-C} + (E_{C-D})_{\text{emf}} + (E_{C-D})_{\text{loss}} + E_{D-A} = 0$$

By applying correct algebraic signs according to the established convention, we get the results given in Table 12.1. Substituting the values from Table 12.1, the equation becomes

$$220 - 0.45(I_2 + 8.0) - 2.0(I_2 + 8.0) - E_m - 0.35(8.0) - 2.0(I_2 + 8.0) = 0$$

Table 12.1

Symbols	Quantities	Notes
$(E_{A-B})_{\text{emf}}$	$+220$ V	Potential of generator
$(E_{A-B})_{\text{loss}}$	$-0.45(I_2 + 8.0)$	Loss in generator
(E_{B-C})	$-2.0(I_2 + 8.0)$	Loss in line
$(E_{C-D})_{\text{emf}}$	$-E_m$	Back-emf of motor
$(E_{C-D})_{\text{loss}}$	$-0.35(8.0)$	Loss in motor
(E_{D-A})	$-2.0(I_2 + 8.0)$	Loss in line

Table 12.2

Symbols	Quantities	Notes
$(E_{D-C})_{emf}$	$+E_m$	Back-emf of motor
$(E_{D-C})_{loss}$	$+0.35(8.0)$	*IR* rise in motor
(E_{C-E})	$-3.0I_2$	Loss in line
$(E_{E-F})_{emf}$	-170 V	Drop across battery
$(E_{E-F})_{loss}$	$-0.75I_2$	Loss in battery
(E_{F-D})	$-3.0I_2$	Loss in line

Simplifying, we get

$$-4.45I_2 - E_m + 181.6 = 0$$

This equation cannot be solved, because there are two unknowns. However, a second equation can be written around a different loop of the circuit.

$$(E_{D-C})_{emf} + (E_{D-C})_{loss} + E_{C-E} + (E_{E-F})_{emf} + (E_{E-F})_{loss} + E_{F-D} = 0$$

From this we can develop Table 12.2 and, therefore, a second equation:

$$E_m + 0.35(8.0) - 3.0I_2 - 170 - 0.75I_2 - 3.0I_2 = 0$$

$$E_m - 6.75I_2 - 167.2 = 0$$

Solving these two equations simultaneously, we obtain the following results:

$$I_2 = 1.3 \text{ A}$$

$$E_m = 1.8 \times 10^2 \text{ V}$$

Then

$$I_1 = I_2 + 8.0 = 1.3 + 8.0 = 9.3 \text{ A}$$

These values can be checked by writing a third equation around the outside loop.

By the procedure above, a set of simultaneous equations may be found that will solve any similar problem, provided the number of unknowns is not greater than the number of circuit paths or loops.

The following general procedure is outlined as a guide to systematically applying Kirchhoff's laws.

1. Sketch a circuit diagram and label all known voltages, resistances, etc. Show + and − signs on potentials.

2. Assume a current direction in each branch of the circuit. If the direction is not known, choose a direction. A negative current solution will indicate that the current is flowing in the opposite direction.

3. Assign symbols to all unknown currents, voltages, etc.

4. Apply Kirchhoff's first law to circuit loops and Kirchhoff's second law at junctions to obtain as many independent equations as there are unknowns in the problem.

5. Solve the resulting set of equations.

Problems

12.1 What is the average current through a conductor that carries 4 375 C during a 2.00-min time period?

12.2 The measured current through a battery is a constant 1.25 A. How many coulombs are supplied by the battery in 24 h?

12.3 Assuming that a current flow through a conductor is due to the motion of free electrons, how many electrons pass through a cross section normal to the conductor in 1 h if the current is 1 650 A?

12.4 Five 1.5-V batteries in series are required to operate a 8.0-W portable radio. What is the current flow? What is the equivalent circuit resistance?

12.5 A small vacuum cleaner designed to be plugged into the cigarette lighter of an auto has a 12.6-V dc motor. It draws 4.0 A in operation. What power must the auto battery deliver? What size resistor would consume the same power?

12.6 When the leads of an impact wrench are connected to a 12.6-V auto battery, a current of 19 A flows. Calculate the power required to operate the wrench. If 75 percent of the power required by the wrench is delivered to the socket, how much energy in joules is produced per impact if there are 1 100 impacts per minute?

12.7 A portable electric drill produces 1.2 hp at full load. If 85 percent of the power provided by the 9.6-V battery pack is useful, what is the current flow? How much power goes into waste heat?

12.8 An ideal 25-V dc power supply is connected to four resistors in series: $R_1 = 7.0\ \Omega$, $R_2 = 12\ \Omega$, $R_3 = 22\ \Omega$, and $R_4 = 120\ \Omega$.

(*a*) Draw the circuit diagram.
(*b*) Determine the equivalent circuit resistance.
(*c*) Calculate the line current.
(*d*) Find the voltage drop across each resistor.
(*e*) Compute the power produced by the 25-V supply and the power consumed by each resistor.

12.9 Three resistors—1.0, 5.0, and 15 MΩ—are connected in series to a 75-V ideal dc voltage source.

(*a*) Draw the circuit diagram.
(*b*) Determine the equivalent circuit resistance.
(*c*) Calculate the line current.
(*d*) Find the voltage drop across each resistor.
(*e*) Compute the power consumed by each resistor.

12.10 An ideal 6.0-V supply is connected to three resistors that are wired in parallel: $R_1 = 12\ \Omega$, $R_2 = 16\ \Omega$, and $R_3 = 1.0\ \text{k}\Omega$.

(*a*) Draw the circuit diagram.

(*b*) Calculate the equivalent circuit resistance.
(*c*) Find the current through the voltage supply.
(*d*) Determine the current through each resistor and the power consumed by each.

12.11 A battery has a measured voltage of 12.6 V when the circuit switch is open. The internal resistance of the battery is 0.25 Ω. The circuit contains resistors of 6.8, 13.7, and 625 Ω connected in parallel.
(*a*) Draw the circuit diagram.
(*b*) Calculate the equivalent circuit resistance as seen by the battery.
(*c*) Find the current flow through the battery when the switch is closed.
(*d*) Compute the current through the 6.8-Ω resistor and the power consumed by it.
(*e*) Calculate the rate at which heat must be removed from the battery if it is to maintain a constant temperature.

12.12 Given the circuit diagram and values in Fig. 12.17, determine the current through the 15-Ω resistor and the voltage drop across it. Find the fraction of the power produced by the battery that is consumed by the 25-Ω resistor.

Figure 12.17

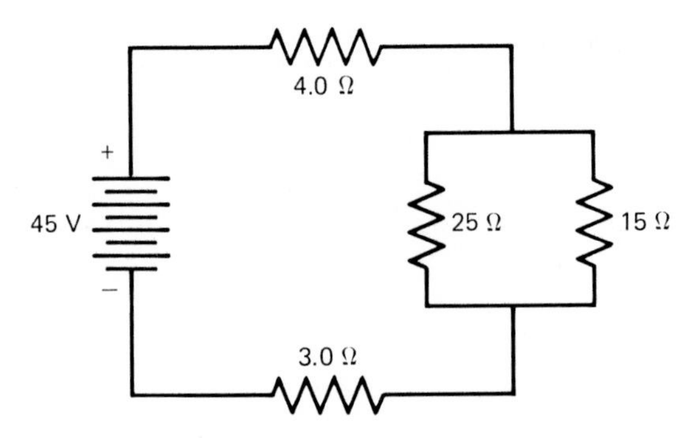

12.13 Add a 1.0-Ω resistor in parallel with the 25- and 15-Ω resistors in the circuit shown in Fig. 12.17 and repeat Prob. 12.12.

12.14 A 24-V battery has an internal resistance of 0.15 Ω. When connected to a load, a current of 95 A flows.
(*a*) What voltage would you expect to measure across the battery terminals when the load is connected?
(*b*) How much power is delivered to the load?
(*c*) At what rate is heat produced in the battery?

12.15 A dc motor has an internal resistance of 0.45 Ω. When connected to a 75-V source, it draws 35 A.
(*a*) What is the back-emf of the motor?
(*b*) How much power is required to drive the motor?
(*c*) At what rate is heat generated in the motor?
(*d*) What power is produced by the motor?

12.16 Refer to Fig. 12.18 and compute the circuit current when $R_1 = 4.5$

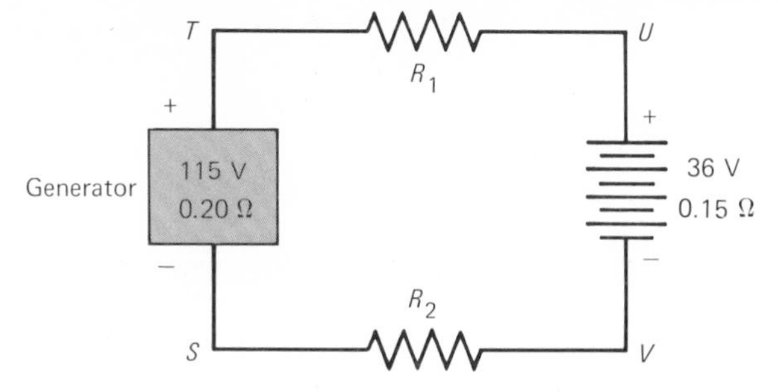

Figure 12.18

Ω and R_2 = 1.5 Ω. Determine the potential at points *T*, *U*, and *V* if the potential at point *S* is 0 V. What percentage of the power available at terminals *S* and *T* is actually charging the battery?

12.17 The 175-V generator in Fig. 12.19 is charging a 90-V battery and driving a motor. Determine the battery charging current, the current through the motor, and the back-emf of the motor. Assume no internal resistance in the generator, motor, and battery.

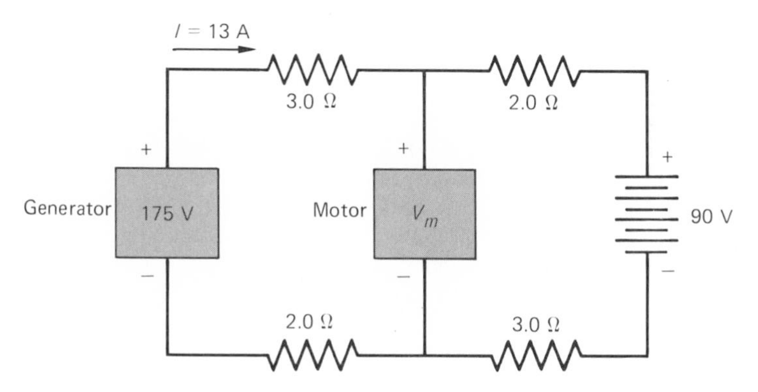

Figure 12.19

12.18 Figure 12.20 shows a resistance circuit driven by two ideal (no internal resistance) batteries. Determine:

(*a*) The currents through each resistor
(*b*) The power delivered to the circuit by the 80-V battery
(*c*) The voltage across the 24-Ω resistor
(*d*) The power consumed by the 12-Ω resistor

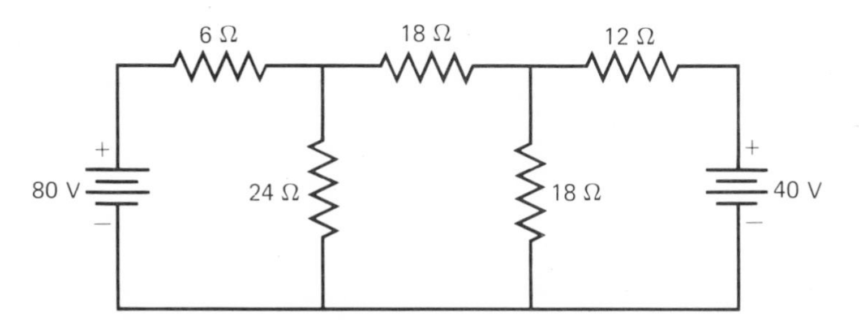

Figure 12.20

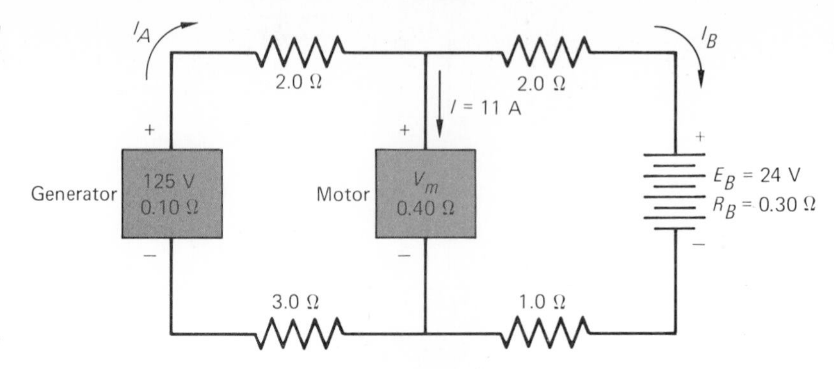

Figure 12.21

12.19 A 125-V generator is driving a motor and charging a 24-V battery (see Fig. 12.21). The measured current through the motor is 11 A. Calculate:

(*a*) Currents I_A and I_B
(*b*) The back-emf of the motor
(*c*) The power delivered by the generator to the circuit
(*d*) The power converted to heat in the motor
(*e*) The voltage across the motor terminals

12.20 An ideal 14-V generator and an ideal 12.6-V battery are connected in a circuit as shown in Fig. 12.22.

(*a*) For $R_A = 2.0\ \Omega$ and $R_B = 3.0\ \Omega$, compute the current through each circuit component (resistors, battery, and generator). Is the battery being charged or discharged?
(*b*) With $R_B = 1.0\ \Omega$, find the current through the battery for $0.50\ \Omega \leq R_A \leq 10.0\ \Omega$, $\Delta R_A = 0.50\ \Omega$. Plot the battery current vs. R_A with R_A as the independent variable. Repeat the process with $R_B = 5.0$ and $10.0\ \Omega$, placing all three curves on the same graph. *Note:* Use a computer program if approved by your instructor.

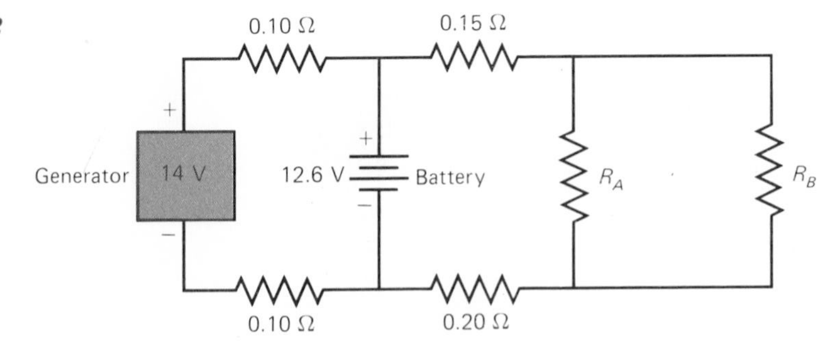

Figure 12.22

12.21 For the circuit shown in Fig. 12.23:

(*a*) Calculate the currents I_1, I_2, and I_3 when the switch is placed at *A*. How much power is consumed by each of the three resistors?
(*b*) Repeat the problem with the switch placed at *B*.

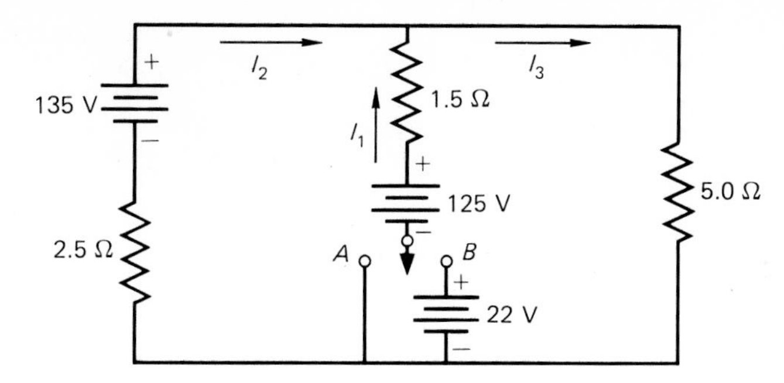

Figure 12.23

12.22 Compute the current through each resistor and the battery shown in Fig. 12.24. Determine the power supplied by the battery and the power consumed by the 25-Ω resistor.

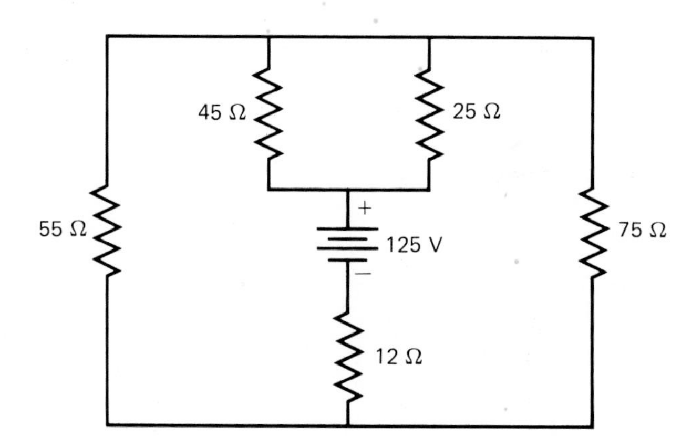

Figure 12.24

12.23 Determine the current through each component of the circuit in Fig. 12.25. Find the power delivered to the 2.0-V battery, the voltage across the 2.0-Ω resistor, and the power consumed by the 6.0-Ω resistor.

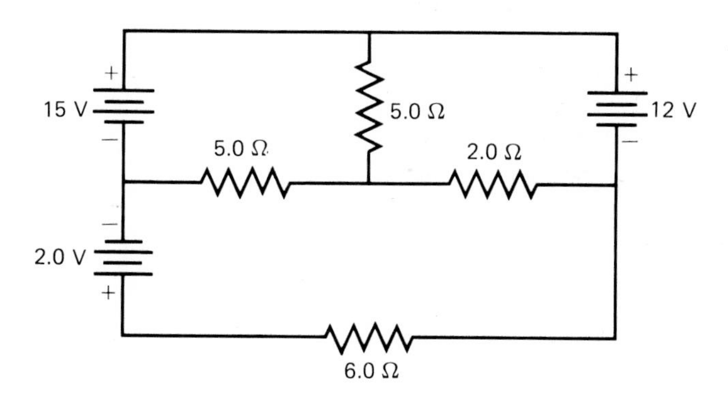

Figure 12.25

13 Energy

13.1 Introduction

Today energy is the world's most important commodity. One way of characterizing the development of society during the past 200 years would be in the substitution of machine power for muscle power. This transformation has been helped by the rapid development of the natural sources of energy, namely, fossil fuels, water, and the atom (see Fig. 13.1). Energy from these natural sources must be converted into forms that can be transported, stored, and applied at the appropriate time and place. The degree of industrial development of society can be determined by the extent of energy usage. There is an excellent correlation between productivity of a nation and its capability to generate energy.

What is energy? Energy cannot be seen; it has no mass or defining characteristics; it is distinguished only by what it can produce. In a broad sense, *energy* may be defined as an ability to produce an effect (change) on matter. Energy may be within an object or may move from one object to another. Thus, energy is usually spoken of as being either stored or in transit.

When supplied, energy can transform natural resources into products and services beneficial to society. We observe the effects of energy when physical changes occur in objects.

Figure 13.1
Coal remains the largest source of fossil energy. Here a giant dragline is used to strip-mine coal in the western United States. (*Sun Company, Inc.*)

13.2 Stored Energy

Stored energy exists in five distinct forms: potential, kinetic, internal, chemical, and nuclear. The ultimate usefulness of stored energy depends on how efficiently the energy is converted into a form that produces a desirable result.

When an object of mass m is elevated to a height h in a gravitational field, a certain amount of work must be done to overcome the gravitational pull. Work is energy associated with a force producing motion of an object. (Work will be considered further in Sec. 13.3.) The object may be said to possess the work required to elevate it to the new position. Its potential energy has been increased. The quantity of work done is the amount of force multiplied by the distance moved in the direction in which the force acts. It is a result of a given mass going from one condition to another. *Potential energy* is thus stored-up energy. It is derived from force and height from a datum plane, not from the means by which the height was attained. Mass m stores up energy when being elevated to height h and does work when it comes back to its starting point. When an object is raised, the force needed is that to overcome the effect of gravity (weight) and the distance is the change in elevation. Thus,

$$\begin{aligned} \text{PE} &= \text{(weight)(height)} \\ &= Wh \\ &= mgh \end{aligned} \qquad 13.1$$

If the units for mass (m) are kilograms, those for acceleration of gravity (g) are meters per second squared, and those for height (h) are meters, then the units for potential energy, or PE, are newton-meters, or joules.

It is unnecessary in most cases to evaluate the total energy of an object, but it is essential to evaluate the energy change when there is a change in the physical state of the object. In the case of potential energy, this is accomplished by establishing a datum plane (see Fig. 13.2) and evaluating the energy possessed by objects in excess of that possessed at the datum plane. Any convenient location, such as sea level, may be chosen as the datum plane. The potential energy at any other elevation is then equal to the work required to elevate the object from the datum plane.

Figure 13.2
Potential energy.

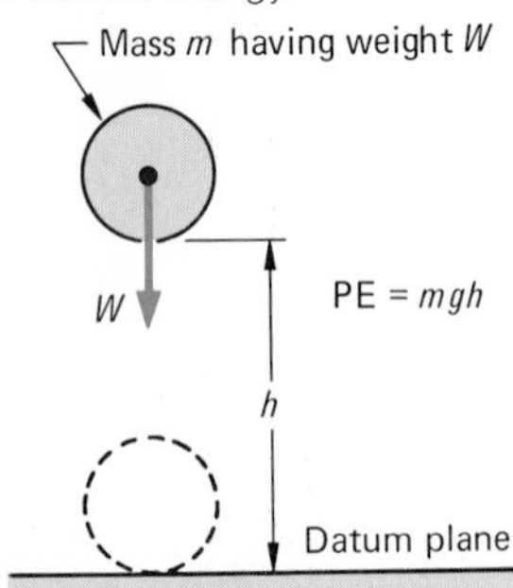

The energy possessed by an object by virtue of its velocity is called *kinetic energy*. It is equal to the energy required to accelerate the object from rest to its given velocity (considering the earth's velocity to be the datum plane). In equation form,

$$\text{KE} = \tfrac{1}{2}mv^2 \qquad 13.2$$

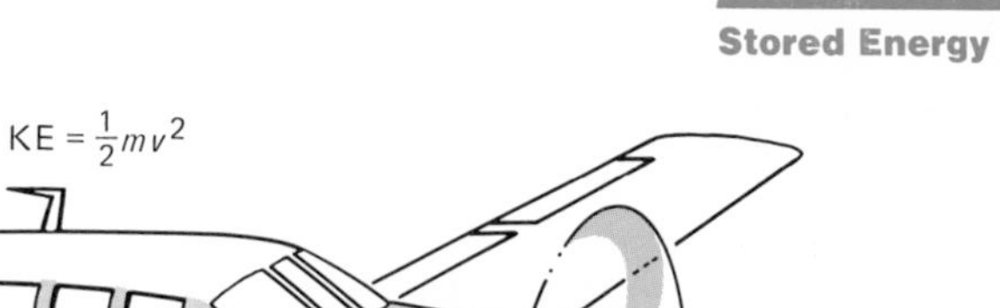

Figure 13.3 Kinetic energy.

where

m = mass, kg

v = velocity, m/s

KE = kinetic energy, J

See Fig. 13.3.

In many situations in nature, an exchange of the forms of energy is common. Consider a ball thrown into the air. It is given kinetic energy when thrown upward. When the ball reaches its maximum altitude, its vertical velocity is zero, but it now possesses a higher potential energy because of the increase in altitude (height). Then as it begins to fall, the potential energy decreases and the kinetic energy increases until it is caught. If other effects such as air friction are neglected and the ball is caught at the same altitude from which it was thrown, there is no change in energy. This phenomenon is called *conservation of energy*, which is discussed in Sec. 13.4.

All matter is composed of molecules that, at finite temperatures, are in continuous motion. In addition, there are intermolecular attractions which vary as the distance between molecules changes. The energy possessed by the molecules as a composite whole is called *internal energy*, designated by the symbol U, which is largely dependent on temperature.

When a fuel is burned, energy is released. When food is consumed, it is converted into energy that sustains human efforts. The energy that is stored in a lump of coal or a loaf of bread is called *chemical energy*. This energy is transformed by a natural process called photosynthesis, that is, a process that forms chemical compounds with the aid of light. The stored energy in combustible fuels is generally measured in terms of heat of combustion, or heating value. For example, gasoline has a heating value of $47.7(10^6)$ J/kg, or $20.5(10^3)$ Btu/lbm.

Certain processes change the atomic structure of matter. During the processes of *nuclear fission* (breaking the nucleus into two parts, which releases high amounts of energy) and *fusion* (combining lightweight nuclei into heavier ones, which also releases energy), mass is transformed into energy. The stored energy in atoms is called *nuclear energy*, which will play an ever-increasing role in the technology of the future.

13.3 Energy in Transit

Energy is transferred from one form to another during many processes, such as the burning of fuels to run engines or the converting of electric energy to heat by passing a current through a resistance. Like all transfer processes, there must be a driving force or potential difference in order to effect the transfer. In the absence of the driving force or potential difference, a state of equilibrium exists and no process can take place. The character of the driving force enables us to recognize the forms of energy in transit, that is, work or heat.

Energy is required for the movement of an object against some resistance. When there is an imbalance of forces and movement against some resistance, mechanical *work* is performed according to the relationship

$$\begin{aligned} W &= (\text{force})(\text{distance}) \\ &= Fd \end{aligned} \qquad 13.3$$

where the force is in the direction of movement. If force has units of newtons and distance is expressed in meters, then work has units of joules.

Electric energy is another form of work. This form of energy is transferred through a conducting medium when a difference in electric potential exists in the medium.

Other forms of energy transfer which are classified as work include magnetic, fluid compression, extension of a solid, and chemical. In each case a driving function exists that causes energy to be transferred during a process.

Heat is energy that is transferred from one region to another by virtue of a temperature difference. The unit of heat is the joule. The large numerical values occurring in energy-transfer computations has led to the frequent use of the megajoule (10^6 J). The symbol used for heat is Q.

The relationship between the energy forms of heat and work during a process is given by the first law of thermodynamics. In addition, every form of work carries with it a corresponding form of friction which may change some of the work into heat. When this happens, the process is irreversible, meaning the energy put into the process cannot be totally recovered by reversing the process. Another way of stating this is that heat is a low-grade form of energy and cannot be converted completely to another form such as work. This concept is basic to the second law of thermodynamics. The first

amples of overall efficiency are 17 to 23 percent for automobile engines (gasoline), 26 to 38 percent for diesel engines, and 20 to 33 percent for turbojet aircraft engines. Thus, in the case of the gasoline automobile engine, for every 80-L (21-gal) tank of gasoline, only 20 L (5.3 gal) ends up moving the automobile.

Care must be exercised in the calculation and use of efficiencies. To illustrate this point, consider Fig. 13.8, which depicts a steam

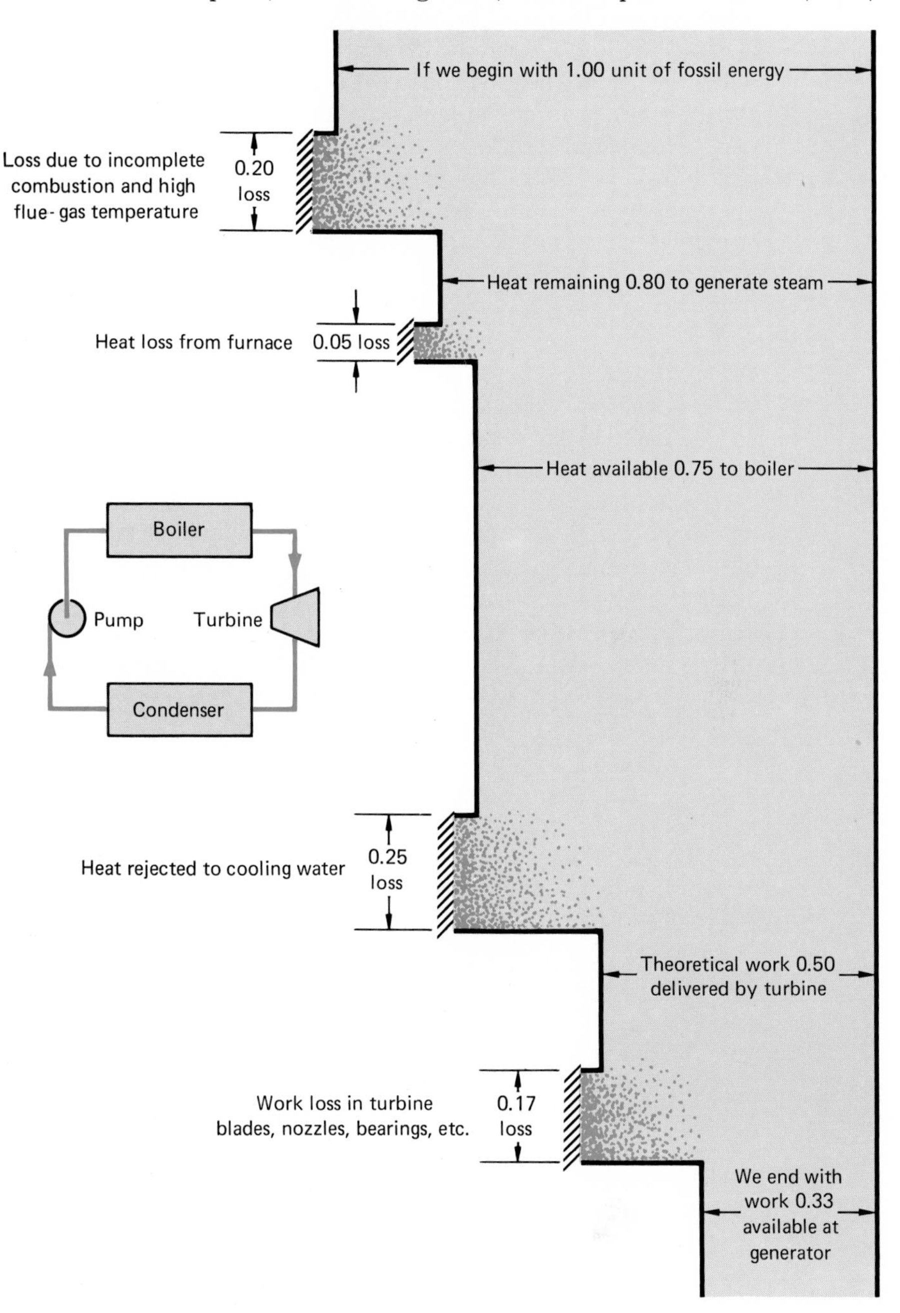

Figure 13.8 Energy losses in a typical steam power plant.

power plant from the burning of fuel for steam generation to driving an electric generator with a turbine. Efficiencies of each stage or combinations of stages in the power plant may be calculated by comparing energies available before and after the particular operations. For example, combustion efficiency can be calculated as 0.80/1.00, or 80 percent. The turbine efficiency is 0.33/0.50, or 66 percent. The overall power plant efficiency up to the electric generator is 0.33/1.00, or 33 percent. By no means is the 33 percent a measure of the efficiency of generation of electricity for use in a residential home. There will be losses in the generator and line losses in the transmission of the electricity from the power plant to the home. The overall efficiency from fuel at the power plant to the electric oven in the kitchen may run as low as 25 to 30 percent. It is perhaps ironic to note that the cycle is complete when the oven converts electricity back into heat, which is where the entire process began.

13.5.4 Power

Power is the rate at which energy is transferred, generated, or used. In many applications it may be more convenient to work with power quantities rather than energy quantities. The SI unit of power is the watt, but many problems will have units of horsepower (hp) and/or foot-pound-force per second (ft·lbf/s) as units, so conversions will be necessary.

Example problem 13.10 A steam turbine at a power plant produces 3 500 hp at the shaft. The plant uses coal as fuel (12 000 Btu/lbm). Using Fig. 13.8 to obtain an overall efficiency, determine how many metric tons of coal must be burned in a 24-h period to run the turbine.

Solution From Fig. 13.8, efficiency is 0.33/1.00, or 33 percent. From Eq. (13.6),

$$\text{Total input} = \frac{\text{useful output}}{\text{overall efficiency}}$$

$$= \frac{3\,500\text{ hp}}{0.33}\left|\frac{2\,545\text{ Btu}}{1\text{ hp}\cdot\text{h}}\right.$$

$$= 2.7(10^7)\text{ Btu/h}$$

The amount of coal needed for 1 day of operation is therefore

$$\text{Coal required} = \frac{2.7(10^7)\text{ Btu}}{1\text{ h}}\left|\frac{1\text{ lbm}}{12\,000\text{ Btu}}\right|\frac{1\text{ kg}}{2.205\text{ lbm}}\left|\frac{1\text{ t}}{10^3\text{ kg}}\right|\frac{24\text{ h}}{1\text{ d}}$$

$$= 24\text{ t/day}$$

air-conditioning fields is called the *energy efficiency ratio*, abbreviated *EER*. In essence, a refrigerating machine is a reversed-heat engine; that is, heat is moved from a low-temperature region to a high-temperature region, requiring a work input to the reversed-heat engine. The expression for efficiency of a reversible-heat engine is

$$\text{Refrigeration efficiency} = \frac{\text{refrigerating effect}}{\text{work input}} \qquad 13.7$$

Refrigeration efficiency is also called *coefficient of performance* (*CP*). The numerical value for CP may be greater than 1.

Industrial-size refrigeration units are measured in tons of cooling capacity. The ton unit originated with early refrigerating machines and was defined as the amount of refrigeration produced by melting 1 ton of ice in a 24-h period. If the latent heat of ice is taken into account, then a ton of refrigeration will be equivalent to 12 000 Btu/h. Home-size units are generally rated in British thermal units per hour of cooling capacity.

The EER takes into account the normal designations for refrigerating effect and work input and expresses these as power rather than energy. Thus, EER is the ratio of a refrigerating unit's capacity to its power requirements:

$$\text{EER} = \frac{\text{refrigerating effect, Btu/h}}{\text{power input, W}}$$

Example problem 13.11 Compute the cost of running a 36 000 Btu/h air-conditioning unit an average of 8 h/d for 1 month if electricity sells for 6.8¢/kWh. The unit has an EER of 8.

Figure 13.9
Cooling towers demonstrating that heat is typically rejected to the atmosphere (or to water).

Figure 13.10
An offshore oil-storage platform assists us in making efficient use of remaining fossil energy sources. (*Phillips Petroleum Company.*)

Solution

$$\text{EER} = \frac{\text{refrigerating effect}}{\text{power input}}$$

$$8 = \frac{36\,000}{\text{power input}}$$

Power input = 4 500 W

$$\frac{\$}{\text{month}} = \frac{4\,500\text{W}}{} \left|\frac{8\text{ h}}{1\text{ d}}\right| \frac{30\text{ d}}{1\text{ month}} \left|\frac{\$0.068}{1\text{ kWh}}\right| \frac{1\text{ kW}}{10^3\text{ W}}$$

$$= \$73.44/\text{month}$$

13.6 Energy Sources in the Future

The engineer will continue to design devices that will perform work by conversion of energy. However, the source of energy used to produce work and the conversion efficiency of the energy into work will play a much greater role in the design procedure.

Much has been written in the popular press concerning the "energy problem" and the development of energy sources to solve the problem. Many opinions are given with regard to how we should proceed to seek new energy sources and how we should use existing sources. The engineer must be able to discern the facts that may not be clear in some of the opinions. The following discussion provides an insight into some energy sources that may be of great benefit in the future.

13.6.1 Wind Power

Heated equatorial air rises and drifts toward the poles. This phenomenon coupled with the earth's rotation results in a patterned air

Figure 13.11
Experimental wind-powered generators.

flow. In the United States, we see this pattern as weather systems moving from west to east across the country. These weather systems possess enormous energy, but the energy is diffused, variable, and difficult to capture.

We are all familiar with the historical use of the wind to drive ships, to pump water, and to grind grain. These uses are still present, even though the advent of inexpensive fossil fuels and rural electrification made them uneconomical for a period of time. Today, a great deal of research is under way in an attempt to capture energy from the wind in an economically feasible manner.

Engineers have learned that the power output from a windmill is approximately proportional to the square of the blade diameter and to the cube of the wind velocity. This suggests that to generate high power output, the windmills must be large and must be located in areas where the average wind velocity is high. Thus, coastal regions and the Great Plains are promising locations.

Efforts continue toward reducing the costs of wind-power installations and toward overcoming the supply-demand problem. For example, we could use excess power generated when the wind is blowing to pump water to an elevated storage area and use the resulting potential energy to drive a water turbine for power generation when the wind is not blowing.

13.6.2 Water Power

There are a number of ways in which water can provide energy to be used by humankind. We can obtain energy, for example, from rivers, tides, ocean currents, waves, and thermal gradients.

Rivers Generation of electricity by use of the energy of falling water provides less than 15 percent of the electricity requirements in the United States (see Fig. 13.12). In most areas of the country, less than one-half of the potential for hydroelectric power has been developed. Hydroelectric power facilities will remain an important factor in energy generation far into the future. However, because of the environmental concerns for the protection of scenic rivers and wildlife, new facility construction will be limited. Hydroelectric power thus will furnish a smaller fraction of the total power production.

Example problem 13.12 How many megawatts of electric power can be obtained from the water of the Zambezi River over Victoria Falls if $1.0(10^5)$ m^3/s of water passes over the falls and drops $1.0(10^2)$ m. Assume a conversion efficiency of 80 percent and that the water turbines can be placed at the bottom of the falls.

Solution Equation (13.6) may be used with an estimated efficiency of 0.80: output = (0.80)(input).

The input may be calculated as the potential energy of the water

Figure 13.12
Electricity generated from water power continues to play a role in our search for energy. The Folsom Dam and power plant provide electricity for parts of California. (*U.S. Bureau of Reclamation.*)

that would be converted to kinetic energy during the 100-m drop. Equation (13.1) is used, with the mass equal to the volume multiplied by the density of the water. Conversion of input energy to power is accomplished by dividing by 1 s.

$$\begin{aligned} PE &= mgh \\ &= V\rho\,(gh) \\ &= (10^5 \text{ m}^3)(10^3 \text{ kg/m}^3)(9.81 \text{ m/s}^2)(100 \text{ m}) \\ &= (9.81)(10^{10}) \text{ kg}\cdot\text{m}^2/\text{s}^2 \\ &= 9.81(10^{10}) \text{ J} \end{aligned}$$

The input power is then

$$\begin{aligned} \text{Input} &= \frac{9.81(10^{10}) \text{ J}}{1 \text{ s}} \\ &= 9.81(10^{10}) \text{ W} \end{aligned}$$

and the output is

$$\begin{aligned} \text{Output} &= (0.80)(9.81)(10^{10}) \text{ W} \\ &= 7.8(10^4) \text{ MW} \end{aligned}$$

Tides The difference between the elevation of the ocean at high tide and low tide varies from 1 to 2 m in most places to 15 to 20 m in some locations. The idea of using this energy source is not new, and any sailor can attest to the power potentially available. The technical feasibility has been proved and plants are in use in France and in the U.S.S.R. The economic feasibility is still in serious doubt and is not yet even claimed by responsible sources. The toughest problems seem to be that vast storage volumes are required and suitable basins are rare.

Ocean currents We are all aware of the warming effect that the Gulf Stream provides for the British Isles and western Europe. This and other ocean currents possess massive amounts of kinetic energy even though they move at very slow velocities. It has been proposed that a series of large (170 m in diameter) turbines be placed in the Gulf Stream off Florida. Ten such turbines could produce power equivalent to that produced by one typical coal-fired power plant. Detractors warn that such installations may reduce the stream velocity to the point that the warming of western Europe would be lost.

Waves No doubt you have watched the surf smash into the beach and have been in awe of the obvious power being displayed. Machines have been made that produce power from the wave action. But to be successful, the installations must be located where the magnitude of the wave action is high and somewhat uniform, and it is difficult and expensive to design such installations against major storms. The best sites in the United States are on the coasts of Washington and Oregon.

Thermal gradients Earlier we discussed the Carnot principle in which the theoretical efficiency of a heat engine is defined by Eq. (13.5). It should be possible to employ the difference in temperature of the warm water at the ocean surface and the cold water at considerable depths to drive a heat engine. It is not uncommon to have temperatures of 75°F at the surface and 40°F at a depth of about 1 500 ft. Such a temperature difference would produce a Carnot efficiency of about 6 percent. An efficiency of about 3 percent might then be realized for an actual engine.

There are major problems to be overcome if a steam cycle is used with these temperatures. Very low pressures (less than 0.5 lbf/in^2 absolute) would need to be provided by a vacuum pump in order to produce steam at 75°F. Even lower pressure, perhaps 0.1 lbf/in^2 absolute, would be needed at the turbine outlet to produce useful work. Because of the low pressures, the steam would occupy a large volume, meaning that all the equipment would have to be much larger than is normally used in a steam power plant. Also, the turbine work output would be very small in comparison to that produced by a conventional system per pound-mass of steam flowing through the turbine.

Perhaps some fluid with a lower boiling temperature, say ammonia, could be employed to better utilize the low temperature of the ocean surface. Other problems to consider include the corrosive action of seawater, protection of the plant from weather, and the difficulties of transmitting the electrical energy produced.

13.6.3 Geothermal Power

It is commonly accepted that the earth's core is molten rock with a temperature of 10 000 to 12 000°F. This source has the potential to provide a large portion of our energy needs. The earth's crust varies in thickness from a few hundred feet to perhaps 20 mi. It is composed mostly of layers of rock—some solid, some porous, and some fractured. Engineers are now exploring the use of the earth's heat (geothermal energy) to produce power from the hot water, steam, and heated rock.

There are many areas throughout the world where large amounts of hot water are available from the earth. The hot water varies in many ways—in quantity, temperature, salinity, and mineral content. You will recall that we earlier discussed the reduced work produced and large volumes required for steam at low temperatures and pressures. These factors explain why much larger geothermal power plants (relative to coal-fired plants) are needed to produce the same amount of power. Since large volumes of hot water must be withdrawn from the earth, there is concern about the possibility

Figure 13.13
Geothermal energy is a potential source of power for areas where this energy exists. (*Phillips Petroleum Company.*)

of subsidence of the earth. Hence, the water used will probably be returned to the formation from which it was withdrawn.

At only a few locations in the world is steam available in sufficient quantities to be used to produce electricity. The only area in the United States producing large quantities of steam is in northern California. Electricity has been produced there for about 25 years at a cost that is less than that produced by fossil-fuel plants. Because of the vast amount of energy available from this source, new plants are going "on line" regularly. This steam is captured by drilling from 500 to 10 000 ft deep, and it has a temperature of about 350°F but at low pressure. Besides the corrosive nature of the steam, it contains several gases, including ammonia and hydrogen sulfide, that have objectionable odors and are poisonous.

In almost all areas we can consider drilling deep enough to tap the heat in the dry, heated rock of the earth. Water could be introduced by one set of pipes and steam removed by another. Since the steam would be at relatively low pressure and temperature, large amounts will be required. This fact points to the need for a large surface area of exposed rock to accomplish the heat transfer. It has been suggested that the rock can be broken by explosions or by hydraulic fracturing to create more exposed surface.

It is clear that using geothermal energy has great potential, but many problems need to be solved and solutions will probably come only after pilot plants have been constructed. Time will tell if this source of energy will become a significant contributor to our needs.

13.6.4 Solar Power

The sun is an obvious source of energy that we have employed in different ways since the beginning of time. It supplies us with many, many times as much energy as we need. Our problems in using this energy lie with collecting, converting, and storing the inexhaustible supply. Our efforts are being directed primarily toward direct heating and steam power plants. We are also experimenting with crop drying and metallurgical furnaces. You are no doubt aware of the increasing use of solar energy in home and building heating and in domestic hot-water supplies. This use will surely continue to increase, particularly in new installations. But we will have to improve on current technology or have tax relief and low-interest loans or other incentives to make solar installations economically feasible.

Solar (photovoltaic or PV) cells have been shown to work on the space vehicles but are extremely costly. We must reduce their cost to between 1 and 2 percent of current costs in order for this application to come into general use.

Example problem 13.13 Estimate the area over which solar energy must be collected to provide the electric energy needed by a small community of 4 500 persons in a 24-h period. Assume a conversion efficiency of 8 percent.

Solution This example requires some assumptions in order to obtain a meaningful result.

Figure 13.14
Solar-heated home.

1. Each home consumes about 13 000 kWh of electricity on the average each year.
2. An average of three persons lives in each home.
3. The sun shines an average of 8 h/d.
4. The typical solar heat transfer rate is 1 kW/m^2.

Equation (13.6) may be used.

$$\text{Input} = \frac{\text{output}}{\text{efficiency}}$$
$$= (\text{area})(\text{solar heat-transfer rate})$$
$$= \text{area } (1 \text{ kW/m}^2)$$

Therefore, the area can be computed as

$$\text{Area} = \frac{1 \text{ m}^2}{0.08 \cdot 1 \text{ kW}} \cdot \frac{13\,000 \text{ kWh}}{\text{year-home}} \cdot \frac{1 \text{ home}}{3 \text{ persons}} \cdot \frac{4\,500 \text{ persons}}{} \cdot \frac{1 \text{ d}}{8 \text{ h}} \cdot \frac{1 \text{ year}}{365 \text{ d}}$$
$$= 84\,000 \text{ m}^2$$
$$= 8.4 \text{ hm}^2$$

13.6.5
Nuclear Power

As mentioned earlier in this chapter, nuclear power may be generated from two processes, fission and fusion. Fission is the splitting of an atom of nuclear fuel, usually uranium 235 (U-235). The splitting process releases a great quantity of heat, which can then be converted to other energy forms. Nearly all reactors on line today producing electric energy are fission reactors using U-235. How-

Figure 13.15
The Duane Arnold Energy Center, Palo, Iowa, is a nuclear-powered electric generating station. Nuclear power is another of the potential sources for additional energy. (*Iowa Electric Light and Power Company.*)

ever, U-235 comprises only 0.7 percent of the natural uranium supply and thus would quickly become a scarce resource if there were total energy dependence on it. Other isotopes, such as U-238 and thorium 232, are relatively abundant in nature. The breeder reactor is designed to use these isotopes that are generally a waste product in current reactors.

The radioactive waste material from nuclear power generation has created some disposal problems. The half-life of radioactive materials can be 1 000 years or more, thus creating a perpetual need for secure disposal of the waste materials from a nuclear-power-generation facility. Radiation leaks and other environmental concerns have led to the shutdown of some facilities. This has made nuclear-power generation a politically sensitive issue. Nuclear engineers and scientists continue to perform research to solve the complex problems associated with this source of energy. As these problems are solved and the general public becomes more aware of the potential afforded by nuclear energy, increased use of nuclear power is likely.

Example problem 13.14 If 1 g of U-235 is capable of producing the same amount of heat as 3 500 kg of coal, how much water would have to be stored $1.0(10^3)$ m above sea level to provide the same energy as 1 kg of U-235? Assume that coal has a heating value of 30.0 MJ/kg.

Solution A kilogram of U-235 would possess the same energy as $3.5(10^6)$ kg of coal and therefore would have a total energy value of

$$(30.0 \text{ MJ/kg})(3.5 \times 10^6 \text{ kg}) = 105(10^6) \text{ MJ}$$

Setting this equivalent to the potential energy of the water,

$$mgh = 105(10^{12}) \text{ J}$$

$$m = \frac{105(10^{12}) \text{ J}}{(9.81 \text{ m/s}^2)(10^3 \text{ m})}$$

$$= 10.7(10^9)(\text{N}\cdot\text{m}\cdot\text{s}^2)/\text{m}^2$$

$$= 10.7(10^9)(\text{kg}\cdot\text{m/s}^2)(\text{m}\cdot\text{s}^2)/\text{m}^2$$

$$= 10.7(10^9) \text{ kg}$$

$$= 10.7 \text{ Tg}$$

Problems

13.1 Determine the amount of kinetic energy possessed by a pickup truck that has a mass of 3 960 lbm and is traveling at 48 mi/h. Express the answer in joules.

13.2 (*a*) Suppose you wish to stop the truck described in Prob. 13.1 in a distance of 75 m. What constant braking force would be required? (*b*) Instead of stopping the truck, you allow it to coast up a grade of

8.6 percent. Assuming no air or roadway friction, what distance in meters will the truck travel along the 8.6 percent grade?

13.3 If an automobile has a kinetic energy of 432 kJ and is traveling at 55 mi/h, what is its mass in kilograms?

13.4 If an automobile weighing 1 875 kg has a kinetic energy of 696 kJ, what is its velocity expressed in kilometers per hour?

13.5 A 2.00-ton automobile coasts down a 10.5 percent grade for 470 m after starting at rest. If we assume no air or road friction, at what speed in meters per second will it be traveling when it crosses the 470-m line?

13.6 A 9.00-Mg airplane is traveling at 450 mi/h at an altitude of 1.00 mi above sea level. What is its total energy in joules?

13.7 A stream flows over a ledge and falls freely 79 m into a pool.
(*a*) What is the water velocity in meters per second just before it enters the pool?
(*b*) What is the kinetic energy in joules in each gallon of water at this point?

13.8 "Speed guns" are regularly employed to measure the velocity of baseballs thrown by pitchers. From what height (in meters) must a ball be dropped to achieve the same speed as a 92-mi/h fastball?

13.9 A crate in a warehouse is moved 42.6 m along a level floor during a period of 13.7 s. A force of 525 N is applied to a cable attached to the crate and the cable makes an angle of 33.6° with the floor.
(*a*) How much work is done (in joules)?
(*b*) What is the average power required (in horsepower)?

13.10 The gravitational field on the surface of the moon is about one-sixth that of the earth at sea level. How much work in joules is done to raise a 198-kg experimental package to an elevation of 82 ft above the moon's surface?

13.11 If the package described in Prob. 13.10, after being put in place, fell back to the surface of the moon, with what velocity in meters per second would it strike the surface?

13.12 In testing a newly designed small cylinder, it is found that a force of 23.7 kN is required to hold the piston against a pressure of 6.38 MPa in the cylinder. What is the diameter in millimeters of the circular piston?

13.13 A piston is 8.23 cm in diameter. The design pressure in the cylinder is 8.00 MPa. To test the cylinder, the cylinder pressure is increased to 2.5 times the design pressure (a safety factor of 2.5). What maximum force in newtons must be applied to the piston to hold it stationary during the test?

13.14 If 33 Btu of heat is added to and 4 650 ft·lbf of work is done on a closed system, what is the change in internal energy (joules) of the system?

13.15 During a process, 1.00 MJ of heat is added to a closed system that contains 9.25 kg of a gas. The internal energy is decreased by 57.5 kJ during the same process. How much work in joules was done on the gas?

13.16 An adiabatic process is carried out on 18.0 lbm of fluid during which

$7.4(10^4)$ ft·lbf of work is done by the fluid. Express the change in energy (in British thermal units) per pound-mass of fluid.

13.17 Heat at the rate of 245 Btu/min is added to a process that delivers work at the rate of 1.75 hp. How much heat (in joules) is lost from the system per minute, and what is the system's overall efficiency?

13.18 A motor burns fuel at the rate of 0.52 L/min, and the fuel has a heating value of 34.2 kJ/mL. What is the output power (in kilowatts) if the overall efficiency is 33.3 percent?

13.19 What is the Carnot efficiency of a heat engine operating between 300 and 40°C?

13.20 Your supervisor asks you to make a quick calculation to determine if it is possible to produce 125 kW of power from an engine if it receives 0.77 MW at 560°C and rejects heat at 245°C. What is your answer and why?

13.21 A low-temperature (37°C) reservoir receives $2.0(10^3)$ Btu/min from a Carnot engine that develops 42 hp. What is the temperature (in degrees Celsius) of the heat source?

13.22 Fuel that has a heating value of 36 MJ/L is burned at the rate of 0.155 L/min in an internal-combustion engine that is used to drive a standby electric generator. The engine has a continuous output of 48 kW, and the generator provides 150 A at 240 V dc.

(*a*) What is the efficiency of the engine?
(*b*) What is the efficiency of the generator?
(*c*) What is the overall efficiency of the system?

13.23 You have a pump and an electric motor that have a combined efficiency of 59 percent. What is the power requirement (in kilowatts) to the motor if you are to pump 1 650 gal of oil to an elevation 43 ft higher than its present location in 23 min? (Oil has a specific gravity of 0.81.)

13.24 A 42-hp pump is used to move a grain-slurry mixture in a processing plant. If the pump is 67 percent efficient, what input horsepower must be supplied? Assuming that the electric motor that drives the pump is 83 percent efficient, what power (in kilowatts) is required?

13.25 A $\frac{1}{2}$-hp electric motor is 91 percent efficient and is used to drive a pump of 73 percent efficiency. How much power (in watts) does the motor require? How much water could this pump-motor combination lift 2.7 m in 52 min?

13.26 Estimate the number of homes that could receive electric energy from fission of 25 kg of U-235. Assume a home requires 13 000 kWh each year, that the conversion efficiency is 72 percent, and that the energy equivalent of U-235 is 10^8 MJ/kg. Determine the surface area (in acres) of solar collectors operating at an efficiency of 9.0 percent that would be needed to supply electric energy to these same homes.

13.27 Using the assumptions (except efficiency) from Example Prob. 13.13, produce a table showing the solar collector area in acres needed to supply electric energy for cities with populations of 5 000, 20 000, 50 000, 100 000, 1 000 000, and 5 000 000. Provide results for conversion efficiencies of 5 to 15 percent in increments of 1 percent. For your city, compare the area of solar collectors required with the total area within the city limits (use an efficiency of 8 percent).

13.28 A 24 000-Btu/h air conditioner is installed in a home. What fraction of the unit's capacity is required to remove the heat generated by six 100-W lamps, four 75-W lamps, a 750-W coffee maker, a 500-W refrigerator, a 800-W microwave oven, and four people watching TV (125 W). The average heat output of a person at rest is about 480 Btu/h. If the EER of the air conditioner is 10.35, what power (in kilowatts) is used by the unit when it is operating? How many kilowatt-hours of electric energy are used during each hour for the conditions described above?

13.29 Compute and plot the monthly cost of operating a series of 36-ton commercial air conditioners vs. EER. Consider values of EER from 7.0 to 11.0, with an increment of 0.5. The electrical rate structure is as follows:

First 200 kWh = 8.66¢/kWh

Next 800 kWh = 7.32¢/kWh

Next 9 000 kWh = 5.66¢/kWh

Over 10 000 kWh = 4.65¢/kWh

Assume that the unit must average 8 h of operation per day during weekdays and 4 h/d on weekends. Use a month beginning on Monday.

13.30 For a 12 500-lbm aircraft, develop a table of altitude, velocity, potential energy, kinetic energy, and maximum altitude that could be attained if all its kinetic energy could be converted to potential energy. Include flight altitudes of 2.0×10^3, 5.0×10^3, 1.0×10^4, 1.5×10^4, and 2.0×10^4 ft and velocities of 125, 150, 175, 200, 225, and 250 mi/h. Plot the maximum attainable altitude for the case of a 5 000-ft flight altitude vs. velocity.

14

Engineering Economy

14.1 Introduction

With the increase in the importance of technology has come a greater role for the engineer in management. Engineers in many cases have become managers or executive officers of businesses and are involved in decision making. When the managers are not engineers, they most often will have engineering advisers who provide reports and analyses that influence their decisions. Steadily over the past decades, the amount of capital investment (money spent for equipment, expansion, etc.) has risen in American business and industry to the point that the average investment now exceeds the annual salary cost. The amount of the investment obviously fluctuates with the type of industry and the degree of automation. This statistic underlies the importance of knowledge of the technical aspects of the enterprise and the increasingly important role of the engineer in financial decision making.

Many examples could be used to illustrate how an engineer might become involved in this decision-making process. Suppose a large manufacturing organization has decided to purchase a computer network. The system will be connecting many aspects of the com-

Figure 14.1
The modern engineering office typically provides the computations hardware and software necessary to analyze several alternative solutions.

pany such as engineering design, purchasing, marketing, manufacturing, field sales, administration and service. Ten different computer vendors are invited to submit bids on the overall network. The bids are to include hardware, software, installation, maintenance, and service costs. The engineer's job then is to do an analysis of alternatives and rank the ten companies based on predetermined criteria.

This assignment is entirely possible once a method of comparison, such as the annual equivalent or the present-worth equivalent, is selected. These methods of analysis are discussed later in the chapter. However they only allow us to look at tangible costs. There are also many intangible items that require accountability.

Few businesses have ever been started with the sole goal of producing a product; the aim of business is to produce a profit. In like manner, an established business is interested only in new ventures or products that will produce a favorable rate of return on the investment. A simple definition of *rate of return* is profit divided by investment, or

$$\text{Annual rate of return} = \frac{\text{annual profit}}{\text{investment}}$$

Since any venture has at least some risk involved, few investors will become interested unless there is a promise of a much greater return than could be realized by deposits in banks and savings and loans or through the purchase of treasury notes or government bonds. These considerations are part of the engineer's concern, either when acting as the decision maker or when attempting to convince the decision maker of the worth of an idea or invention.

14.2 Simple Interest

The idea of interest on an investment is certainly not new. The New Testament refers to banks, interest, and return. History records business dealings involving interest at least 40 centuries ago. Early business was largely barter in nature with repayment in kind. It was common during the early years of the development of the United States for people to borrow grain, salt, sugar, animal skins, etc., from each other to be repaid when the commodity was again available. Since most of these items depended on the harvest, annual repayment was the normal process. Likewise, since the lender expected to be repaid after no more than a year, simple annual interest was the usual transaction. When it became impossible to repay the loan after a year, the interest was calculated by multiplying the principal amount by the product of the interest rate and the number of periods (years):

$$I = Pni \qquad 14.1$$

where I = interest accrued

P = principal amount

n = number of interest periods

i = interest rate per period (as a decimal, not as a percent)

Therefore, if $1 000 were to be loaned at 7 percent annual interest for 5 years, the interest would be

$$I = Pni$$
$$= (1\ 000)(5)(0.07)$$
$$= \$350$$

and the amount S to be repaid is

$$S = P + I \qquad 14.2$$
$$= 1\ 000 + 350$$
$$= \$1\ 350$$

It can be seen that

$$S = P + I = P + Pni \qquad 14.3$$
$$= P(1 + ni)$$

14.3 Compound Interest

As time progressed and business developed, the practice of borrowing became more common, and the use of money replaced the barter system. It also became increasingly more common that money was loaned for longer periods of time. Simple interest was relegated to the single-interest period, and the practice of compounding developed. It can be shown using Eq. (14.3), $n = 1$, that the amount owed at the end of one period is

$$P + Pi = P(1 + i)$$

The interest generated during the second period is then $(P + Pi)i$. It can be seen that interest is being calculated not only on the principal but upon the previous interest as well. The sum S at the end of two periods becomes

P	principal amount
$+ Pi$	interest during first period
$+ Pi + Pi^2$	interest during second period
$P + 2Pi + Pi^2$	sum after two periods

This can be factored as follows:

$$P(1 + 2i + i^2) = P(1 + i)^2$$

The interest during the third period is

$$(P + 2Pi + Pi^2)i = Pi + 2Pi^2 + Pi^3$$

and the sum after three periods is

$P + 2Pi + Pi^2$	sum after second period
$+ Pi + 2Pi^2 + Pi^3$	interest during third period
$P + 3Pi + 3Pi^2 + Pi^3 = P(1 + i)^3$	sum after three periods

This procedure can be generalized to n periods of time and will result in

$$S_n = P(1 + i)^n \qquad 14.4$$

where S_n is the sum generated after n periods.

Therefore, based on interest compounded annually, the preceding example becomes (5 years at 7 percent)

$$S_n = P(1 + i)^n$$

$$S_5 = (1\ 000)(1.07)^5$$

$$S_5 = \$1\ 402.55$$

The same result can be obtained by the following arithmetic:

Principal amount	=	\$1 000.00
Interest during first year: (1 000)(0.07)	=	70.00
Sum after 1 year	=	1 070.00
Interest during second year: (1 070)(0.07)	=	74.90
Sum after 2 years	=	1 144.90
Interest during third year: (1 144.90)(0.07)	=	80.14
Sum after 3 years	=	1 225.04
Interest during fourth year: (1 225.04)(0.07)	=	85.75
Sum after 4 years	=	1 310.79
Interest during fifth year: (1 310.79)(0.07)	=	91.76
Sum after 5 years	=	\$1 402.55

Care must be exercised in using interest rates and payment periods to make sure that the interest rate used is the rate for the period selected.

Consider calculating the sum after 1 year. If the annual interest rate is 12 percent compounded annually, then $i = 0.12$ and $n = 1$. However, when the annual rate is 12 percent, but it is to be compounded every 6 months (semiannually), then $i = 0.12/2$ and $n = 2$. This idea can be extended to a monthly compounding period, with $i = 0.12/12$ and $n = 12$, or a daily compounding period, with $i = 0.12/365$ and $n = 365$.

Example problem 14.1 What lump sum must be paid at the end of 4 years if \$8 000 is borrowed from a bank at a 12 percent annual interest rate compounded (*a*) annually, (*b*) semiannually, (*c*) monthly, and (*d*) daily?

Solution

(*a*) Compounded annually:

$S_n = P(1 + i)^n$

where $i = 0.12$

$n = 1 \times 4$ years

$P = \$8\,000$

$S_4 = 8\,000(1.12)^4$

$= \$12\,588.15$

Note: Always calculate the answer to the nearest penny.

(*b*) Compounded semiannually:

$S_n = P(1 + i)^n$

where $i = 0.12/2$

$n = 2 \times 4$ years

$P = \$8\,000$

$S_8 = 8\,000(1.06)^8$

$= \$12\,750.78$

(*c*) Compounded monthly:

$S_n = P(1 + i)^n$

where $i = 0.12/12$

$n = 12 \times 4$ years

$P = \$8\,000$

$S_{48} = 8\,000(1.01)^{48}$

$= \$12\,897.81$

(*d*) Compounded daily:

$S_n = P(1 + i)^n$

where $i = 0.12/365$

$n = 365 \times 4$ years

$P = \$8\,000$

$$S_{1460} = 8000\left(1 + \frac{0.12}{365}\right)^{1460}$$

$= \$12\,927.57$

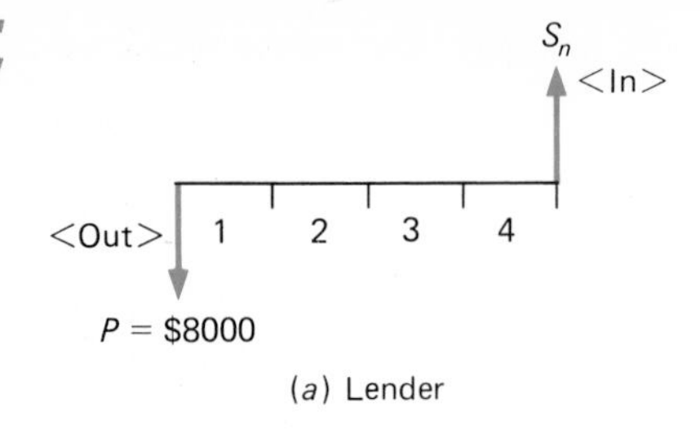

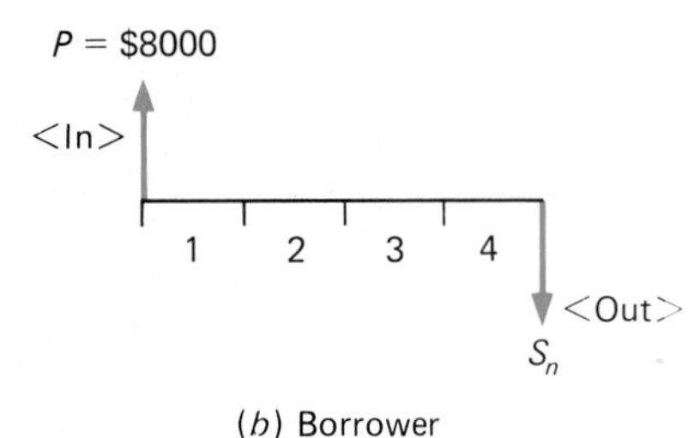

Figure 14.2
Cash-flow diagram: (*a*) lender; (*b*) borrower.

14.4 Cash-Flow Diagram

The transaction described in Example Prob. 14.1 can be graphically illustrated in a cash-flow diagram. Since a cash-flow diagram is very useful in the visualization of any transaction, it will be used throughout this chapter. The following general rules apply:

1. The horizontal line is a time scale. Normally, years are given as the interval of time.

2. The arrows signify cash flow. A downward arrow means money out, and an upward arrow means money in.

3. The diagram is dependent upon the point of view from which it is constructed, i.e., upon whether it is the lender's or the borrower's point of view (see Fig. 14.2).

14.5 Present Worth

"Present worth" is a term that is encountered when trying to fix the worth of a given sum at the current time. It is the amount of money that must be invested now to produce a prescribed sum at another date. (It could also be the worth today of a previous investment.)

In other words, if you were guaranteed an amount of money S_n 4 years from today, then P would be the present worth of S_n where the interest is i and $n = 4$ (assuming annual compounding). Since this analysis is exactly the inverse of finding a future sum, we have

$$P = S_n(1 + i)^{-1} \qquad 14.5$$

As an example, if you could convince a lending institution (i.e., a banker) that you have a guaranteed amount of money available 4

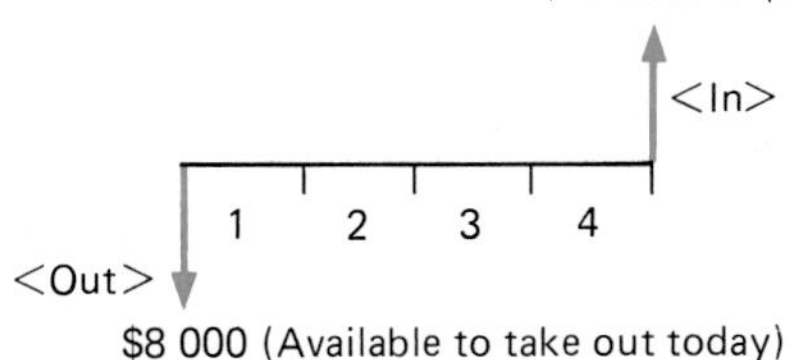

Figure 14.3
Banker's cash-flow diagram.

years from today, it may be possible to borrow the present worth of that amount. If the guaranteed sum (4 years later) is equal to $12 588.16, then the present worth at 12 percent annual interest (compounded annually) is $8 000.00. See Fig. 14.3.

$$P = S_n(1 + i)^{-n}$$

where $S = \$12\ 588.15$

$i = 0.12$

$n = 4$

$$P = \frac{12\ 588.15}{(1.12)^4}$$

$$= \$8\ 000.00$$

In situations that involve economic decisions, the following questions may arise:

1. Does it pay to make an investment now?
2. What is the current benefit of a payment that will be made at some other date?
3. If a series of payments is made over specific intervals throughout a designated time span, what is this worth now?

In such cases, the answer can be examined by finding the present worth of the transaction.

Many businesses calculate their present worth each year, since the change in their present worth is a measure of the growth of the company. The following example problem will help demonstrate the concept of present worth.

Example problem 14.2 Listed below are five transactions. Determine their present worth if money is currently valued at 10 percent annual interest compounded annually. Determine the current net-cash equivalent assuming no interest has been withdrawn or paid. Draw a cash-flow diagram for each.

Figure 14.4
Owner's cash-flow diagram.

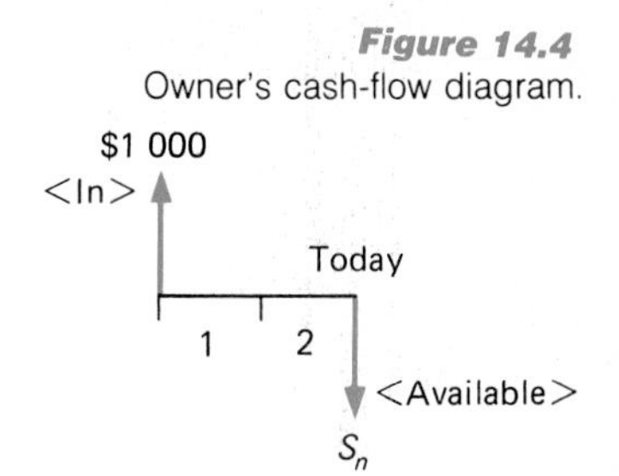

Solution

(*a*) $1 000 deposited 2 years ago (Fig. 14.4):

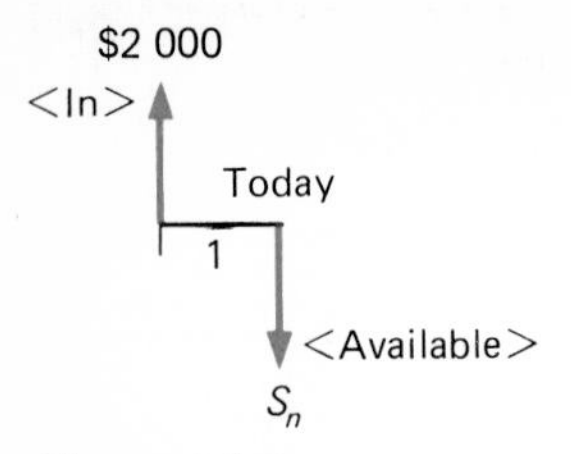

Figure 14.5
Owner's cash-flow diagram.

$S_n = 1\,000(1.10)^2 = \$1\,210.00$

(*b*) \$2 000 deposited 1 year ago (Fig. 14.5):

$S_n = 2\,000(1.10)^1 = \$2\,200.00$

(*c*) \$3 000 to be received 1 year from now (Fig. 14.6):

$P = 3\,000(1.10)^{-1} = \$2\,727.27$

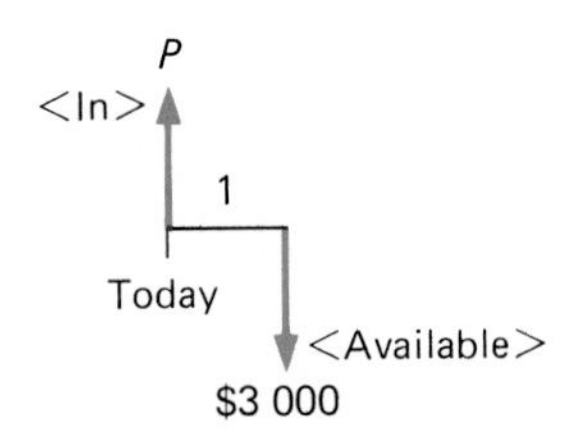

Figure 14.6
Owner's cash-flow diagram.

(*d*) \$4 000 to be paid 2 years from now (treated as negative for the owner since it must be paid) (Fig. 14.7):

$P = -\,4\,000(1.10)^{-2}$

$\quad = -\,\$3\,305.79$

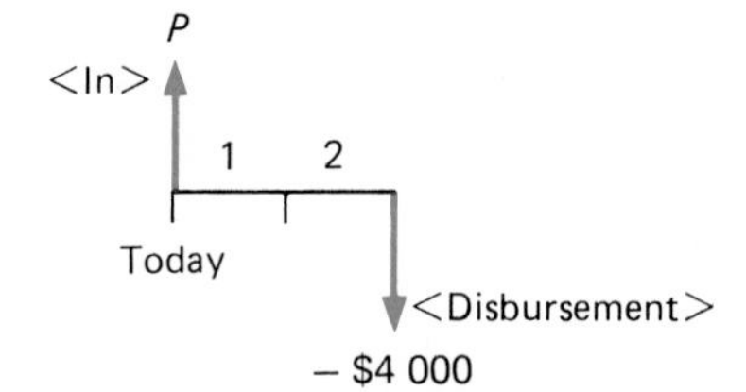

Figure 14.7
Owner's cash-flow diagram.

(*e*) \$5 000 to be received 4 years from now (Fig. 14.8):

$P = 5\,000(1.10)^{-4} = \$3\,415.07$

Present worth of the five transactions:

	\$1 210.00
	2 200.00
	2 727.27
	−3 305.79
	3 415.07
Present worth =	\$6 246.55

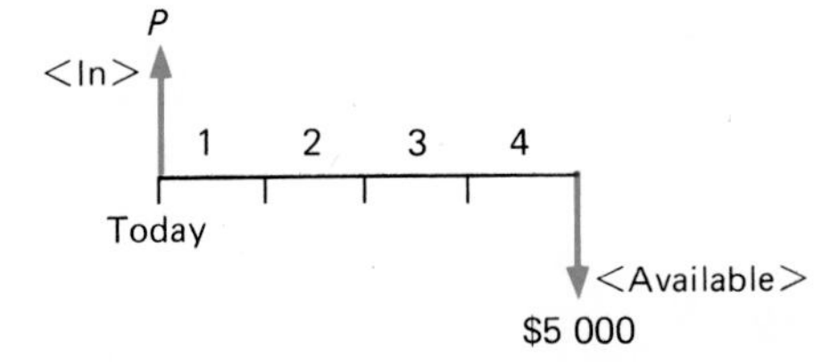

Figure 14.8
Owner's cash-flow diagram.

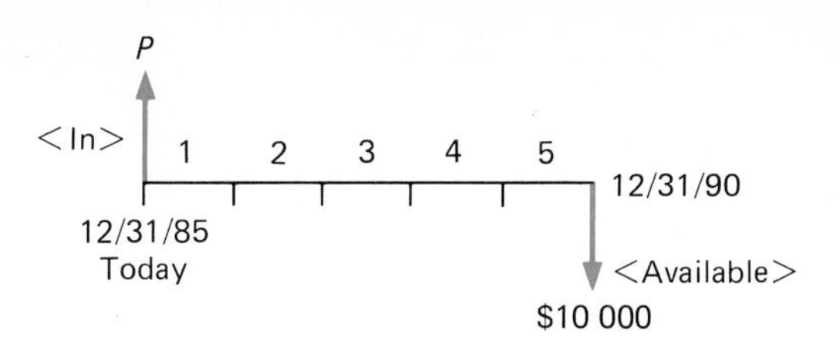

Figure 14.9
Owner's cash-flow diagram.

Example problem 14.3 On December 31, 1990, there will be $10 000 available from the settlement of an estate. What sum of money does that represent on December 31, 1985, at the annual interest of 12 percent compounded annually?

Solution (Fig. 14.9)

$$P = 10\,000(1.12)^{-5}$$
$$= \$5\,674.27$$

14.6 Annuities

The concept of an annuity can be developed from the idea of compound interest. Since it involves a series of equal payments at regular intervals, the point in time at which compounding begins must be examined.

Annuities are used by financial agencies in several ways; each form of annuity has been given a name to distinguish one from another.

14.6.1 Sinking Fund

A sinking fund is an annuity that is established in order to produce an amount of money at some future time. It might be used to develop cash for an expenditure that one knows is going to occur—for instance, a Christmas fund to pay for presents. If you wish to trade cars at regular intervals, it is much more economical to pay in advance. A cash-flow diagram for the sinking fund is shown in Fig. 14.10.

If A is deposited at the end of each period and interest is compounded at i, the sum S_n will be produced after n periods.

It can be seen that the last payment will produce no interest, the payment at period $n - 1$ will produce interest equal to A times i,

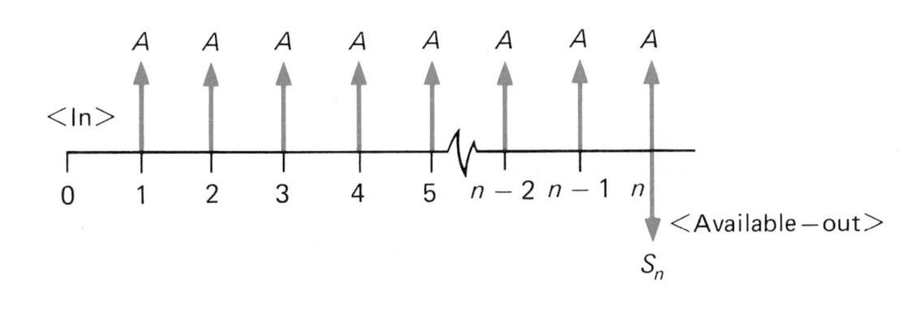

Figure 14.10
Saver's cash-flow diagram.

the payment at period $n - 2$ will produce $A(1 + i)^2$, and so forth. Hence, the sum produced will be as follows:

Deposit at end of period	Interest generated	Sum due to this payment
n	None	$A(1)$
$n - 1$	$A(i)$	$A(1 + i)$
$n - 2$	$A(1 + i)i$	$A(1 + i)^2$
$n - 3$	$A(1 + i)^2 i$	$A(1 + i)^3$

$$S_4 = \text{sum for four payments} = A(4 + 6i + 4i^2 + i^3)$$

If we multiply this expression by i, add and subtract 1, and divide by i, then

$$S_4 = A\left[\frac{(4i + 6i^2 + 4i^3 + i^4 + 1) - 1}{i}\right]$$

$$= A\left[\frac{(1 + i)^4 - 1}{i}\right]$$

It can be shown that

$$S_n = A\left[\frac{(1 + i)^n - 1}{i}\right] \qquad 14.6$$

Therefore, if you want to accumulate S_n during n periods, A must be deposited at the end of each period at i interest rate, compounded at each period.

Sinking funds are often used to accumulate sufficient money to replace worn-out or obsolete equipment.

Example problem 14.4 How much money would be accumulated by a sinking fund if \$90 is deposited at the end of each month for 3 years with an annual interest rate of 10 percent?

Solution (Fig. 14.11)

$$S_n = A\left[\frac{(1 + i)^n - 1}{i}\right]$$

where $A = \$90$

$i = 0.10/12$ (monthly compounding)

$n = 12 \times 3$

$$S_n = 90\left[\frac{(1 + 0.10/12)^{36} - 1}{0.10/12}\right]$$

$$= \$3\,760.36$$

Example problem 14.5 It is desired to accumulate \$10 000 to replace a piece of equipment after 8 years. How much money must

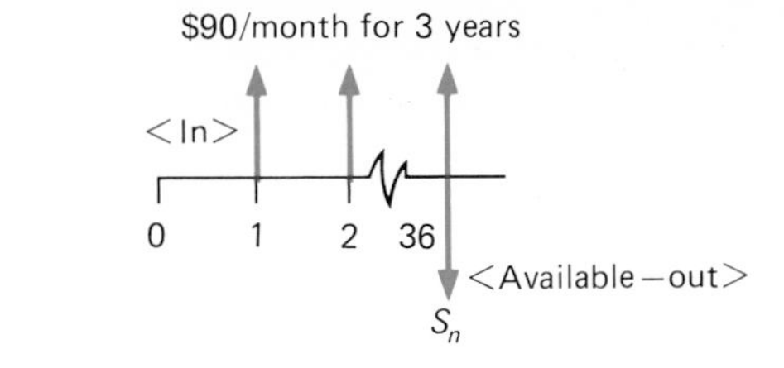

Figure 14.11
Saver's cash-flow diagram.

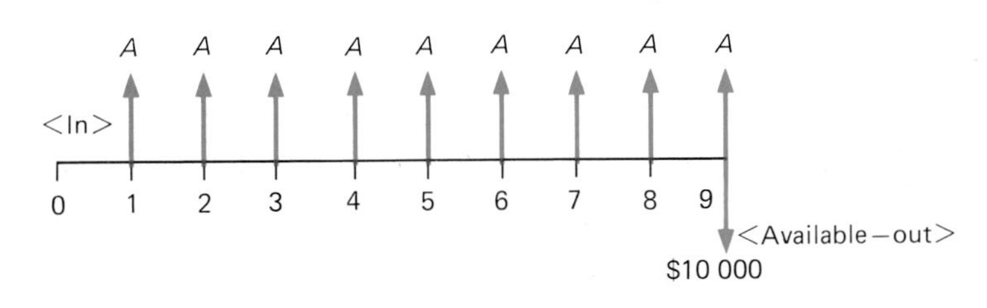

Figure 14.12
Company's cash-flow diagram.

be placed annually into a sinking fund that earns 7 percent interest? Assume that the first payment is made today and the last one 8 years from today with interest compounded annually.

Solution (Fig. 14.12)

$$S_n = A\left[\frac{(1+i)^n - 1}{i}\right]$$

$$A = \frac{S_n(i)}{(1+i)^n - 1}$$

$$= \frac{0.07(10\,000)}{(1.07)^9 - 1}$$

$$= \$834.86$$

Note that $n = 9$ in this example because the first payment must occur at the end of the first period so that the sinking-fund formula will be valid.

14.6.2
Installment Loan

A second and very popular way that annuities are used to retire a debt is by making periodic payments instead of a single large payment at the end of a given time period. This is the time-payment plan used by most retail businesses and lending institutions. It is also used to amortize bond issues. A cash-flow diagram for this scheme is illustrated in Fig. 14.13.

In this case the principal amount P is the size of the debt and A is the amount of the periodic payment that must be made with interest compounded at the end of each period. It can be seen that if P were removed from the time line and S_n placed at the end of the nth period, it would be a sinking fund. Furthermore, it can be

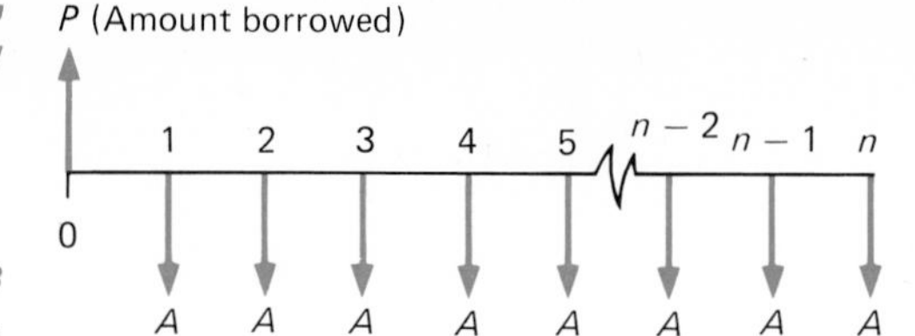

Figure 14.13
Buyer's cash-flow diagram.

shown that S_n would also be the value of P placed at compound interest for n periods [$S_n = P(1 + i)^n$]. Likewise, P can be termed the present worth of the sinking fund that would be accumulated by the deposits. Therefore, since

$$S_n = P(1 + i)^n \qquad \text{and} \qquad S_n = A\left[\frac{(1 + i)^n - 1}{i}\right]$$

the present worth becomes

$$P = A\left[\frac{(1 + i)^n - 1}{i(1 + i)^n}\right] \tag{14.7}$$

The term within the brackets is known as the *present worth* of a sinking fund, or the *uniform annual payment present-worth factor*.

It follows that

$$A = P\left[\frac{i(1 + i)^n}{(1 + i)^n - 1}\right] \tag{14.8}$$

where the term in brackets is most commonly called the *capital recovery factor*, or the *uniform annual payment annuity factor*, and is the reciprocal of the uniform annual payment present-worth factor.

To clarify the concept of installment loans, consider the following example problem.

Example problem 14.6 Suppose you borrow $1 000 from a bank (lending agency) for 1 year at 12 percent annual interest compounded monthly. Consider paying the loan back by three different methods:

1. You keep the $1 000 for 1 year and pay the bank back at the end of the year in a lump sum. What would you owe? (See Fig. 14.14.)

$$S_n = 1\,000(1 + 0.01)^{12}$$
$$= \$1\,126.83$$

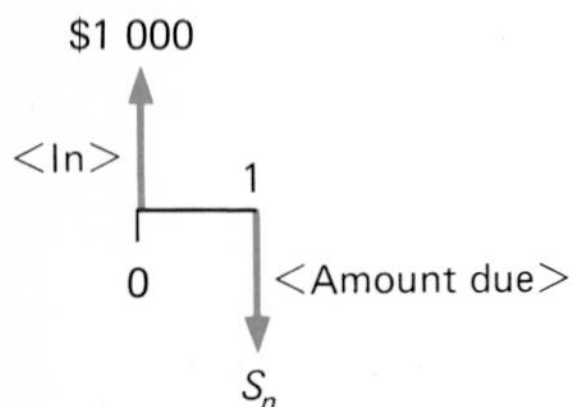

Figure 14.14
Borrower's cash-flow diagram.

2. Since you will owe the bank $1 126.83 at the end of 1 year, what if you decide to set up a sinking fund with the credit union and accumulate $1 126.83 in 12 payments? How much would each payment be (assuming the same 12 percent interest)? (See Fig. 14.15.)

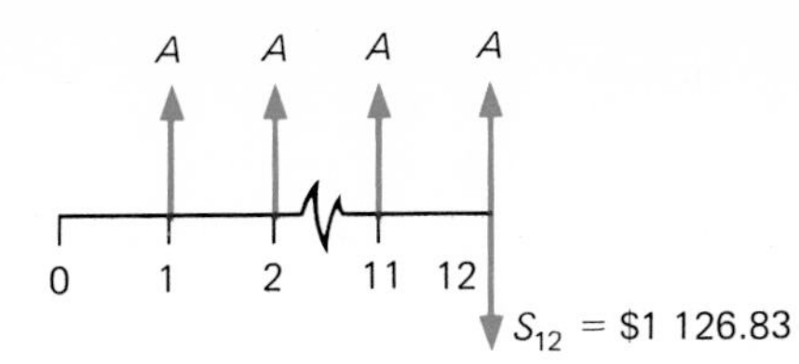

Figure 14.15
Cash-flow diagram for a savings account.

$$S_n = A\left[\frac{(1+i)^n - 1}{i}\right]$$

where $S_n = \$1\ 126.83$

$i = 0.12/12$

$n = 1 \times 12$

$$A = \frac{S_n(i)}{(1+i)^n - 1}$$

$$= \$88.85$$

So if we made payments to the credit union of $88.85 per month for a year into a sinking fund we could take $1 126.83 from the credit union to the bank at the end of 12 months.

3. A third method is the installment loan. You borrow $1 000 from the bank, make payments of $88.85 per month to the bank, and at the end of a year you have retired the debt. (See Fig. 14.16.) In other words, the amount of money due at the end of the year is

$$S_n = P(1+i)^n$$

which is the same amount derived from

$$S_n = A\left[\frac{(1+i)^n - 1}{i}\right]$$

so $P(1+i)^n = A\left[\frac{(1+i)^n - 1}{i}\right]$

Typical applications of the installment-loan concept are demonstrated in the following two example problems.

Example problem 14.7 An automobile that has a total cost of $5 760 is to be purchased by trading in an older car for which $1 320

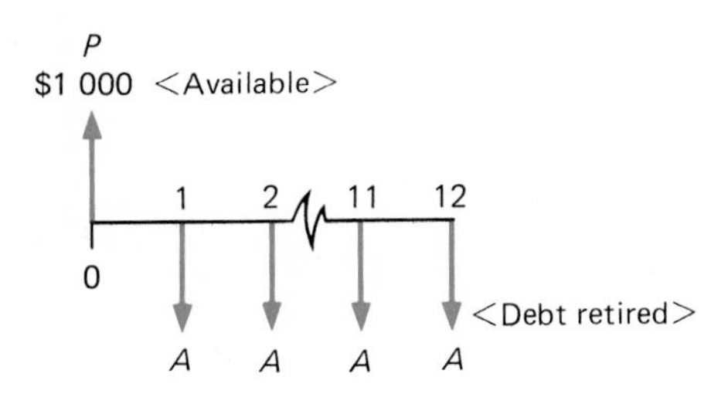

Figure 14.16
Borrower's cash-flow diagram.

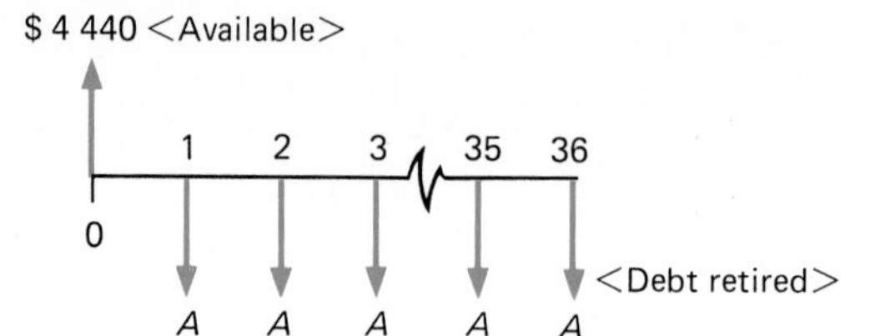

Figure 14.17 Car owner's cash-flow diagram.

is allowed. If the interest rate is 9 percent a year and the payments will be made monthly for 3 years beginning at the end of the first month, what are the monthly payments? (See Fig. 14.17.)

Solution

$$A = (5\,760 - 1\,320)\left[\frac{(0.007\,5)(1.007\,5)^{36}}{1.007\,5^{36} - 1}\right]$$

$$= \$141.19$$

If 36 monthly payments of \$141.19 had been placed in a sinking fund at 9 percent interest, they would generate

$$141.19\left(\frac{1.007\,5^{36} - 1}{0.007\,5}\right) = \$5\,810.35$$

Another way of expressing the relationship is by saying that \$4 440 (that is, 5 760 − 1 320) is the present worth of 36 monthly payments of \$141.19, beginning in 1 month at $\frac{3}{4}$ percent per month.

Example problem 14.8 Assume that the auto-purchase arrangement in Example Prob. 14.7 is the same except that no payments are to be made until 6 months after purchase and then a total of 36 monthly payments are to be made. What is the amount of the monthly payment?

Solution See Fig. 14.18.

$$\text{Balance due after trade-in} = 5\,760 - 1\,320 = \$4\,440$$

$$\text{Balance due after 5 months} = 4\,440(1.007\,5)^5 = \$4\,609.02$$

$$A = 4\,609.02\left[\frac{(0.007\,5)(1.007\,5)^{36}}{1.007\,5^{36} - 1}\right] = \$146.57$$

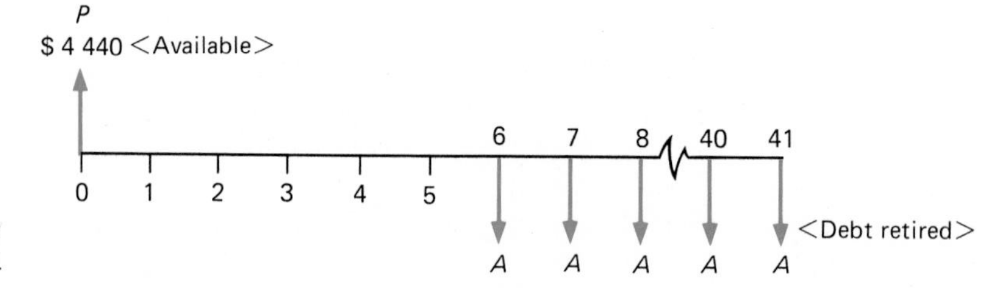

Figure 14.18 Car owner's cash-flow diagram.

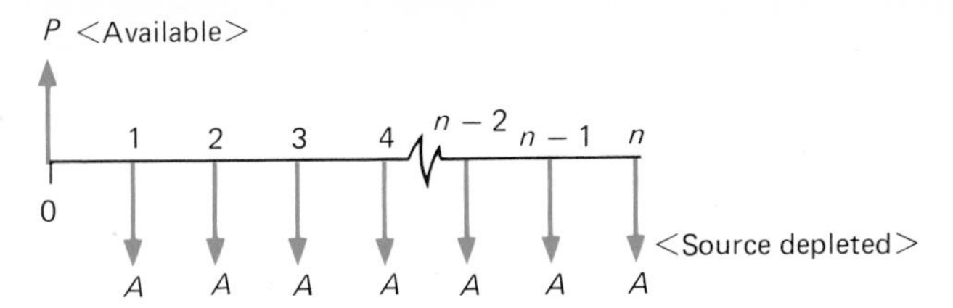

Figure 14.19
Series of monthly withdrawals.

Note that the unpaid balance is compounded for only 5 months because the first payment marks the end of the first period of the annuity. This is normally referred to as a *deferred annuity.*

14.6.3
Retirement Plan

A third way to consider annuities is the classic way, that is, the time when a sum of money is returned in monthly installments at retirement. The formula that applies is Eq. (14.7), and the cash-flow diagram is shown in Fig. 14.19.

The problem could be restated as follows: How much money P must be available at retirement so that A can be received for n periods, assuming i interest rate? Equation (14.7) can be solved for the amount of money P that must be accumulated by retirement if an amount A is to be withdrawn for n periods at a given interest rate.

To extend this idea, how much money P would need to be available if monthly amounts A are to be withdrawn forever? If we look at the equation, do we solve for $n = \infty$?

Perhaps we could restate the problem by asking what amount of money P must be available at retirement so that an amount A can be withdrawn each month and never affect the principal amount P?

If we use the compound interest formula, i.e., $S_n = P(1 + i)^n$ and find the amount of interest generated for 1 month (here that is equal to A), then $S_n = P + A = P(1 + i)^1$. If we solve this equation for P, then

$$P = \frac{A}{(1 + i)^1 - 1}$$

and given an interest rate (say, $i = 0.10/12$) and a fixed monthly income (say, $A = \$100$), then $P = \$12\ 000$.

14.7
Analysis of Alternatives

An engineer makes use of the material presented in this chapter in a very practical way. It is used to analyze a situation so that an intelligent decision can be made. It is normally true that several alternatives are available, each having some strong attributes. The task is to compare each alternative and select the one that appears superior, all things considered.

Before examining some practical examples, a few definitions must be reviewed.

First cost This is the initial cost of purchase and includes items such as freight, sales tax, and installation.

Life This refers to the number of years of service the user expects from the item or property.

Salvage This is the net sum to be realized from the disposal of an item or property after service. It normally includes removal costs, freight out, etc.

It is not always possible to determine these values with certainty, so the engineer must often work with data that are inexact. Even though the data are not perfect, an engineer can make better decisions with economic comparisons than without.

The most obvious method of comparing costs is to determine the total cost of each alternative. An immediate problem arises in that the various costs occur at intervals, so the total number of dollars spent is not a valid method of comparison. It has been shown that the present worth of an expenditure can be calculated. If this is done for all costs, the present worth of buying, operating, and maintaining two or more alternatives can then be compared. The present-worth method is referred to as the *capitalized cost*. Simply stated, it is the sum of money necessary to buy, maintain, and operate a facility. The alternatives must obviously be compared for the same length of time, and replacements due to short-life expectancies must be considered.

A second method, preferred by those who work with annual budgets, is to calculate the *average annual cost* of each. The approach is similar to the capitalized cost method, but the numerical value is in essence the annual contribution to a sinking fund that would produce a sum identical to the capitalized cost placed at compound interest.

Many investors approach decisions on the basis of the profit a venture will produce in terms of percent per year. The purchase of a piece of equipment, a parcel of land, a new product line, etc., is thus viewed favorably only if it appears that it will produce an annual profit greater than a certain percent. The amount of an acceptable percent return is not constant but fluctuates with the money market. Since there is doubt about the amount of the profit and certainly there is a chance of a loss, it would not be prudent to proceed if the prediction of return was not considerably above "safe" investments such as bonds.

The example that follows illustrates the use of these two methods (capitalized cost and average annual cost) and includes a third technique called *future worth* that provides a check.

Each method compares money at the same point in time or over the same time period. Each method is different yet each yields the same conclusion.

Example problem 14.9 Consider the purchase of two computer-aided-design (CAD) systems. Assume the annual interest rate is 12 percent.

	System 1	System 2
Initial cost	$100 000	$65 000
Maintenance and operating cost	4 000/year	8 000/year
Salvage	18 000 after 5 years	5 000 after 5 years

Using each of the three methods below compare the two computer-aided-design systems and offer a recommendation:

1. Annual cost
2. Present worth
3. Future worth

Solution

1. *Annual cost* (Fig. 14.20)

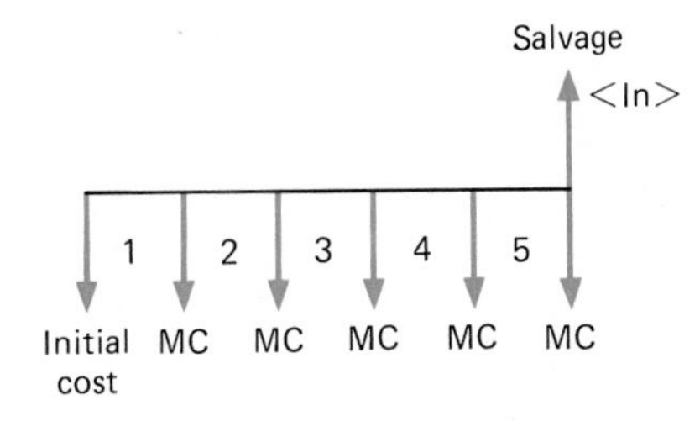

Figure 14.20
Company's cash-flow diagram (MC = maintenance cost).

System 1

a. Initial cost

$$A = P\left[\frac{i(1+i)^n}{(1+i)^n - 1}\right]$$

$P = 100\ 000$

$i = 0.12$

$n = 5$

$A = \$27\ 740.97/\text{year}$

b. Maintenance and operating costs

MC = 4 000/year

c. Salvage

$$S_n = A\left[\frac{(1+i)^n - 1}{i}\right]$$

System 2

a. Initial cost

$$A = P\left[\frac{i(1+i)^n}{(1+i)^n - 1}\right]$$

$P = 65\ 000$

$i = 0.12$

$n = 5$

$A = \$18\ 031.63/\text{year}$

b. Maintenance and operating costs

MC = 8 000/year

c. Salvage

$$S_n = A\left[\frac{(1+i)^n - 1}{i}\right]$$

$$A = \frac{S_n(i)}{[(1 + i)^n - 1]} A$$

$S = 18\,000$

$i = 0.12$

$n = 5$

$A = \$(-)2\,833.38/\text{year}$

$$A = \frac{S_n(i)}{[(1 + i)^n - 1]}$$

$S = 5\,000$

$i = 0.12$

$n = 5$

$A = \$(-)787.05/\text{year}$

System 1 (annual-cost analysis)

+	27 740.97
+	4 000.00
(−)	2 833.38
	$28 907.59

System 2 (annual-cost analysis)

+	18 031.63
+	8 000.00
(−)	787.05
	$25 244.58

Conclusion: System 2 is less expensive (Fig. 14.21).

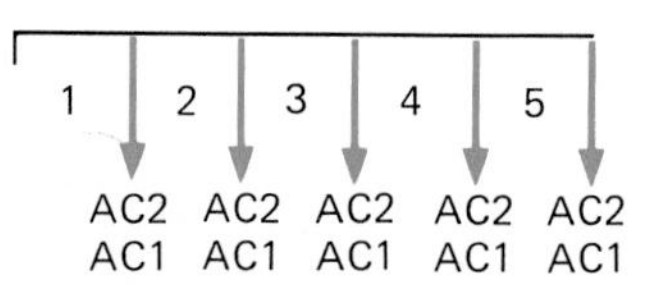

Figure 14.21
Company's cash-flow diagram.

System 2 annual cost =
AC2 = $ 25 244.58

System 1 annual cost =
AC1 = $ 28 907.59

2. *Present worth*

System 1:

a. Initial cost = $100 000
b. Maintenance and operating costs

$$P = A\left[\frac{(1 + i)^n - 1}{i(1 + i)^n}\right]$$

$A = \$4\,000$

$i = 0.12$

$n = 5$

$P = \$14\,419.10$

c. Salvage

$P = S_n(1 + i)^{-n}$

$S_n = 18\,000$

$i = 0.12$

$n = 5$

$P = \$(-)10\,213.68$

System 2:

a. Initial cost = $ 65 000
b. Maintenance and operating costs

$$P = A\left[\frac{(1 + i)^n - 1}{i(1 + i)^n}\right]$$

$A = \$8\,000$

$i = 0.12$

$n = 5$

$P = \$28\,838.21$

c. Salvage

$P = S_n(1 + i)^{-n}$

$S_n = 5\,000$

$i = 0.12$

$n = 5$

$P = \$(-)2\,837.13$

System 1 (present-worth analysis)

$ 100 000.00
14 419.11
(−)10 213.68
$ 104 205.43

System 2 (present-worth analysis)

$ 65 000 .00
28 838 .21
(−)2 837.13
$ 91 001 .08

Conclusion: System 2 is less expensive (Fig. 14.22).

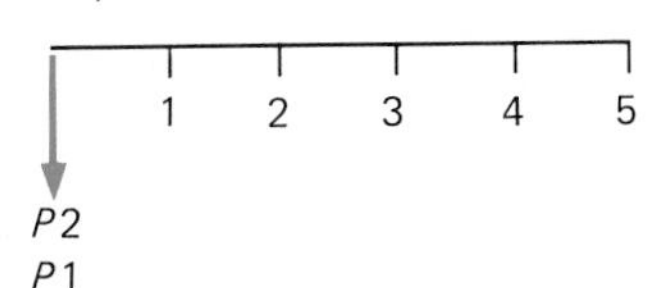

System 2 present worth =
$P2$ = $91 001.08

System 1 present worth =
$P1$ = $104 205.43

Figure 14.22
Company's cash-flow diagram.

3. *Future worth*

System 1:

a. Initial cost = $100 000

$S_n = P(1 + i)^n$

$P = \$100\ 000$

$i = 0.12$

$n = 5$

$S_n = \$176\ 234.17$

b. Maintenance and operating costs

$$S_n = A\left[\frac{(1 + i)^n - 1}{i}\right]$$

$A = \$4\ 000$

$S_n = \$25\ 411.39$

c. Salvage = $(−)18 000

System 1 (future-cost analysis)

176 234.17
25 411.39
(−) 18 000.00
$183 645.56

System 2:

a. Initial cost = $ 65 000

$S_n = P(1 + i)^n$

$P = \$65\ 000$

$i = 0.12$

$n = 5$

$S_n = \$114\ 552.21$

b. Maintenance and operating costs

$$S_n = A\left[\frac{(1 + i)^n - 1}{i}\right]$$

$A = \$8\ 000$

$S_n = \$50\ 822.78$

c. Salvage = $(−)5 000

System 2 (future-cost analysis)

114 552.21
50 822.78
(−) 5 000 .00
$160 374.99

Conclusion: System 2 is less expensive (Fig. 14.23).

Figure 14.23
Company's cash-flow diagram.

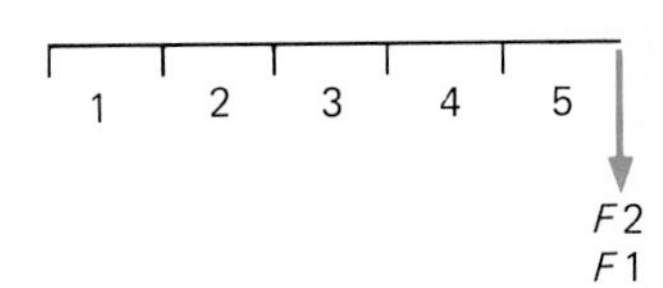

System 2 future cost =
$F2$ = $160 374.99

System 1 future cost =
$F1$ = $183 645.56

14.8 Depreciation

With the passage of time, the value of most physical property suffers a reduction, which is called *depreciation*. The price that a willing buyer will pay a willing seller for a property is called the *market value*. The amount that the property can be sold for at a later date when it is no longer wanted by the owner is called the *salvage value*. Therefore, the market value at purchase minus the salvage value at discard is the amount of depreciation. The concept is simple and obvious; however, determination of depreciated value depends on an estimation of salvage value. The fact that most property does depreciate with time indicates that the amount of the depreciation is part of the cost of doing business. If the property in question is a machine that produces a given number of items per year and has a limited life expectancy, then a portion of the cost of the machine must be charged to each item produced and sold. A second reason for calculating depreciation can be to create a fund from which the cost of replacing the worn-out property can be taken. A third reason is that the Internal Revenue Service (IRS) accepts depreciation as a cost of doing business, and the taxable profits can legally be reduced by the amount of the calculated depreciation. There are several methods acceptable to the IRS of calculating depreciation, four of which will be examined. Many accepted depreciation methods fail to account for the time-value of money.

14.8.1 Straight-Line Depreciation

One general requirement of methods of depreciation is that they not be overly complex. The least complex is the straight-line method. It assumes uniform depreciation over the life of the property and can be calculated from the formula

$$d = \frac{C - C_\ell}{\ell} \tag{14.9}$$

where d = annual amount of depreciation

C = cost, new

C_ℓ = value at age ℓ, years

ℓ = age, years

Example problem 14.10 If a machine costs $8 000 new and can be traded in for $1 600 after 16 years, what is the annual straight-line depreciation?

Solution

$$d = \frac{C - C_\ell}{\ell} = \frac{8\,000 - 1\,600}{16}$$

$$= \$400$$

This method has an added advantage in that during times of inflation, if the annual depreciation is invested, it will produce more than the original cost of the machine and will tend to overcome the increased cost of replacement.

14.8.2 Sinking-Fund Method

This method assumes that it is desirable to produce a fund to replace the property. Moreover, it also assumes that an amount equal to the new cost minus the salvage cost as shown by the formula

$$d = (C - C_\ell)\left[\frac{i}{(1 + i)^\ell - 1}\right] \quad 14.10$$

will replace the property.

Example problem 14.11 Using the data from Example Prob. 14.10, compute the first year's depreciation by the sinking-fund method. Assume 8 percent annual interest.

Solution

$$d = (8\,000 - 1\,600)\left(\frac{0.08}{1.08^{16} - 1}\right)$$

$$= \$211.05$$

It can be seen that this method requires much less money to be put aside each year in order to accumulate the \$6 400 (8 000 − 1 600), which is an advantage if funds are limited and \$6 400 is truly the sum that is required. However, if the computation is to be used for tax purposes or to replace property whose cost will probably escalate, the sinking fund will not produce the most desirable results. It must be understood that the total depreciation at the end of any year is the amount of the sinking fund that is established. The total depreciation at the end of 2 years may be found, using Eq. (14.6), as follows:

$$211.05\left(\frac{1.08^2 - 1}{0.08}\right) = \$438.98$$

so that depreciation during the second year is \$227.93.

14.8.3 Sum of Years' Digits

A third method of calculating depreciation has been given the name "sum of years' digits." It is calculated by assigning depreciation in decreasing amounts as follows (example is for an expected life of 8 years):

Year of life	Reverse order of years of life	Fractional depreciation
1	8	8/36
2	7	7/36
3	6	6/36
4	5	5/36
5	4	4/36
6	3	3/36
7	2	2/36
8	1	1/36
	Sum = 36	Sum = 36/36 = 1

In this method, the analyst must first determine the total depreciation as before and then assign annual depreciation, the largest amount during the first year with decreasing amounts each year thereafter. The method produces a rapid depreciation in the early years and thereby has tax advantages.

Example problem 14.12 Calculate the depreciation for each year in Example Prob. 14.10.

Year of life	Reverse order of years of life	Fractional depreciation	Depreciation during the year
1	16	16/136	752.94
2	15	15/136	705.88
3	14	14/136	658.82
4	13	13/136	611.76
5	12	12/136	564.71
6	11	11/136	517.65
7	10	10/136	470.59
8	9	9/136	423.53
9	8	8/136	376.47
10	7	7/136	329.41
11	6	6/136	282.35
12	5	5/136	235.29
13	4	4/136	188.24
14	3	3/136	141.18
15	2	2/136	94.12
16	1	1/136	47.06
	Sum = 136		

14.8.4 The Matheson Formula

This formula produces what is commonly called the *declining balance method*. It is sometimes called the *constant percentage method* inasmuch as it assumes that the depreciation is a percentage of the remaining value. This produces a greater depreciation the first year and smaller amounts in each succeeding year, similar to the sum of years' digits method. The declining balance method can never depreciate a property to zero—not a serious problem because the depreciation in later years becomes very small.

If k represents the rate of depreciation, the depreciation in the first year is Ck. The value at the start of the second year is $C - Ck$, or $C(1 - k)$. The depreciation during the second year is $C(1 - k)k$ and the total depreciation in 2 years is $2Ck - Ck^2$, so that $C(1 - k)^2$ is the value after 2 years.

It can be seen that the value remaining after n years is $C_n = C(1 - k)^n$, where C_n is the depreciated value.

$$\frac{C_n}{C} = (1 - k)^n \quad \text{and} \quad \left(\frac{C_n}{C}\right)^{1/n} = 1 - k$$

so that for ℓ years

$$k = 1 - \left(\frac{C_\ell}{C}\right)^{1/\ell}$$

Substitution of this expression for k into $C_n = C(1 - k)^n$ yields

$$C_n = C\left(\frac{C_\ell}{C}\right)^{n/\ell} \qquad 14.11$$

Example problem 14.13 What is the depreciated value at the end of 4 years of a property that costs \$15 000 new and has an estimated value of \$3 000 at the end of 9 years? Use the Matheson formula.

Solution

$$C_n = C\left(\frac{C_\ell}{C}\right)^{n/\ell}$$

$$= 15\,000\left(\frac{3\,000}{15\,000}\right)^{4/9}$$

$$= \$7\,335.64$$

The total depreciation during these 4 years is clearly $15\,000 - 7\,335.64 = \$7\,664.36$. This method produces a very rapid depreciation and is widely used for that reason.

Table 14.1 summarizes some of the data given above and compares the four methods of depreciation that have been discussed. (See also Fig. 14.24.) It can be seen that these methods produce values that vary considerably. The choice of the method employed will depend on the purpose of the calculation and the convention established by a particular industry or company.

Table 14.1 Comparison of Depreciation Methods*

Age years	Depreciation during the year				Depreciated value remaining			
	Straight line	Sinking fund	Sum of years' digits	Matheson	Straight line	Sinking fund	Sum of years' digits	Matheson
1	400	211.05	752.94	765.57	7600.00	7788.95	7247.06	7234.43
2	400	227.93	705.88	692.31	7200.00	7561.02	6541.18	6542.12
3	400	246.17	658.82	626.05	6800.00	7314.85	5882.36	5916.07
4	400	265.86	611.76	566.15	6400.00	7048.99	5270.60	5349.92
5	400	287.14	564.71	511.96	6000.00	6761.85	4705.89	4837.96
6	400	310.10	517.65	462.98	5600.00	6451.75	4188.24	4374.98
7	400	334.91	470.59	418.67	5200.00	6116.84	3717.65	3956.31
8	400	361.70	423.53	378.60	4800.00	5755.14	3294.12	3577.71
9	400	390.64	376.47	342.37	4400.00	5364.50	2917.65	3235.34
10	400	421.89	329.41	309.61	4000.00	4942.61	2588.24	2925.73
11	400	455.64	282.35	279.98	3600.00	4486.97	2305.89	2645.75
12	400	492.05	235.29	253.19	3200.00	3994.92	2070.60	2392.56
13	400	531.56	188.24	228.96	2800.00	3463.36	1882.36	2163.60
14	400	573.98	141.18	207.05	2400.00	2889.38	1741.18	1956.55
15	400	619.90	94.12	187.23	2000.00	2269.49	1647.06	1769.32
16	400	669.50	47.06	169.32	1600.00	1600.00	1600.00	1600.00

*Table is based on an original cost of $8000.00, salvage value of $1600.00, life expectancy of 16 years and an interest rate of 8 percent.

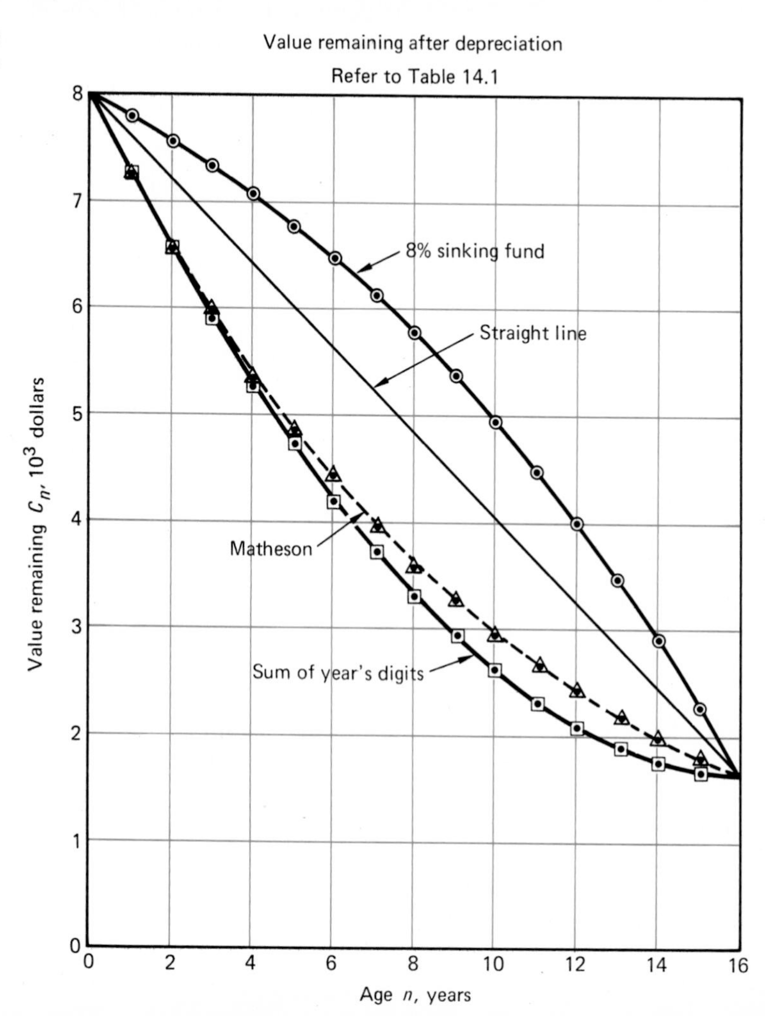

Figure 14.24 Typical depreciation schedule methods.

Problems

14.1 What is the present worth of $10 000.00 to be paid to you in 4 years if the money is thought to be worth (*a*) 6 percent, (*b*) 12 percent, and (*c*) 24 percent? Assume annual compounding.

14.2 How much must be invested now to grow to $20 000 in 8 years if the annual interest rate is 7.0 percent compounded (*a*) annually, (*b*) semi-annually, (*c*) monthly?

14.3 The taxes on a home are $1 800.00 per year. The mortgage stipulates that the owner must pay $\frac{1}{12}$ of the annual taxes each month (in advance) to the bank so that the taxes can be paid on March 31 of the next year (taxes are due on April 1). Assuming that money is worth 11.5 percent annual interest, compounded monthly, determine how much profit the bank makes on the owner's tax money.

14.4 Your hometown has been given $400 000 from the estate of a citizen. The gift stipulates that the money cannot be used for 3 full years but may be invested. If the money can be invested at 10.2 percent annual interest, how much will be available in 3 years if it is compounded (*a*) annually, (*b*) semiannually, (*c*) daily?

14.5 If the interest rate is 9.8 percent per year, how long will it take for an investment to double in value, assuming monthly compounding?

14.6 You have just made an investment that will repay $728.21 at the end of each month; the first payment is 1 month from today and the last one 5 years from today.

(*a*) If the interest rate is 13.2 percent per year compounded monthly, what amount did you invest?

(*b*) If you can refuse all the payments and allow the money to earn at the same rate as stated, how much will you have at the end of 5 years?

14.7 You have made an investment that will yield $8 000.00 exactly 6 years from today. If the current interest rate is 8.3 percent compounded annually, what is your investment worth today?

14.8 You just borrowed $2 000.00 and have agreed to repay the bank $500.00 at the end of each of the next 5 years. What is the annual interest rate of the loan?

14.9 Today you have a savings account of $16 240. Based on an annual interest rate of 8 percent, what equal amount can you withdraw from the account at the end of each month for 4 years and leave $4 000 in the account?

14.10 On January 1 this year you borrowed $20 000 toward component parts for a product you hope to have on the market in September of this year. You have agreed to an annual interest rate of 11.6 percent, compounded monthly. You also have agreed to begin repaying the debt on October 1, making equal monthly payments until the loan is repaid on February 1 next year. What are your monthly payments?

14.11 Referring to Prob. 14.10, suppose that you are unable to make the monthly payments and your creditor agrees to allow you to make a single payment on February 1 next year. How much will you owe at that time?

14.12 If sales at your company are doubling every 4 years, what is the annual rate of increase?

14.13 You have $25 000 to invest, and you have decided to purchase bonds that will mature in 6 years. You have narrowed your choices to two types of bonds. The first class pays 14.2 percent annual interest. The second class pays 9.6 percent annual interest but is tax-free, both federal and state. Your income bracket is such that your highest income tax rate is 36 percent federal and the state income tax is 12 percent of the federal. Which is the better investment for you at this time? Assume that all conditions remain unchanged for the 6-year period.

14.14 On your twenty-fifth birthday you open an IRA (individual retirement account). At that time and on each succeeding birthday up to and including your sixtieth, you deposit $2 000. During this period the interest rate on the account remains constant at 9.8 percent. No further payments are made and beginning 1 month after your sixty-fifth birthday you begin withdrawing equal payments (the annual interest rate is the same as before). How much will you withdraw if the account is to be depleted with a last check on your eighty-fifth birthday?

14.15 You wish to retire at age 60 and at the end of each month thereafter, for 25 years, to receive $3 000. Assume that you begin making monthly payments into an account at age 23. You continue these payments until age 60. If the interest rate is constant at 10.4 percent, how much must be deposited monthly between ages 23 and 60?

14.16 Two machines are being considered to do a certain task. Machine A costs $22 000 new and $2 600 to operate and maintain each year. Machine B costs $30 000 new and $1 200 to operate and maintain each year. Assume that both will be worthless after 10 years and that the interest rate is 15.0 percent. Determine by the annual-cost method which alternative is the best buy.

14.17 Assume you needed $10 000 on April 1, 1982, and two options were available:

1. Your banker would loan you the money at an annual interest rate of 24.0 percent, compounded monthly, to be repaid on September 1, 1982.
2. You could cash in a certificate of deposit (CD) that was purchased earlier. The cost of the CD purchased September 1, 1981 was $10 000.00. If left in the savings and loan company until September 1, 1982, the CD's annual interest is 15.6 percent, compounded monthly. If the CD is cashed in before September 1, 1982, you lose all of the interest for the first 3 months and the interest rate is reduced to 6 percent, compounded monthly, after the first 3 months.

Which option is better and by how much? (Assume an annual rate of 15 percent, compounded monthly, for any funds for which an interest rate is not specified.)

14.18 Two machines are being considered for the same task. Machine A costs $18 000.00 new and is estimated to last 6 years. The cost to replace machine A will be 10.0 percent more each year than it was the year before. It will cost $1 200.00 per year to operate and maintain machine A. It will have a trade-in (salvage) value of $1 500.00. Machine B costs $38 000.00 to buy, will last 12 years, and will have a trade-in value of $2 000. The cost of operation and maintenance is $700.00 per year.

Compare the two machines and state the basis of your comparison. Include a cash-flow diagram for each alternative. Assume all interest rates at 12.0 percent per year unless otherwise stated.

14.19 Two machines are being considered for the same task. Machine A costs $21 000.00 new and is estimated to last 6 years. It will then have a sale or trade-in value of $1 500.00. The cost to replace machine A will be 10 percent more each year than it was the year before. It will cost $1 400.00 per year to operate and maintain machine A, payable at the end of each year. Machine B costs $40 000.00 to buy and will last 12 years. It will have a sale or trade-in value of $2 000.00. The cost to operate and maintain machine B will be $800.00 per year, payable at the end of each year.

Assume that the task will be performed for 12 years. Compare the two machines, state your basis of comparison, and include a cash-flow diagram. All interest rates are 14.00 percent per year unless otherwise stated.

14.20 You have borrowed as follows:

January 1, 1982	$24 000
July 1, 1983	$33 000
January 1, 1984	$39 000

The agreed-upon annual interest rate was 11 percent, compounded semi-annually. How much did you owe on July 1, 1984?

You agreed to begin making monthly payments on October 1, 1984. Assume that the interest rate was still 11 percent but was compounded monthly. How much was each of the 15 payments?

14.21 You have reached agreement with an auto dealer regarding a new car. He has offered you a trade-in allowance of $2 250 on your old car for a new one that he has "reluctantly" reduced to only $8 830 (before trade-in). He has further agreed to a contract that requires you to pay $180.44 each of the next 48 months, beginning 1 month from today. What is your interest rate, expressed as an annual percentage? Give your answer to the nearest 0.01 percent. With your instructor's approval, write a computer program to solve the problem.

14.22 You have been assigned the task of estimating the annual cost of operating and maintaining a new assembly line in your plant. Your calculations indicate that during the first 4 years the cost will be $100 000 per year; the next 5 years will cost $130 000 per year; and the following 6 years will cost $155 000 per year. If the interest rate is constant at 13.8 percent over the next 15 years, what will be the average annual cost of operation and maintenance?

14.23 Your parents are planning to buy a new home for $115 000. They have an agreement with the savings and loan company to borrow the needed money if they pay 25 percent in cash and monthly payments for 25 years at an interest rate of 13.2 percent, compounded monthly.

(*a*) What monthly payments will be required?

(*b*) How much principal reduction will occur in the first payment?

14.24 If your parents had started a savings account that paid them 8.8 percent, compounded monthly, and their payments into the account were the same as in Prob. 14.23, how long would they have had to make payments in order to purchase the home for cash? (Assume the same down payment amount was available as in Prob. 14.23.)

14.25 Your bank pays 5.25 percent on Christmas Club accounts. How much must you put into an account weekly beginning on January 2 in order to accumulate $800 on December 4? Assume weekly compounding.

14.26 You can purchase a treasury note today for 94.2 percent of its face value of $10 000. Every 6 months you will receive an interest payment at the annual rate of 9.88 percent of face value. You can then invest your interest payments at the annual rate of 8.0 percent, compounded semiannually. If the note matures 6 years from today, how much money will you receive from all of the investments? Express this also as an annual rate of return.

14.27 What is the present worth of each of the following list of assets and liabilities and your net present worth?

(*a*) You deposited $3 000 exactly 3 years ago.

(*b*) You have a checking account with a current balance of $1 127.19.

(*c*) You must pay $4 150 exactly 4 years from now.

(*d*) Today you just made the twenty-ninth of 36 monthly payments of $29.52.

(*e*) You will receive $8 000 exactly 5 years from now.

Assume that all annual interest rates are 7.5 percent. Assume monthly compounding in figuring (*d*).

14.28 Plant revisions are necessary to put a new product into production. These revisions could be done at one time or they could be done in several increments. It is estimated that a single project completed on July 1, 1985, would cost $210 000. If completed over several years, the costs are estimated to be as follows:

1st phase	July 1, 1985	$100 000
2nd phase	Jan. 1, 1988	80 000
3rd phase	July 1, 1992	60 000
4th phase	July 1, 1994	40 000
5th phase	July 1, 1997	50 000

If money can be borrowed at 11.2 percent annual interest, compounded quarterly, which project is most economical? By how much? What other factors might affect your decision?

14.29 EJMN Engineering has estimated that the purchase of computer-graphics work stations costing $150 000 will reduce the firm's drafting expenses by $12 500 per month during a 2-year period. If the work stations have zero salvage value in 2 years, what is the firm's expected annual rate of return on investment? (*Rate of return* is the equivalent interest rate that must be earned on the investment to produce the same income.) If approved by your instructor, write a computer program.

14.30 Engineers at Specialty Manufacturing are writing a justification report to support purchase of a DNC milling center (mill, controller, microcomputer, installation, etc.). They have learned that the total initial cost will be $70 000. The labor savings and improved product quality will result in an estimated benefit to the company of $2 100 each month over a 10-year time period. If the salvage value of the center is about $10 000 in 10 years, what annual rate of return (annual percentage) on investment did the engineers calculate? Write a computer program if assigned by your instructor.

14.31 Many new engineering graduates will finance new cars. Automobiles are typically financed for 3 years with monthly payments to the loan agency. Assume you borrowed $10 000.00 for 36 monthly payments at 15 percent annual interest.

Mortgage table

Payment date	Payment number	Monthly payment	Monthly principal	Monthly interest	Total interest per calendar year
9/1/85	1	$XXX.XX	$XXX.XX	$XXX.XX	
10/1/85	2				
11/1/85	3				
12/1/85	4				
1/1/86	5				$XXX.XX
2/1/86	6				
⋮					

(*a*) Prepare the table below (write a computer program if approved by your instructor).

Payment number	Monthly payment	Monthly principal	Monthly interest
1	$XXX.XX	$XXX.XX	$XXX.XX
2			
⋮			

(*b*) If you decided to pay the loan off at the end of 10 months, what would be the balance? At the end of 20 months? At the end of 30 months?
(*c*) What was the total interest paid in the first 12 payments? The second 12 payments? The last 12 payments?
(*d*) If the lending institution puts your monthly payments into a sinking fund, what would the total value be at the end of 3 years at the above interest rate?

14.32 Many of you will eventually purchase a house. Few will have the total

cash on hand so it will be necessary to borrow the money from some home-loan agency. Typically the money P is borrowed at a fixed annual interest rate I for N years. As a home owner you can deduct interest paid on both state and federal tax returns. From the loan information given below, write a computer program to produce the mortgage table.
Loan: $80 000.00, August 1, 1985

P = $80 000 P = $80 000

$I = 12\%$ $i = 0.12/12$

N = 25 years $n = 25 \times 12$

Foundations of Design

PART FIVE

15

Engineering Design—A Process

15.1 Introduction

In Chap. 1 we alluded to design in engineering as an iterative process involving decisions at each step (see Fig. 15.1). The purpose of this chapter is to expand on these introductory remarks by considering the design process in more detail. It is possible to subdivide the entire design process in many different ways. In this chapter, it has been broken down into nine steps, so that each step can be explained separately. To make each phase as understandable and practicable as possible, we relate the progress of an actual student design project at the end of each step. We recount, in other words, how a team of beginning engineering students accomplished each phase of the process from start to finish.

Before we begin our discussion, we must first examine engineering design in general and its place in the activities of the technology team that was described in Chap. 1. Many practicing engineers hold the opinion that design is *the* distinguishing feature of engineering. They feel that most of our efforts are directed toward producing systems and devices that use our natural resources in the most effective, efficient manner to satisfy human needs. The real test of

1. Identification of a need
2. Definition of the problem
3. Search
4. Criteria and constraints
5. Alternative solutions
6. Analysis
7. Decision
8. Specification
9. Communication

Figure 15.1
The design process.

Figure 15.2
Engineers use a computer-aided-design (CAD) work-station to design new and improved products. Without the assistance of design software and computer-graphics enhancement, an industry may well find itself no longer competitive. (*International Business Machines Corporation.*)

Figure 15.3
A design team discusses the components in an assembly. (*Allen-Bradley.*)

the systems and devices is found in their use. That is, do they truly fit into our society and improve it?

The application of computer technology to engineering design has compressed the life cycle of many products; that is, new and improved products are designed, manufactured, and marketed in a much shorter time period than in the past. We may expect this trend to continue into the future. Many items we consider necessities today will be superseded by innovations in just a few years. We will find our standard of living increasingly improved and our needs efficiently served by engineers applying technology.

15.1.1 The Design Process

A simple definition of *design* is "to create according to a plan." A *process*, on the other hand, is a phenomenon identified through step-by-step changes that lead toward a required result. Both these definitions suggest the idea of an orderly, systematic approach to a desired end. Figure 15.4 shows the design process as continuous and cyclic in nature. This idea has validity in that many problems arise during the design process that generate subsequent designs. You should not assume that each of your design experiences will necessarily follow the sequential steps without deviation. Experienced designers will agree that the steps as shown are quite logical; but on many occasions, designers have had to repeat some steps or perhaps have been able to skip one or more.

Before beginning an overview of the entire design process, we must state that limits are always placed on the amount of time available. Normally we establish a time frame or a series of deadlines for ourselves before we begin the process. It is almost impossible

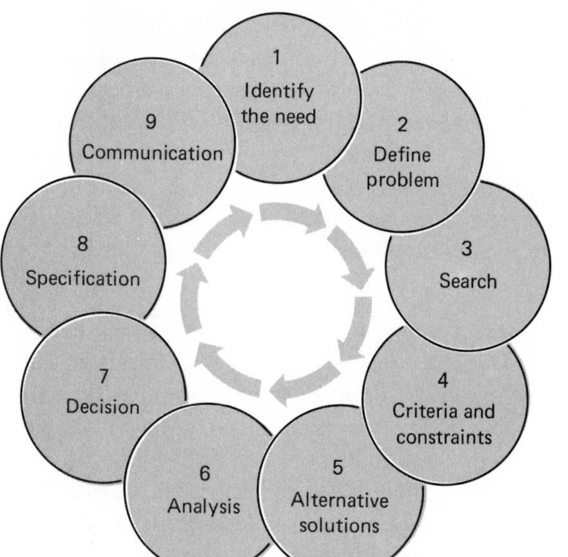

Figure 15.4
The design process is cyclic in nature.

for us to tell you how much time should be allocated for each step, because the problems are so varied. A sample of a time frame is shown in Fig. 15.5.

The whole process begins when a need is recognized: Put simply, someone feels that something must be done. Oftentimes it is not the designers who are involved at step 1; but they usually assist in defining the problem (step 2) in terms that allow it to be scrutinized. Information is gathered at step 3, and then the nature of the solution to the problem is determined and boundary conditions (constraints) are established (step 4). At step 5, several possible solutions are entertained and the creative, innovative talents of the designer come

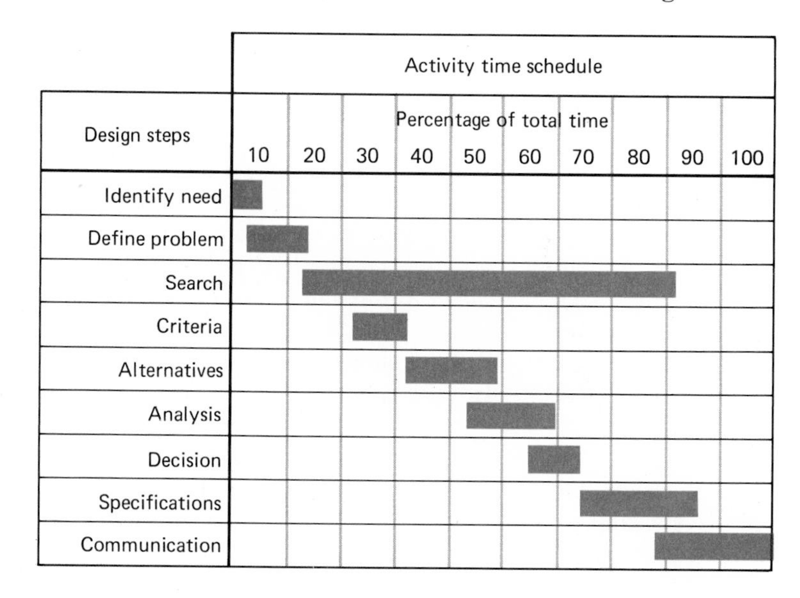

Figure 15.5
In order to control the design process, a time schedule must be developed early.

into play. This is followed by detailed analysis of the alternatives (step 6), after which a decision is reached regarding which one should be completely developed (step 7). Specifications of the chosen concept are prepared (step 8), and its merits are explained to the proper people or agencies (step 9) so that implementation (construction, production, etc.) can be accomplished. A more complete explanation of each step as well as a reporting of the actions of the student design team constitutes the remainder of this chapter.

15.2 Identification of a Need—Step 1

Before the process can begin, someone has to recognize that some constructive action needs to be initiated. This may sound vague, but understandably so, because such is the way the process normally begins. Engineers do not have supervisors who tell them to identify a need. You might be asked to do so in the classroom, because some professors may have you work on a project that you choose rather than one that is assigned. When most of us speak of a need, we generally refer to a lack or shortage of something we consider essential or highly desirable. Obviously, this is an extremely relative thing, for what may be a necessity to some could be a luxury to others.

More often than not, then, someone other than the engineer decides that a need exists. In private industry, it is essential that products sell for the company to survive. Most of the products have a life cycle that goes from the development stage, when the expenditures by the organization are high and sales are low, to the peak demand period, when profits are high, and eventually to the point where the product becomes obsolete. Even though a human need may still exist, the economic demand does not, because a more attractive alternative has become available. With obsolescence of a product, the company perceives a need to phase out the product and to develop one that is profitable. Inasmuch as most companies exist to make a profit, profit can be considered to be the basic need.

A bias toward profit and economic advantage should not be viewed as a selfish position, because products are purchased by people who feel that what they are buying will satisfy a need that they perceive as real. Society appreciates anyone who provides essential and desirable services, as well as goods that we use and enjoy. The consumers are ultimately the judges of whether there is truly a need. In like manner, the citizens of a community decide whether or not to have paved streets, parks, libraries, adequate police and fire protection, and scores of other things. City councils vote on the details of the programs. However, during the period when citizens and decision makers are formulating their plans, engineers are involved in supplying factual information to assist them. After the policy decisions have been made, engineers conduct studies, surveys, tests, and computations that allow them to prepare the detailed design plans, drawings, etc., that shape the final project.

15.2.1 The Chapter Example—Step 1

Throughout the remainder of the chapter, we will trace the steps five beginning engineering students* followed to produce a design for their class project. As a starting point, a professor may assign students the task of identifying a need. It usually is easier to approach such an assignment by beginning with a very broad area of technology, such as energy, for example.

The five students who were chosen were informed that all the student teams would be involved in some area dealing with the energy problem. Their professor began with construction of a decision tree, shown as Fig. 15.6. The class discussed sources of energy and jointly added the first level of subproblems: fossil, wind, geothermal, solar, nuclear, and organic. The class was then divided into

*The five students were John W. Benike, Douglas L. Carper, Patrick J. Grablin, Rick Sessions, and David L. White.

Figure 15.6 A decision tree pertaining to energy.

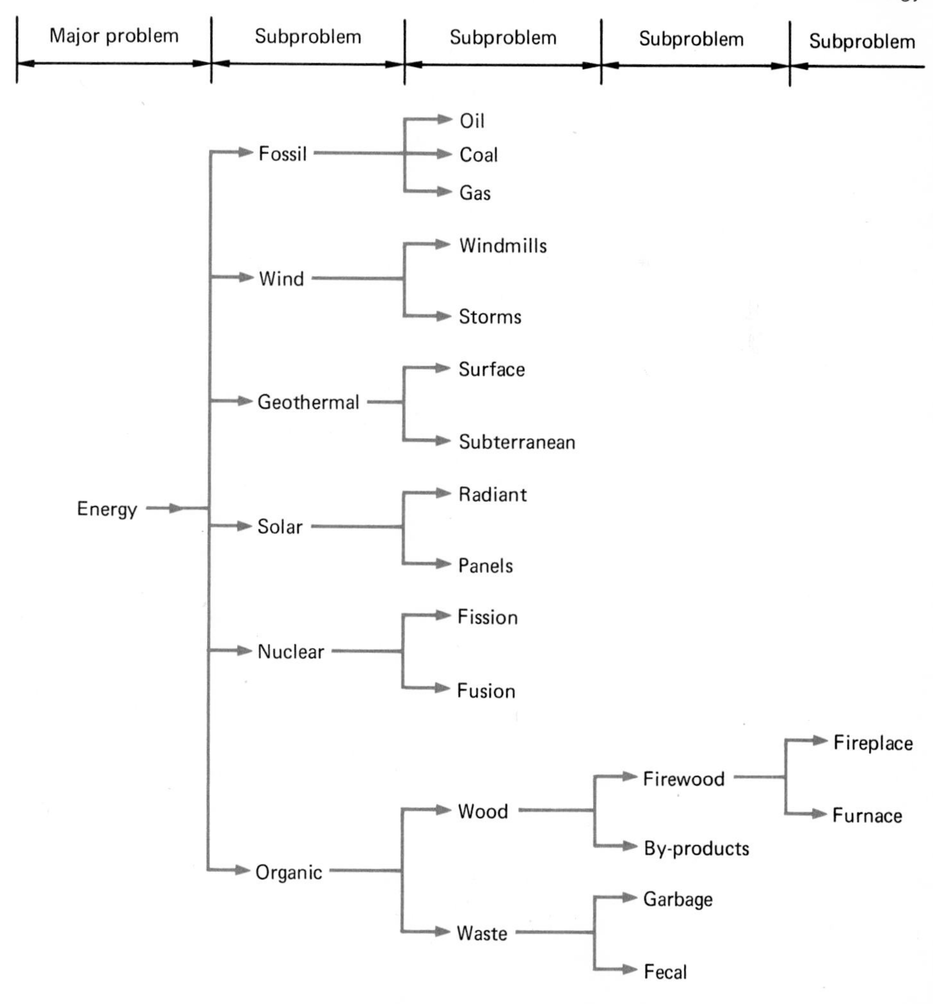

groups; the groups began to further subdivide *one* of the energy sources listed above. There probably is no end to this procedure, but it does provide quickly a wide range of topics from which needs may be more easily recognized. Our student design team developed Fig. 15.6, as shown, and thus began their discussion of the general topic of firewood for use in fireplaces. They recorded statements such as the following:

1. More and more people are deciding to have fireplaces in their homes and apartments.
2. Firewood is not as commercially available as it used to be.
3. The price of firewood has risen significantly.
4. People are now more willing to cut and split their own firewood than they were previously.
5. The small, inexpensive chain saw has made the cutting portion of the task more acceptable, but splitting the wood is still a major problem.

After a reasonable period of discussion involving these topics as well as others related to firewood, they agreed upon the following initial statement of need:

There is a need for an inexpensive supply of firewood for use in the home.

(We will see later that this statement was changed by the students in much the same way that professional engineers refine and redefine problems during a design process.)

15.3 Problem Definition—Step 2

There is often a temptation to construct quickly a mental picture of a gadget that if properly designed and manufactured will satisfy a need. In the general case of a need for firewood, we obviously know that we can call a supplier and have firewood delivered, but the cost factor is equally obvious. We could get a bit facetious and decide to burn our furniture. There have been emergency situations during extreme storms when this was the best solution available. If some friendly neighbor will supply us with firewood at no cost or effort, a problem does not exist. An important point to realize at this time is had we allowed ourselves to focus on a specific piece of equipment or a single method of obtaining firewood in the very beginning, we would never have considered the statements mentioned above.

15.3.1 Broad Definition First

The need as previously stated does not point to any particular solution and thereby leaves us with the opportunity to consider a wide range of alternatives before we agree on a specific problem statement. Consider for a moment a partial array of possibilities that will satisfy the original statement that there is a need for an inexpensive supply of firewood for use in the home.

1. Purchase firewood from a supplier.
2. Use something other than firewood (coal or rolled newspaper).
3. Make use of existing equipment.
4. Hire a portion of the work done (probably the splitting of logs).
5. Design improved equipment.

This is not an exhaustive list. You may want to take a few minutes to add to it. The first item is the current solution for many people. The other items show some promise of being an improvement over simply purchasing the firewood. You may think that item 1 should not be listed because it offers no change. If so, you are mistaken, because the status quo is the solution that is selected most often, at least as a temporary measure.

15.3.2 Symptom versus Cause

If you cough and do nothing but suck on a cough drop, you may be treating the symptom (the tickle in your throat) but doing little to alleviate the cause of the tickle. This approach may be expedient; but it can many times result in a repetition of the problem if the tickle is caused by a virus or a foreign object of some sort. Engineers seldom tell a client to take two aspirins and call back tomorrow, but they can sometimes be guilty of failing to see the real problem.

For many years residential subdivisions were designed so that the rainfall would drain away quickly, and expensive storm sewer systems were constructed to accomplish the task. Not only were the sewers expensive, but they also resulted in transporting the water problem downstream for someone else to handle. In recent years, perceptive engineers have designed land developments so that the rainfall is temporarily held and released over a longer period of time. This approach employs smaller, less expensive sewers and reduces the likelihood of flooding downstream. The real problem was not how to get rid of the rainfall as originally assumed, but what should be done to control the water.

15.3.3 Solving the Wrong Problem

In the 1970s, the problem of increasing fatalities in auto accidents was clearly recognized. It was shown that the fatality rate could be significantly reduced if the driver and front-seat passenger used lap and shoulder belts. The solution technique that was implemented was to build in an interlock system that required the belts to be latched before the auto could be started. That solution certainly should have solved the problem but it didn't. It attacked the problem of requiring that the belts be physically used, but it did nothing to solve the real problem—that of driver attitude. The driver and passenger still did not wish to use belts and did everything possible to avoid it, even to having the interlock system removed.

15.3.4 The Chapter Example—Step 2

The students whose progress we are following considered the possibilities outlined in Sec. 15.3.1. Their discussions covered a range of topics; and from their notebook we have listed a few of their pertinent recorded thoughts.

1. The range of possible solutions must be reduced before the problem can be solved.
2. People really like the smell of burning wood (in preference to paper or coal).
3. There already exists an adequate supply of wood in most areas.
4. Chain saws are already well developed; hence, cutting the wood into proper lengths is not a pertinent problem.
5. Time is very short, so a problem must be chosen that we (the students) can solve.

Most assuredly these young people had other thoughts, many of which were not recorded. The result of their consideration was a slightly revised problem definition:

There needs to be available to the average household an inexpensive, efficient method of splitting a small quantity of firewood.

In practice, you and other engineers will face restrictions that will affect the quality of your solutions. Many times your solution will have to meet governmental regulations in order to qualify for grants of money; or perhaps safety requirements by some agency cannot be met if certain materials are used. In almost all your projects, there will be cost and time constraints that force you to make decisions that are not what you really want to do. Such decisions, once made, then control many of your subsequent actions on that project.

This situation was faced by our student design team. By limiting the range of possible solutions and accepting the present method of cutting wood, they eliminated even the consideration of other burning materials. We are not being critical of their decision because we have experienced similar time and resource constraints.

15.4 Search—Step 3

Most of your productive professional time will be spent locating, applying, and transferring information—all sorts of information. This is not the popular opinion of what engineers do, but it is the way it will be for you. Engineers are problem solvers, skilled in applied mathematics and science, but they seldom, if ever, have enough information about a problem to begin solving it without first gathering more data. This search for information may reveal facts about the situation that result in redefinition of the problem.

15.4.1 Types of Information

The problem usually dictates what types of data are going to be required. The one who recognized that something was needed (step 1) probably listed some things that are known and some things that need to be known. The one or ones who defined the problem had to have knowledge of the topic or they could not have done their part (step 2). Generally, there are several things that we look for in beginning to solve most problems. For example:

1. What has been written about it?
2. Is something already on the market that may solve the problem?
3. What is wrong with the way it is being done?
4. What is right with the way it is being done?
5. Who manufactures the current "solution"?
6. How much does it cost?
7. Will people pay for a better one if it costs more?
8. How much will they pay (or how bad is the problem)?

15.4.2 Sources of Information

If anything can be said about the last half of the twentieth century, it is that we have had an explosion of information. The amount of

Figure 15.7 Research is ongoing throughout the design process. Manufacturers' catalogs are often a source of information. (*Stanley Consultants.*)

data that can be uncovered on most subjects is overwhelming. People in the upper levels of most organizations have assistants who condense most of the things that they must read, hear, or watch. When you begin a search for information, be prepared to scan many of your sources and catalog them so that you can find them easily if the data subsequently appears to be important.

Some of the sources that are readily available include the following:

1. Your library. Many universities have courses that teach you how to use your library. Such courses are easy when you compare them with those in chemistry and calculus, but their importance should not be underestimated. There are many sources in the library that can lead you to the information that you are seeking. You may find what you need in an index such as the *Engineering Index,* but don't overlook the possibility that a general index, such as *The Reader's Guide* or *Business Periodicals Index,* may also be useful. The *Thomas Register of American Manufacturers* may direct you to a company that makes a product that you need to know more about. *Sweets Catalog* is a compilation of manufacturer's information sheets and advertising material about a wide range of products. There are many other indexes that provide specialized information. The nature of your problem will direct which ones may be helpful to you. Don't hesitate to ask for assistance from the librarian.

2. Government documents. Many of these are housed in special sections of your library, but others are kept in centers of government—city halls, county court houses, state capitols, and Washington, D.C. The agencies of government that regulate, such as Interstate Commerce Commission, Environmental Protection Agency, regional planning agencies, make rules and police them. The nature of the problem will dictate which of the myriad of agencies can fill your needs.

3. Professional organizations. The American Society of Civil Engineers is a technical society that will be of interest to students majoring in civil engineering. Each major in your college is associated with not one but often several such societies. The National Society of Professional Engineers is an organization that most engineering students will eventually join, as well as at least one technical society such as the American Society for Mechanical Engineers (ASME), the Institute of Electrical and Electronics Engineers (IEEE), or any one of dozens that serve the technical interests of the host of specialties with which professional practices seem most closely associated. Many engineers are members of several associations and societies. Other organizations, such as the American Medical Association and the American Bar Association, serve various professions, and all have publications and referral services.

4. Trade journals. They are published by the hundreds, usually specializing in certain classes of products and services.

5. Vendor catalogs. Perhaps your college subscribes to one of the several information services that gather and index journals and catalogs. These data banks may have tens of thousands of such items available to you on microfilm. You need only learn how to use them.

6. Individuals that you have reason to believe are somewhat expert in the

field. Your college faculty has at least several, maybe many. There are, no doubt, some practicing engineers in your city.

15.4.3
Recording Your Findings

The purpose of a bibliography is to direct you to more information than is included in the article you are reading. The form of the bibliography makes it easy to find the reference. So it seems reasonable for you to record your information sources in proper form so that if that reference is to be cited in your report, you are ready to do it properly. By so doing, you are ensuring that it can be found again quickly and easily. Few things are more discouraging than to be unable to locate an article that you found once and know will be helpful if you could locate it again.

It is usually a good procedure to record each reference on a card or sheet of paper. English teachers usually recommend the use of file cards, but engineers seem to prefer information put in a bound notebook. Whatever your choice, Fig. 15.8 is recommended as a reasonable form of record.

As we are looking at something like a piece of equipment we oftentimes have thoughts and ideas that should be recorded for future reference. At such times, our ability to sketch is an invaluable tool because so many details can be graphically shown but are very difficult to describe in words.

15.4.4
The Chapter Example—Step 3

The team of engineering students whose project we are following realized the importance of the research phase but were also aware of the overall time constraints on the design process. They decided that a detailed research plan was needed so that specific assignments could be made to avoid conflict or overlap of effort. After considerable discussion, specific research areas were assigned to each team member. They consulted home builders about the demand for fireplaces, suppliers of firewood, several manufacturers of chain saws, companies that sell chain saws, a landscape architect, city govern-

TA 152.17 Inganere, M. E.
G273 *Heat, Air, and Gas Power*. McGraw-Hill, New York, 1973.

Has good tables in appendix and formulas on pages 52–55 covering heat transfer cases that may occur on project.

Figure 15.8
Documentation of research findings is essential if the findings are to be useful later. Recording pertinent information on a card will permit easy retrieval of the findings when needed.

ment, the library, a county extension service, an engineer, and a company that sells commercial log-splitters.

One of the team members was assigned the task of checking the library to determine what products were currently available. The librarian explained about the various indexes, so the student selected the *Thomas Register*. After failing to find anything listed under "Logs," he tried "Splitters." Figure 15.9 is a reproduction of his notes. (The appendix of the student report includes copies of letters sent by the design team along with the responses received.)

With this information the student went to the *Yellow Pages* of the local telephone book, wherein he learned that one of the products

Figure 15.9
A record of manufacturers who produce log-splitters.

Source: Thomas Register of American Manufacturers
Thomas Register Catalog File

Listing: Splitters: Wood, Firewood, Kindling, etc.

Location: Engineering Library – Reference Tables

Manufacturers

1) Gordon Corporation
P.O. Box 244-TR
Farmington, CT
(Hydraulic)

2) H. L. Diehl Co.
South Windham, CT

3) Vermeer Mfg. Co.
3804 New Sharon Rd.
Pella, IA
(Powered, log, hydraulic, trailer)
(515) 628-3141

4) Tree King Mfg. & Engineering, Inc.
North St.
Showhegan, ME
(Hydraulic)

5) Lindig Mfg. Corp.
1831 West County Rd.
St. Paul, MN
(612) 633-3072

6) Carthage Machine Co., Inc.
571 West 3rd Ave.
Carthage, NY
(Wood for pulp mills)

7) Equipment Design & Fabrication, Inc.
722 N. Smith St.
Charlotte, N.C.
(Log)

8) Pabco Fluid Power Co.
5752 Hillside Ave.
Cincinnati, OH
(Log)
(513) 941-6200

9) Piqua Engineering, Inc.
234-52 First Street
Piqua, OH
(513) 773-2464

10) Henke Mfg. Co., Inc.
433 W. Florida St.
Milwaukee, WI
(Log)

11) Didier Mfg. Co.
1652 Phillips Ave.
P.O. Box 806
Racine, WI
(Hydraulic, log)

12) Murray Machinery, Inc.
104 Murray Road
Wausau, WI
(Hydraulic, paper roll)

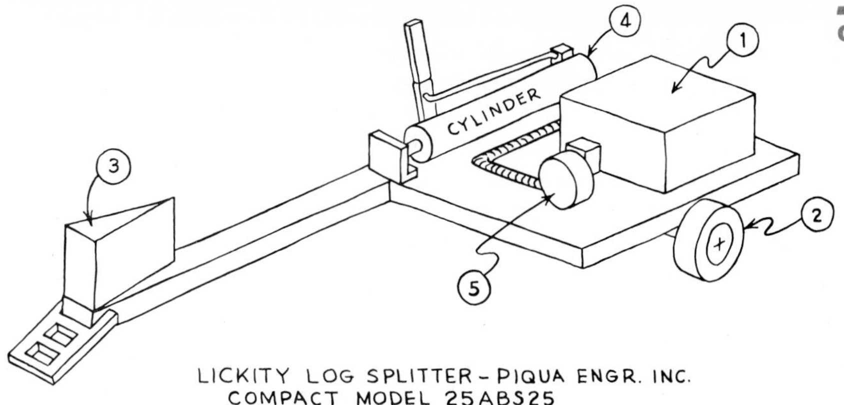

Figure 15.10
A sketch is an excellent method of recording certain types of information.

was sold locally; so a team member was assigned to visit the dealer. The dealer was temporarily out of advertising pamphlets, so the team member sketched the floor model and recorded pertinent data about it, the notes for which are shown as Fig. 15.10. Their procedure is to be commended and highly recommended in that they adequately documented the information in sufficient detail so that it could be used or the manufacturer could be contacted for additional data.

The initial research stage has provided us with added information about the problem, so that we are now ready to begin to describe the design in terms of things it must be or must have and what attributes are most important.

15.5 Criteria and Constraints—Step 4

When we approach a problem, we always know some things about the answer that tend to describe what the final solution will be like. Please don't misread that statement; we did not say that we know what the best solution is, rather that we know some things *about* it. We may know, for instance, that it has to fit into a certain place, a fact that gives us an idea about its size. We may know that it can't cost more than so many dollars or that certain colors simply will not work. These three items tend to put limits or restrictions—what we commonly call constraints—on the solution. Such an idea is not new to you, and it certainly is a routine experience for the practicing engineer. We face such a situation in almost every decision we make, even those that are not really important. When you arose this morning, you had to choose some clothes to wear. You probably limited the choice to those hanging in your closest (or maybe in your roommate's closet). This was a constraint *you* placed on the solution, not one that really existed until you made it so. In most fields of engineering, formulas have been developed and are used in designs of various kinds. Many, and probably most, of them are valid in a certain range of physical conditions. For instance, the hydraulic conditions of the flow of water are not valid below 0°C or above 100°C and are restricted to normal pressure ranges. Figure 9.23 shows a stress-strain curve. You can easily see that the relationship above and below the yield point is quite different. We normally refer to these constraints or limits as *boundary conditions*, and they occur in many different ways.

As we view a problem, we may know that some attributes of a solution will be good and some will be detrimental. Perhaps lower cost, less weight, or greater reliability will be desirable. Usually a considerable list of attributes can be made that help us evaluate the ideas that will come to us as we analyze the problem and consider alternatives. These attributes, called *criteria*, allow a comparison to be made among alternative solutions. Unlike constraints, they are not quantifiable.

The criteria will be based on your background knowledge and the research that you have conducted, but they should not be based on ideas about what the solution should be. *Always* establish the criteria before you attempt to generate alternative solutions.

15.5.1 Design Criteria

Whereas each project or problem has a personality all its own, there are certain characteristics that occur in one form or another in a great many projects. We should ask ourselves, "What characteristics are most desirable and which are not applicable?" Typical design criteria are listed below.

1. Cost—almost always a heavily weighted factor
2. Reliability
3. Weight
4. Ease of operation and maintenance
5. Appearance
6. Compatibility
7. Safety features
8. Noise level
9. Effectiveness

10. Durability
11. Feasibility
12. Acceptance

There will be other criteria and perhaps some of those given are of little or no importance in some projects, so a design team in industry or in the classroom must decide which criteria are important to the design effort. Since value judgments have to be made later, it probably makes little sense to include those criteria that will be given relatively low weights. There are oftentimes mild disagreements at this point, not about which criteria are valid, but rather about how much weight should be assigned to each. It is often better if the team members make their assignments of weight independently and then compile all the results. This tends to dampen the effect of the more persuasive members at the same time that it forces all team members to contribute consciously. Usually there are not many instances where one of the members strongly disagrees with the mean value of the weight assigned to each criterion. Some negotiation may be required, but it is seldom a difficult situation to resolve.

15.5.2 The Chapter Example—Step 4

Most people feel comfortable when they talk in general terms about a great many things, because as long as they do not have to get specific, an avenue of escape from their position is left open. The students whose progress we are following were not so fortunate, because they were facing a real problem. A review of their time

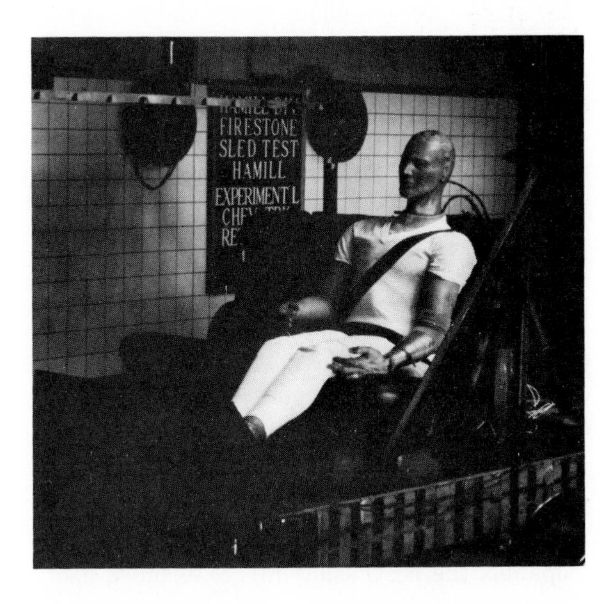

Figure 15.11
A safety criterion may be evaluated by model testing. (*Firestone Tire and Rubber Company.*)

schedule indicated that it was time to make some decisions. They therefore agreed on the following assigned weights:

1. Cost: 30 percent
2. Portability: 20 percent
3. Ease of operation: 15 percent
4. Safety: 15 percent
5. Durability: 10 percent
6. Use of standard parts: 10 percent

What areas of agreement and disagreement do you see between our list of 12 criteria and their list of 6?

The most obvious difference is that they included the use of standard parts as an important criterion, but it was not listed at all in Sec. 15.5.1. Just why this is important is not clear unless you try to place yourself in their position. If the team has plans to manufacture the log-splitter, then it will be much easier and less expensive to begin operations if many of the components can be purchased rather than manufactured in their own plant.

They agreed that cost, weight (portability), ease of operation, and safety were important. The others on our list of 12 were either not considered or were considered to be of low importance (less than 5 percent).

They did not list any specific constraints, but we must assume that they are practical individuals who would assign a very low rating to a concept that exceeded some level of performance. For instance, if one of their ideas had an estimated cost of $500, it would no doubt be rated at near zero on a 10-point scale. They did not, however, tell us at what cost the zero rating begins; is it $200 or $100 or $75? In like manner they did not say at what weight they would consider a concept to be too heavy to be portable. They did give us a little help by restricting the projected users to adults (excluding children).

It is our conclusion that at this point they have decided that the solution to their problem will be a portable log-splitter that can be operated by a single adult. If this is true, then we can conclude that a loop has been installed in the design process. This is not unusual. Figure 15.4 might very well have a number of arrows to show that problem solvers do return to steps in the design process that have supposedly been completed. In this particular case, we can assume that our students have redefined their problem even though they don't say so and don't report having undertaken any additional research. Again, this is not unusual.

15.6 Alternative Solutions—Step 5

We are now ready to see if we can think of a good solution. The best way that we can be reasonably sure that we arrive at a good (hopefully the best) solution is to examine many possibilities.

Suppose that you are chief engineer for a manufacturing company and are faced with appointing someone to the position of director of product testing. This is an important position, because all the company's products are given rigorous testing under this person's direction before they are approved. You must compare all the candidates with the job description (criteria) to see who would do the best job. This seems to be a ridiculously simple procedure, doesn't it? Well, we think it does too, but many times such a process is not followed and poor appointments are made. In the same way that the list of candidates for the position has to be made, so must we produce a list of possible answers to our problem before we can go about the job of selecting the best one.

15.6.1 The Nature of Invention

The word "invention" strikes fear into the minds of many people. They say, "Me, an inventor?" The answer is "Why not?" One reason why we don't fashion ourselves as inventors is that some of our earliest teaching directed us to be like the other boys and girls. Since much of our learning was by watching others, we learned how to conform. We also learned that if we were like the other kids, no one laughed at us. We can recall in our early days, even preschool, that the worst thing that could happen would be if people laughed at us. We'll bet that most of you have a similar feeling when you say something that is not too astute, and it is followed by smiles and polite laughs. Moreover, we don't like to experiment because most experiments fail. It is a very secure person who never has to try something that he or she hasn't already done well. Think about it: When you were in the first few grades at school, didn't you feel great when you were called on by the teacher and you knew the answer? Don't you, even today, try to avoid asking your professors a question because you don't want the professor or your classmates to know that you don't know the answer? Most of us like to be in the majority. Please don't assume that we are saying that the majority of the people *are* wrong or that it is bad to be like other people. However, if we dwell on such behavior, then we will never do anything new. A degree of inventiveness or creativity is essential if we are to arrive at solutions to problems that are better than the way things are being done now. If we can remove the blocks to creativity, then we have a good chance of being inventive.

The father of one of the authors had a motto above his desk that read as follows:

Life's greatest art,
Learned through its hardest knocks,
Is to make
Stepping stones of stumbling blocks.

He doesn't know whether this was original with his father or not, but he remembers it after not having seen it for over 20 years, and it surely applies to the process of developing ideas.

15.6.2 Building the List

There are a great number of techniques that can be used to assist us in developing a list of possible solutions. Three of the more effective methods will be briefly discussed.

Checkoff lists, designed to direct your thinking, have been developed by a number of people. Generally the lists suggest possible ways that an existing solution to your problem might be changed and used. Can it be made a different color, a different shape, stronger or weaker, larger or smaller, longer or shorter, of a different material, reversed or combined with something else? It is suggested that you write your list down on paper and try to conceive of how the current solution to the problem might be if you changed it according to each of the words on your list. Ask yourself: Why is it like it is; will change make it better or worse; did the original designers have good reason for doing what they did or did they simply follow the lead of their predecessors?

Morphological listing gives a visual conception of the possible combinations that might be generated. These listings are usually shown as grids or diagrams. It is easy to visualize a rectangular prism as shown in Fig. 15.12. This example indicates that we are considering the log-splitting problem as composed of three subdivisions: power source, power delivery, and splitting principles. These are then subdivided, as shown in Fig. 15.12. The prism produces 72 different combinations such as the one indicated by the shaded volume. Here we would have the logs split by torsion applied by a

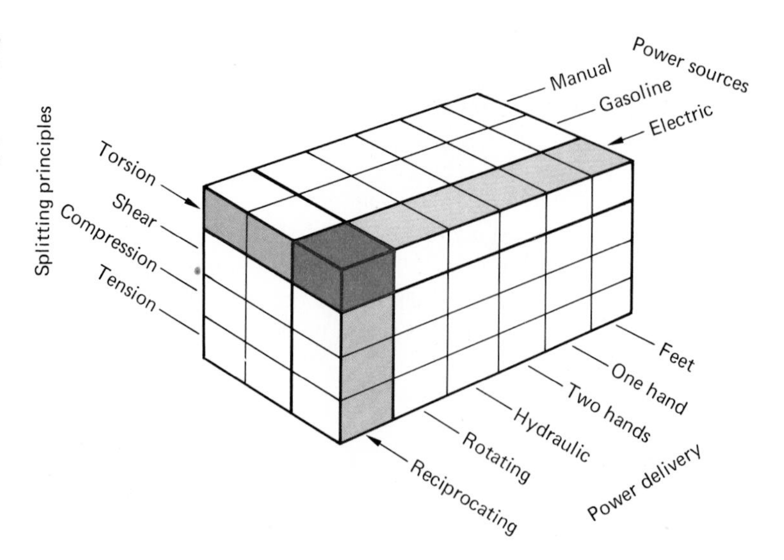

Figure 15.12
A morphological chart. There are 72 combinations of the three attributes. The shaded segment, for example, would employ an electric reciprocating device that splits by torsion.

reciprocating motion generated by an electric power source. Surely you can think of more than the three major subdivisions shown in Fig. 15.12 and can provide additional ideas for each of them.

Brainstorming is a technique that has received wide discussion and support. The mechanics of a brainstorming session are rather simple. The leader states the problem clearly and ideas about its solution are invited. The length of productive sessions varies, but it is usually in the half-hour range. Often it takes a few minutes for a group to rid itself of its natural reserved attitude. But brainstorming can be fun, so choose a problem area and try it with some friends. Be prepared for a surprise at the number of ideas that will develop.

There are many descriptions of this process, most of which can be summed up as follows:

1. The size of the group is important. We have read of successful groups that range from three to fifteen; however, it is generally agreed that six to eight is an optimum number for a brainstorming session.
2. Free expression is essential. This is what brainstorming is all about. Any evaluation of the exposed ideas is to be avoided. Nothing should be said to discourage a group member from speaking out.
3. The leader is a key figure, even though free expression is the hallmark. The leader sets the tone and tempo of the session and provides a stimulus when things begin to drag.
4. The members of the group should be equals. No one should feel any reason to impress or support any other member. If your supervisor is also a member, you must steer clear of concern for his or her feelings or support for his or her ideas.
5. Recorders are necessary. Everything that is said should be recorded, mechanically or manually. Evaluation comes later.

We have discussed a few techniques that are recommended to stimulate your thought processes. You may choose one of the free-

Figure 15.13
Design teams meet frequently to discuss problems and to be updated on the total project. (*Bourns, Inc.*)

wheeling techniques or perhaps a well-defined method. Regardless of your preferences, we think you will be pleased and even surprised at the large list of ideas that you can develop in a short period of time.

15.6.3 The Chapter Example—Step 5

Our team of engineering students approached their task of generating ideas by setting up a brainstorming session. Their minds were already tuned in on the problem; they had produced a list of candidates for the ultimate decision. They admit that they erred by giving preliminary evaluation to some of the ideas, which is a hard rule not to violate. The following is a list of ideas as it appeared in their written report. It is not, however, their total list.

Ideas for splitting wood

1. Hydraulic cylinder (vertical or horizontal) used as a method to apply force.
2. Auto-jack or fence-tightener concept in order to apply pressure through a mechanical advantage. (See Fig. 15.14.)
3. Use of compressed air to force wedge through log.
4. Adaptations of conventional hand tools such as the axe, mall, or wedge.
5. Power or manual saws.
6. Heavy pile driver with block and tackle for raising weight.
7. High-voltage arc between electrodes; similar to lightning.
8. Spring-powered wedge using either compression or tension.
9. Sliding mass that drives wedge into wood.
10. Drop wedge from elevated position onto the log.
11. Electronic sound that produces compression waves strong enough to split logs.
12. Wedge driven by explosive charge.
13. Spinning hammermill that breaks by shearing and concussion like a rock crusher.
14. Separate or split with intensive concentrated high energy such as laser beam.

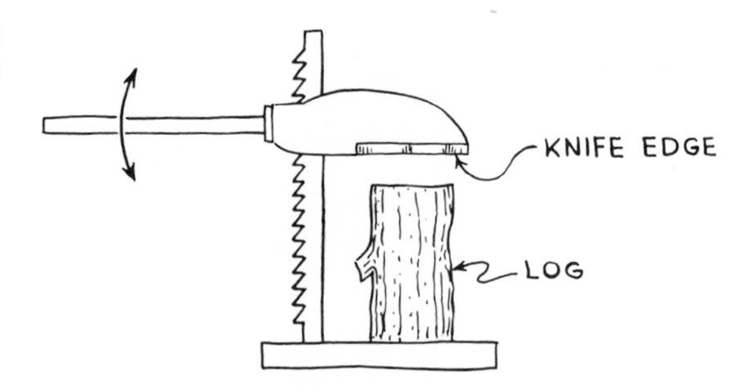

Figure 15.14 Thumbnail sketches are oftentimes helpful in describing ideas.

15. Force a conical wedge into log and apply a torsional force.
16. Use a large mechanical vice with one jaw acting as a wedge.
17. Drill core (hole) in wood, fill with water, cap, and freeze.
18. Cut wood into slabs rather than across the grain.
19. Apply couple to ends of log causing a shearing action.
20. Drop log from elevated position onto fixed wedge.

15.7 Analysis—Step 6

At this point in the design process we have defined the problem, expanded our knowledge of the problem with a concentrated search for information, established constraints for a solution, selected criteria for comparing solutions, and generated alternative solutions. In order to determine the best solution in light of available knowledge and criteria, the alternative solutions must be analyzed to determine performance capability. Analysis thus becomes a pivotal point in the design process. Potential solutions which do not prove out during the analysis phase may be discarded or, under certain conditions, may be retained with a redefinition of the problem and a change in constraints. Thus one may need to repeat segments of the design process (Fig. 15.4) after completing the analysis.

Analysis involves the use of mathematical and engineering principles to determine the performance of a solution. Consider a system, such as the cantilever beam in Fig. 15.15, constrained by the laws of nature. When there is an input to the system (the applied load P), analysis will determine the performance of the system (beam)—that is, the deflection, stress buildup, and so forth. Keep in mind that the objective of design is to determine the best solution (system) to a need.

The process of analysis is shown in the following example problem.

Example problem 15.1 Determine the deflection of the beam in Fig. 15.15 under the following conditions. Assume the beam is structural steel.

Figure 15.15

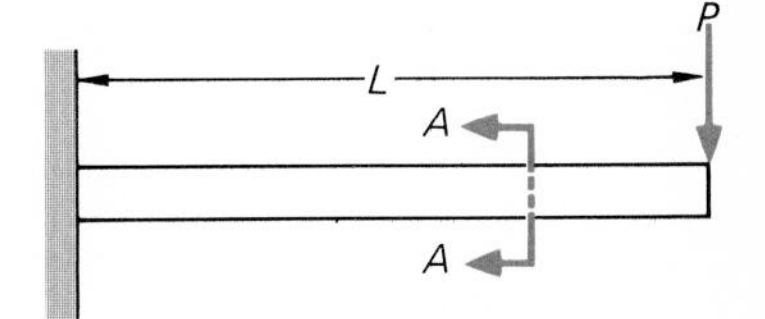

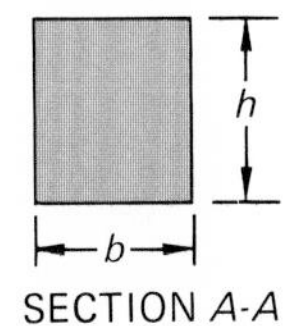

$L = 4.0$ m

$h = 4.0 \times 10^1$ cm

$b = 2.0 \times 10^1$ cm

$P = 1.0 \times 10^5$ N

Solution The deflection of the end of a cantilever beam for the configuration shown is given by

$$d = \frac{PL^3}{3EI} \text{ (constraint equation)}$$

where d = deflection, m

$$E = \text{modulus of elasticity, a material constant, Pa}$$
$$= 2.07(10^{11}) \text{ Pa for structural steel}$$
$$I = \text{moment of inertia, m}^4$$

For a rectangular cross section,

$$I = \frac{bh^3}{12}$$
$$= \frac{(0.2)(0.4)^3}{12}$$
$$= 1.067(10^{-3}) \text{ m}^4$$

Therefore,

$$d = \frac{(10^5)(4)^3}{3(2.07)(10^{11})(1.067)(10^{-3})}$$
$$= 9.66(10^{-3}) \text{ m}$$
$$= 0.97 \text{ cm}$$

This result would be forwarded to the designer for incorporation into the decision phase. The time required to produce an analysis is critical to the design process. If it takes longer to do an analysis than the schedule (Fig. 15.5) permits, then the results are somewhat meaningless. The engineer must exercise some judgment in selecting the method of analysis in order to assure results within the time limit. You can visualize the potential of computers in the analysis effort. With modern, high-speed machines coupled with computer graphics, the engineer can model and analyze systems much faster than in the past. In addition, many alternatives can be investigated in a brief amount of time. Thus if the constraint equation in Example Prob. 15.1 had been programmed, it would have been a simple task to change any of the parameters—b, h, P, L, or E—and see the effect on the deflection immediately. It would have been possible to produce plots of each parameter vs. deflection using computer graphics. It is obvious that the more possibilities one can investigate, the better the problem is understood and the better the design will be.

We will discuss the beam of Example Prob. 15.1 in more detail in Sec. 15.8 to further illustrate the value of analysis in engineering design.

Analysis performed by engineers in most design projects is based on the laws of nature, the laws of economics, and common sense.

15.7.1 The Laws of Nature

You have already come into contact with many of the laws of nature, and you will no doubt be exposed to many more. At this point in your education, you may have been exposed to the conservation principles: the conservation of mass, of energy, of momentum, and of charge. From chemistry you are familiar with the laws of Charles, Boyle, and Guy-Lussac. In mechanics of materials, Hooke's law is a statement of the relationship between load and deformation. Newton's three principles serve as the basis of analysis of forces and the resulting motion and reactions.

Many methods exist to test the validity of an idea against the laws of nature. We might test the validity of an idea by constructing a mathematical model, for example. A good model will allow us to vary one parameter many times and examine the behavior of the other parameters. We may very well determine the limits within which we can work. Other times we will find that our boundary conditions have been violated and, therefore, the idea must be discarded.

Results of an analysis of a mathematical model are frequently presented as graphs. Very often the slopes of tangents to curves, points of intersection of curves, areas under or over or between curves, or other characteristics provide us with data that can be used directly in our designs.

Computer graphics enables a mathematical model to be displayed on a screen. As parameters are varied, the changes in the model and its performance can also be quickly displayed to the engineer.

The preparation of scale models of proposed designs is often a necessary step (see Fig. 15.17). This can be a simple cardboard cutout or it can involve the expenditure of great sums of money to test the model under simulated conditions that will predict how the real thing will perform under actual use. A prototype or pilot plant is sometimes justified because the cost of a failure is too great to

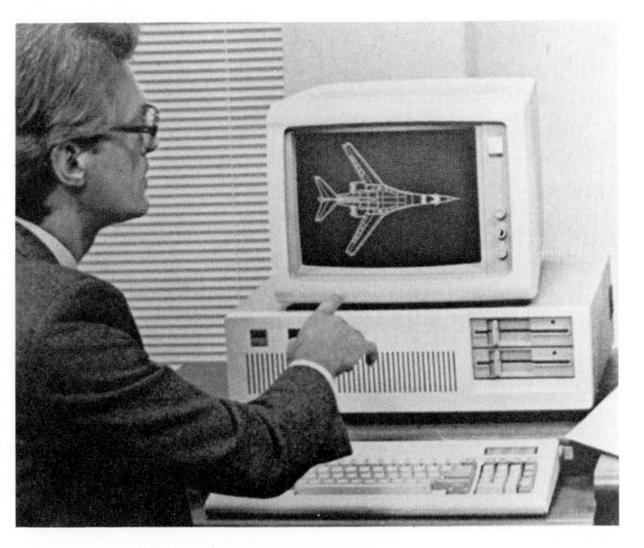

Figure 15.16
Computer graphics enables complex designs to be modeled, verified, and analyzed without the expense of constructing a scale model or prototype. (*International Business Machines Corporation.*)

Figure 15.17
Engineers study a model of a hydraulic structure used to predict the performance for much larger structures prior to construction. (*Stanley Consultants.*)

chance. Such a decision usually comes only after other less-expensive alternatives have been shown to be inadequate.

You probably have surmised that the more time and money that you allot to your model, the more reliable is the data that you receive. This fact is often distressing because we want and need good data but have to balance our needs against the available time and money.

15.7.2
The Laws of Economics

Section 15.7.1 introduced the idea that money and economics are part of engineering design and decision making. We live in a society that is based on economics and competition. It is no doubt true that many good ideas never get tried because they are deemed to be economically infeasible. Most of us have been aware of this condition in our daily lives. We started with our parents explaining why we could not have some item that we wanted because it cost too much. Likewise, we will not put some very desirable component into our designs because the value gained will not return enough profit in relation to its cost.

Industry is continually looking for new products of all types. Some are desired because the current one is not competing well in the marketplace. Others are tried simply because it appears that people will buy them. How do manufacturers know that a new product will be popular? They seldom know with certainty. Chapter 6 dealt with statistics, an important consideration in market analysis. Some of you may find that probability and statistics are truly fascinating and get involved with sampling popular opinion. The techniques of this area of mathematics allow us to make inferences about how large groups of people will react based on the reactions of a few. It is beyond our study at this time to discuss the techniques, but industry routinely employs such studies and invests millions of dollars based on the results.

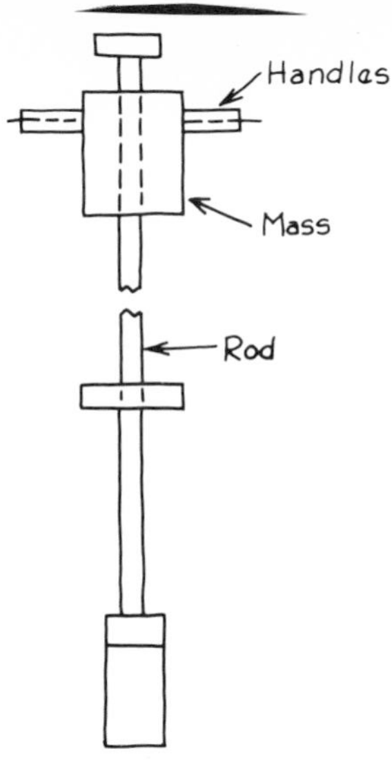

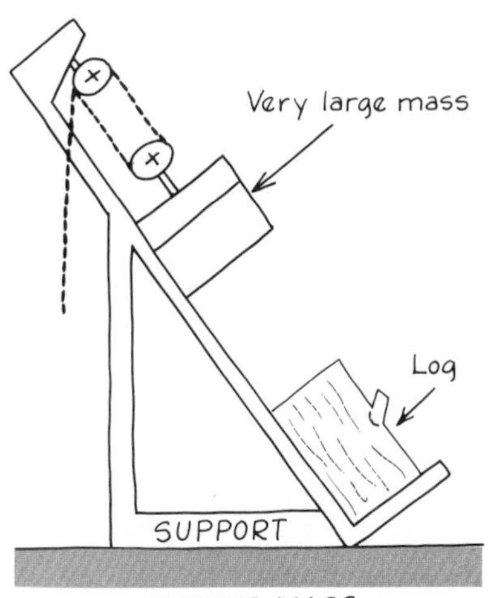

SLIDING MASS
IDEA SKETCH #3

Figure 15.20
(continued)

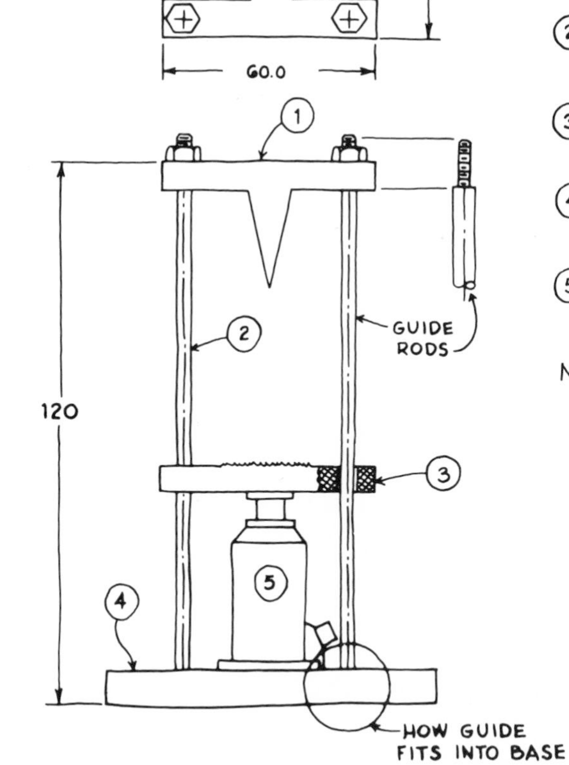

ASSEMBLY DRAWING

(1) WEDGE – 1 EA.
REMOVABLE FOR TRANSPORT AND SHARPENING. MADE OF STEEL.

(2) GUIDE RODS – 4 EA.
HOLDS ASSEMBLY SOLID, ALLOWS DISASSEMBLY

(3) SLIDER – 1 EA.
SLIDES ALONG RODS, BASE FOR WOOD.

(4) BASE – 1 EA.
LARGE ENOUGH TO PREVENT TIPPING.

(5) HYDRAULIC CYLINDER – 1 EA.
COMMERCIALLY AVAILABLE.

NOTE: DIMENSIONS IN CENTIMETERS

Figure 15.21
Concept development of the pressure-wedge idea, sketch #3 from Fig. 15.19.

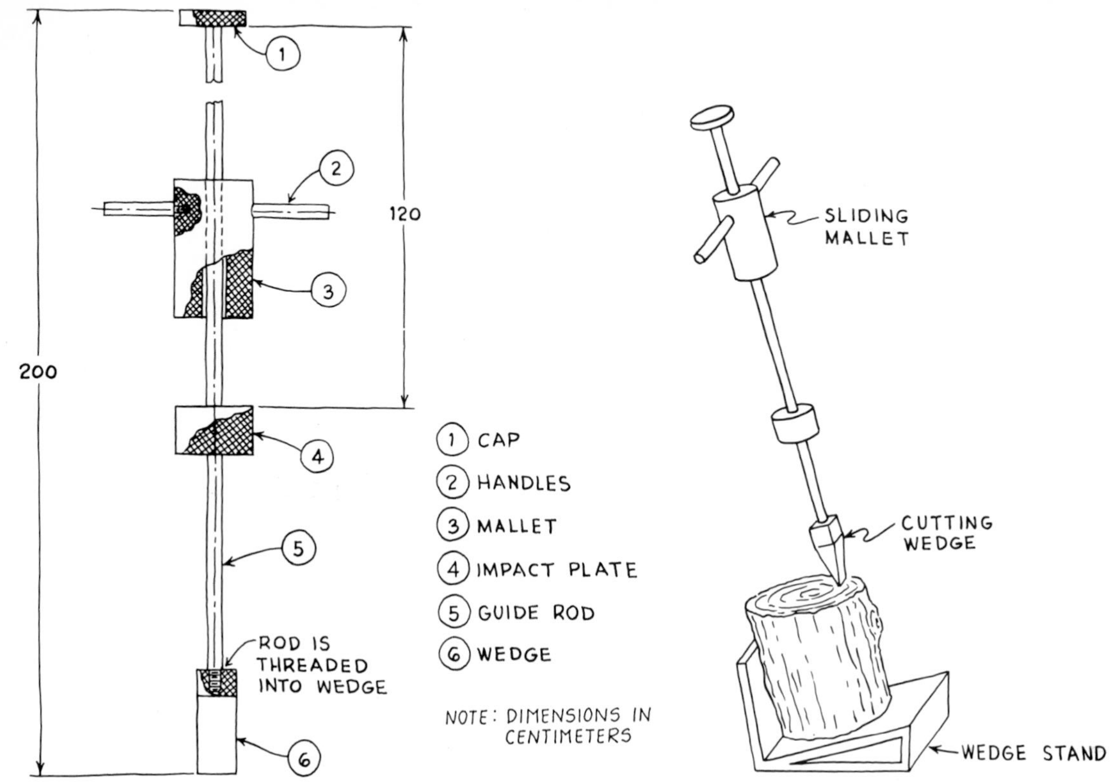

Figure 15.22
Concept development for the sliding-mass idea, sketch #2 from Fig. 15.20. This concept was selected as the final team project.

essary to evaluate each idea sketch. This phase involves considerable detail and is often referred to as "concept development." The final concepts must be developed to the point that comparative judgments can be intelligently made in evaluating each concept in light of the criteria.

Concept sketches are shown for both the pressure wedge (Fig. 15.21) and the sliding mass (Fig. 15.22).

15.8.3
Criteria in Decision

The objective of the entire design process is to choose the best solution for a problem. The steps that precede the decision phase are designed to give information that leads to the best decision. It should be quite obvious by now that poor research, a less-than-adequate list of alternatives, or inept analysis would reduce one's chances of selecting a good, much less the best, solution. Decision making, like engineering itself, is both an art and a science. There have been significant changes during the past few decades that have changed decision making from being primarily an art to what it is at the present, with probability, statistics, optimization, and utility theory all routinely used. It is not our purpose to explore these

Figure 15.23
Three members of a design team are trying to reach a decision concerning the side slopes of an earth-filled dam. (*Stanley Consultants.*)

topics, but simply to note their influence and to consider for a moment our task of selecting the best of the proposed solutions to our problem. The term "optimization" is almost self-explanatory in that it emphasizes that what we seek is the best, or optimum, value in light of a criterion. As you study more mathematics, you will acquire more powerful tools through calculus and numerical methods for optimization.

In order to illustrate optimization, we will return to the beam problem illustrated in Fig. 15.15 and Example Prob. 15.1. Our objective will be to determine the least mass of the beam for prescribed performance conditions. You will recall in our discussion of analysis that a system, the laws of nature and economics, an input to the system, and an output are involved. Analysis gives the output if the system, laws, and input are known.

If we consider the inverse problem—that is, if we were looking for a system, given the laws, input, and output—we would be using *synthesis* rather than analysis. Synthesis is not as straightforward as analysis, since it is possible to have more than one system that will perform as desired. But if we specify a criterion for selecting the best solution, then a unique solution is possible. Performing synthesis with a criterion for a best solution, often called a *payoff function,* is a meaningful and realistic part of modern-day design.

Example problem 15.3 (Refer to Fig. 15.15.) Determine the dimensions b, h for the least beam mass under the following conditions:

The deflection cannot exceed 4.0 cm.

The height h cannot be greater than three times the base b.

$E = 2.0 \times 10^{11}$ Pa
$L = 4.0$ m
$P = 1.0 \times 10^{5}$ N

The mass is a minimum when the cross-sectional area $A = bh$ is a minimum.

Procedure The system we are after is the beam shape $b \times h$ within the conditions specified above; the law is the deflection equation from Example Prob. 15.1; the inputs are L, P, and E, and the output is range of permissible deflection. The deflection equation becomes

$$d = \frac{PL^3}{3EI}$$

$$0.04 = \frac{10^5(4)^3}{3(2)(10^{11})I}$$

$$\text{or } I = 2.667(10^{-4})\ \text{m}^4$$

Then

$$\frac{bh^3}{12} = 2.667(10^{-4})$$

Thus

$$b = \frac{3.2(10^{-3})}{h^3}$$

This equation is a relationship for the beam under the condition that the deflection is a constant 4.0 cm. The expression is plotted in Fig. 15.24. Note that values to the right of the curve represent beam dimensions for which the deflection would be less than 4.0 cm. Those to the left would cause the deflection to exceed 4.0 cm; thus that portion of the "design space" for b and h is invalid.

Next we demonstrate the effect of the required relationship between b and h by plotting the line $b = h/3$, as shown in Fig. 15.24. Points above this line represent valid geometric configurations; those below do not.

Now we have a better picture of the "design space," or solution region for our problem. A point $h = 30$, $b = 30$ represents a satisfactory solution since it falls within all conditions *except* possibly minimum mass. Many designs stop at this point where a nominal solution has been found. These are the designs that may not survive in the marketplace because they are not optimum. In fact, to get a nominal solution we could have guessed values for b and h and very quickly had an answer without going to all the effort we have up to this point.

The constraint equation is the area:

$$A = bh$$

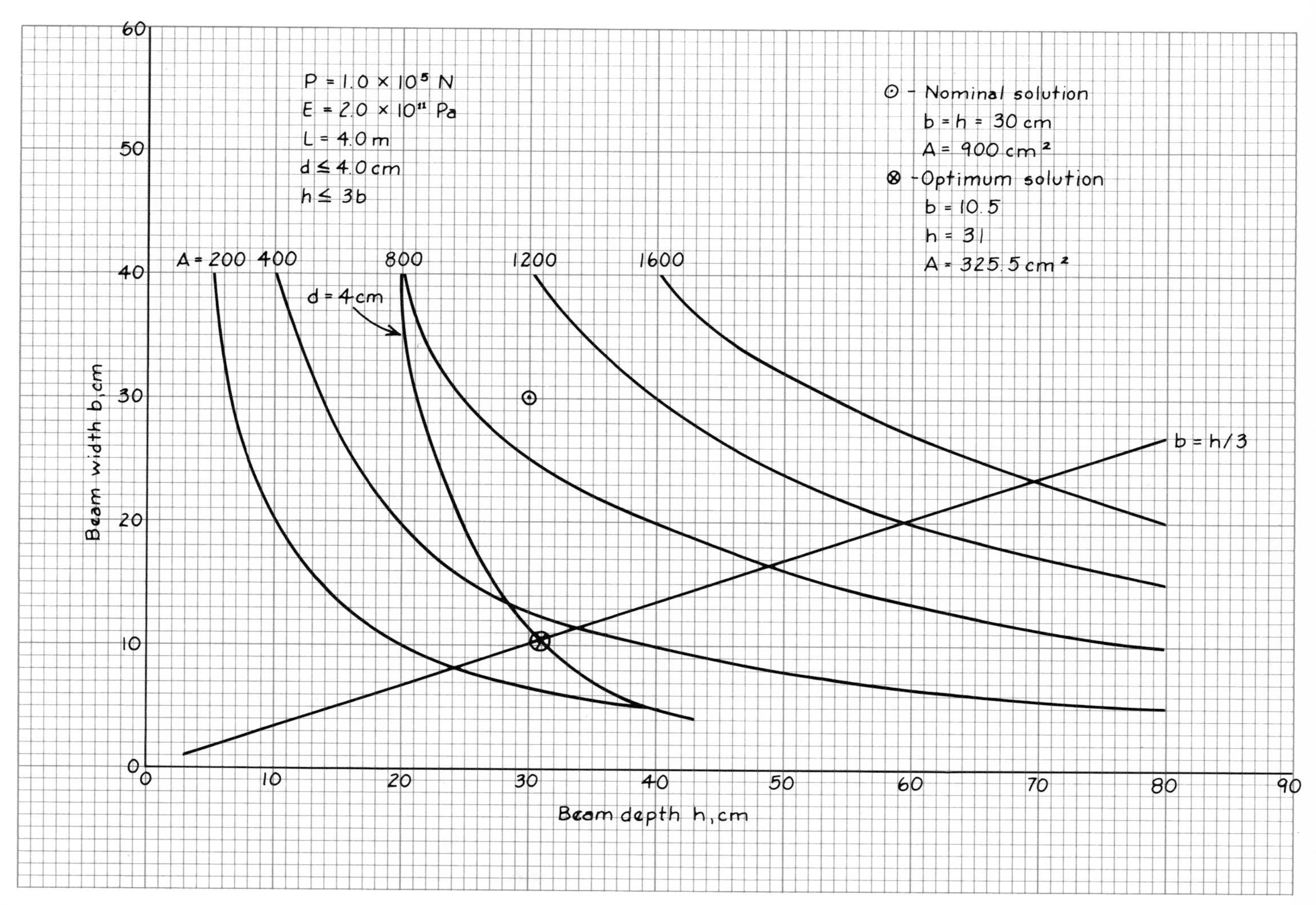
P = 1.0 × 10⁵ N
E = 2.0 × 10¹¹ Pa
L = 4.0 m
d ≤ 4.0 cm
h ≤ 3b
⊙ - Nominal solution
b = h = 30 cm
A = 900 cm²
⊗ -Optimum solution
b = 10.5
h = 31
A = 325.5 cm²
A = 200
400
800
1200
1600
d = 4 cm
b = h/3
Beam width b, cm
Beam depth h, cm
0
10
20
30
40
50
60
70
80
90

Figure 15.24

To illustrate how the area varies with respect to the other parameters, we plot lines of constant area in Fig. 15.24. To get started we note that for $b = 30$, $h = 30$, $A = 900$ so we begin near this with $A = 800$ and move to the left and right from there by incrementing A. The choices for the curves A = constant are completely arbitrary.

Now the complete problem has been illustrated in Fig. 15.24. The optimum solution is indicated and in this instance falls on the border of the design space, which is often the case. You might ask at this point why the lines A = constant were plotted. In this example you can very quickly see that the function $A = bh$ does not have a relative minimum because it is asymptotic to the axes. However, in general the constraint equation will be of a complex nature such that the plot characteristics are not obvious.

Another value of this plot, or alternatively a computer data base of the information, is the ease of extension to similar problems. For example, if the beam were lengthened to 6 m this would affect the deflection equation. A new equation could be quickly plotted on Fig. 15.24, yielding another optimum solution. Now that you have followed through this procedure, you are capable of using a computer program to very quickly investigate a multitude of cantilever beam designs for the general conditions given.

15.8.4 The Chapter Example—Step 7 Completed

The method of decision employed by the student design team is one that has considerable merit, and one that you may find simple to use. As mentioned in Sec. 15.5.3, they established six criteria and assigned weights to them. They later examined each of the five surviving concepts and graded them on a 0-to-10 scale. The grade was multiplied by the weight in percent and the points were recorded. A total of 1 000 points would be perfect. The concept with the greatest number of points is considered the best alternative. The results of their evaluation are shown in matrix form in Fig. 15.25.

Note that the winning alternative did not receive the highest rating for safety, ease of operation, and durability, and it tied for highest in portability. So, our team must report that this alternative has some shortcomings but that it is the best they could find.

15.9 Specification—Step 8

After progressing through the design process up to the point of reaching a decision, many feel that the romance has gone out of the project. The suspense and uncertainty of the solution are over, but much work still lies ahead. Even if the new idea is not a breakthrough in technology but simply an improvement in existing technology, it must be clearly defined to others. Many very creative people are ill-equipped to convey to others just exactly what their proposed solution is. It is not the time to use vague generalities

Decision matrix

Criteria	Weight W, percent	Selected concepts (see below) 1	2	3	4	5	6
Cost	30	6 / 180	7 / 210	7 / 210	7 / 210	9 / 270	
Ease of operation	20	10 / 200	7 / 140	9 / 180	10 / 200	7 / 140	
Safety	15	9 / 135	7 / 105	6 / 90	5 / 75	8 / 120	
Portability	15	6 / 90	5 / 75	4 / 60	10 / 150	10 / 150	
Durability	10	8 / 80	9 / 90	10 / 100	9 / 90	9 / 90	
Use of standard parts	10	7 / 70	8 / 80	8 / 80	6 / 60	9 / 90	
Total	100	755	700	720	785	860	

Rating scale R	
Excellent	9–10
Good	7– 8
Fair	5– 6
Poor	3– 4
Unsatisfactory	0– 2

Rating → 6 / 180 ← $R \times W$

Selected concepts

1. Auto-jack principle (item #2)
2. Drop wedge from elevation (item #10)
3. Spring-powered wedge (item #8)
4. Wedge driven by explosion (item #12)
5. Sliding mass (item #9)
6. Additional concepts

Figure 15.25
Each concept was rated by the team on a scale of 0 to 10 for each criterion. The rating was multiplied by the criterion weight and then summed. Concept 5 was chosen as the optimum even though it did not receive the highest rating for three of the six criteria.

about the general scope and approximate size and shape of the chosen concept. One must be extremely specific about all of the details, regardless of the apparent minor role that each may play in the finished product.

15.9.1 Specification by Words

One medium of communication that the successful engineer must master is that of language, written and spoken. Your problems in professional practice would be considerably less complicated if you were required to defend and explain your ideas only to other engineers. Few engineers have such luxury; most must be able to write and speak clearly and concisely to people who do not have comparable technical competence and experience. They may be officials of government who are bound so tightly by budgets and the need for public acceptance that only the best explanation is good

Figure 15.26
Detail drawings of each component must be prepared unless standard (stock) parts are used. (*Stanley Consultants.*)

enough to pierce their protective armor. They may be people in business who know that capital is limited and that they cannot defend another poor report to stockholders. Without appropriate documentation, they will not be nearly as certain as you are that your idea is a good one. Therefore, this phase of specification—communication—is so important that we have assigned it as the final step in the design process. Before discussing it, however, we will discuss another means of communication.

15.9.2 Graphical Specification

Appendix C gives a brief overview of some of the basic principles of graphical procedures. As an engineer, you will not be expected to be an accomplished drafter, but you will be expected to understand graphical techniques. You will probably have many occasions in which to work closely with technicians and drafters as they prepare the countless drawings essential to the manufacture of your design. You will not be able to do your job properly if you cannot sketch well enough to portray your idea or to read drawings well enough to know whether the plans that you must approve will actually result in your idea being constructed as you desire.

A lathe operator in the shop, an electronics technician, a contractor, or someone else must produce your design. How is the person to know what the finished product is to look like, what materials are to be used, what thicknesses are required, how it is to work, what clearances and tolerances are demanded, how it is to be assembled, how it is to be taken apart for maintenance and repair, what fasteners are to be used, etc.?

Typical drawings that are normally required include:

1. A sufficient number of detail drawings describing the size and shape of each part

2. Layouts to delineate clearances and operational characteristics
3. Assembly and subassemblies to clarify relationship of parts
4. Written notes, standards, specifications, etc., concerning quality and tolerances
5. A complete bill of materials

Included with the drawings are almost always written specifications, although certain classes of engineering work refer simply to documented standard specifications. Most cities have adopted one of several national building codes, so all structures constructed in that city must conform to the code. It is quite common for engineers and architects to refer simply to the building code as part of the written specifications and to write detailed specifications for only items that are not covered in the code. This procedure saves time and money for all by providing uniformity in bidding procedures. Many groups have produced standards that are widely recognized. For instance, there are standards for welds and fasteners for the obvious reasons—ease of specification and economy of manufacture. Moreover, there are such standards for each discipline of engineering.

15.9.3 The Chapter Example—Step 8

Reproductions of two drawings prepared by the student team are given as Figs. 15.27 and 15.28. The following drawings were part of their completed report:

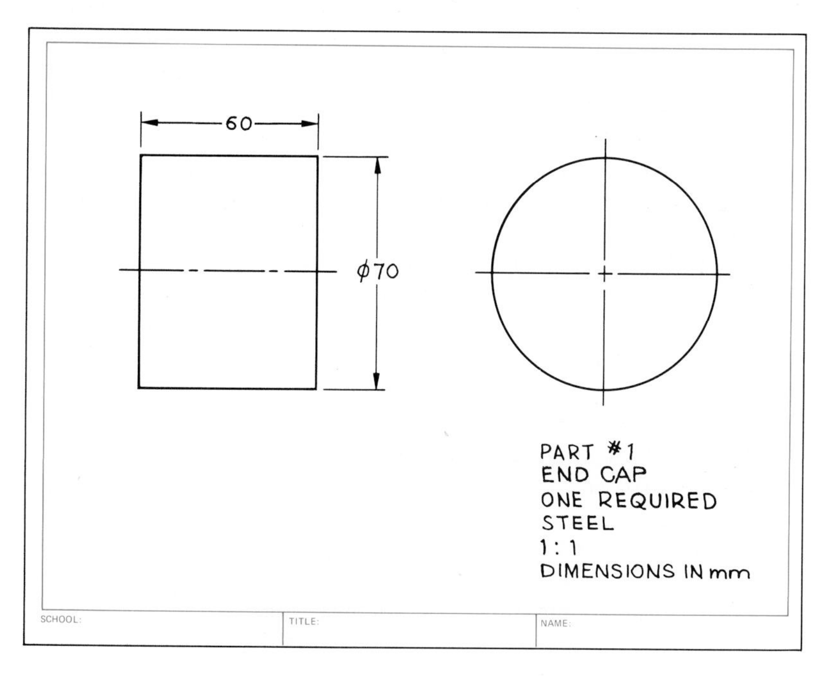

Figure 15.27 Detail drawing of part 1 (see Fig. 15.28).

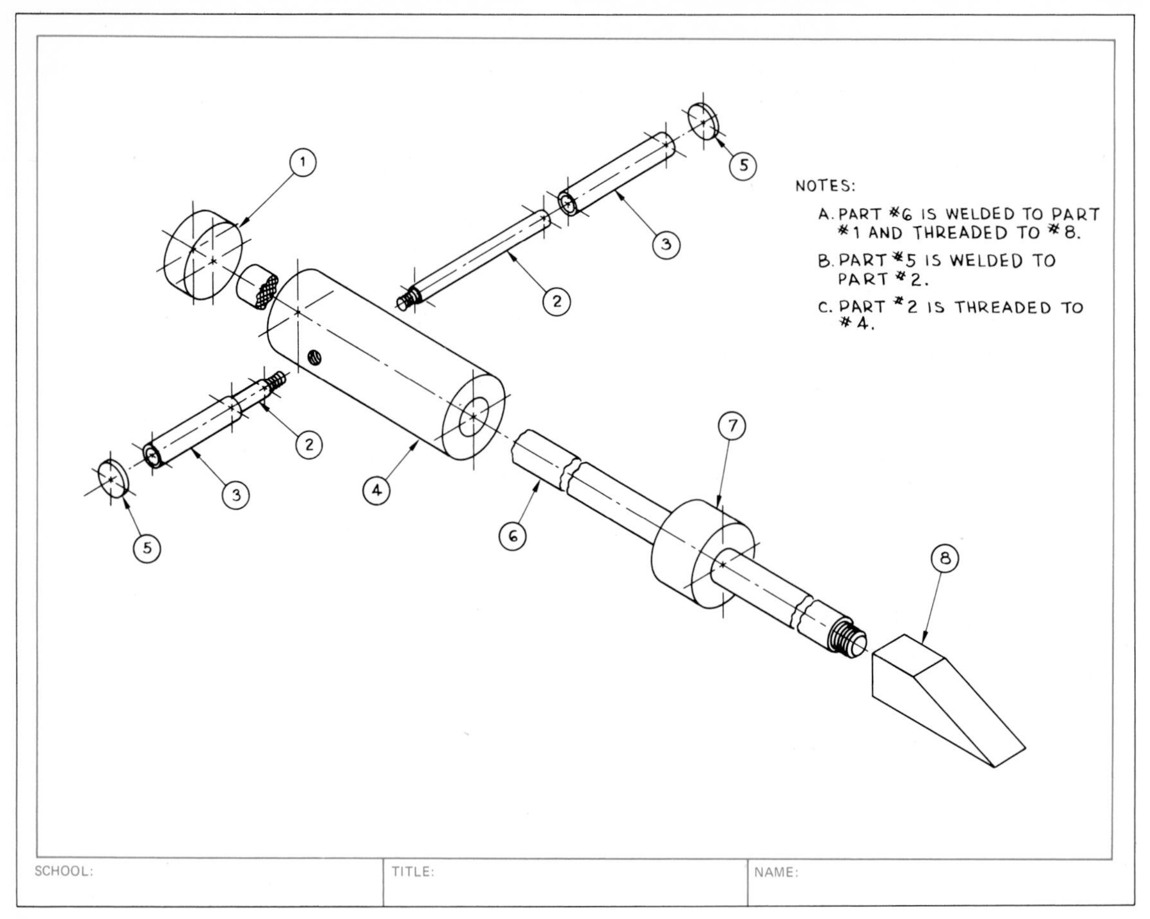
1
2
3
4
5
6
7
8
NOTES:
A. PART #6 IS WELDED TO PART #1 AND THREADED TO #8.
B. PART #5 IS WELDED TO PART #2.
C. PART #2 IS THREADED TO #4.
SCHOOL:
TITLE:
NAME:

Figure 15.28
Exploded pictorial drawing of the log-splitter.

1. Detail drawings of all eight parts of the log-splitter (one shown)
2. Exploded pictorial of the log-splitter
3. Detail drawing of the wedge stand (see Fig. 15.22 for pictorial)
4. Welding assembly of the log-splitter
5. Welding assembly of the wedge stand
6. Complete parts list

The drawings were accompanied by a cost analysis, weight summary, and description of the operating characteristics.

None of the team worked in a plant that produces such items, so no doubt an experienced detailer or drafter would find reason to be critical of their drawings, but we feel that this phase of the process was performed quite well and that there will be few misunderstandings as a result of omissions on their part.

15.10 Communication—Step 9

During the 1960s, the word "communication" seemed to take on a very high priority at conferences and in the professional journals. The need for conveying information and ideas had not changed, but there was an awareness of too much incomplete and inaccurate rendering of information. At most of the conferences the authors have attended over the past 20 years, one or more papers either discussed the need for engineers to develop greater skills in communication or demonstrated a technique for improving the skills. Students at most colleges are required to complete freshman English courses and, at some schools, a technical writing course in their junior or senior year; but many professors and employers feel that not enough of these types of courses are required.

For our purposes here, however, we will discuss only the salient points involved in design step 9.

15.10.1 Selling the Design

It is certainly the responsibility of any profession to inform people of findings and developments. Engineering is no exception in this regard. Our emphasis here will be on a second type of communication, however: selling, explaining, or persuading.

Selling takes place all the way through the design process. Individuals who are the most skillful at it will see many of their ideas develop into realities. Those who are not so good at it will no doubt become frustrated with their supervisors for not exploring what they feel is a perfectly good idea in more depth.

If you are working as a design engineer for an industry, you cannot simply decide on your own that you will try to improve Model X of the product line. Industry is anxious for their engineers to initiate ideas, but they won't necessarily approve all of them. As an engineer with a company, you must convince those who decide what assignments you get that the idea is worth the time and money required to develop it. Later, after the design has developed to the point where it can be produced and tested, you must again persuade management to place it into production.

It is a natural reaction to feel that your design has so many clear advantages that selling it should not be necessary. Such may be the case; but in actual situations, things seldom work so smoothly and simply. You will be selling or persuading or convincing others almost daily in a variety of ways. Among the many forms of communication are written and oral reports.

15.10.2 Written Reports

The types of reports that you will write as an engineer will be varied, so a precise outline that will serve for all of the reports cannot be supplied. The two major divisions of reports are those used by individuals within the organization and those used primarily by clients or customers. Many times the in-house reports follow a strict form prescribed by the organization, whereas those intended for the client are usually designed for the particular situation. The nature of the project and the client usually determine the degree of formality employed in the report. Clients often state that they wish to use the report in some particular manner, which may direct you to the style of report to use. For instance, if you are a consultant for a city and have studied the needs for expansion of a power plant, the report may be very technical, brief, and full of equations, computations, etc., if it is intended for the use of the city's engineer and public utilities director. However, if the report is to be presented to citizens in an effort to convince them to vote for a bond issue to finance the expansion, the report will take a different flavor.

Reports generally will have the following divisions or sections:

1. Appropriate cover page
2. Abstract
3. Table of contents
4. Body
5. Conclusions and recommendations
6. Appendixes

Abstract. A brief paragraph indicating the purpose and results of the effort being reported, the abstract is used primarily for archiving so that others can quickly decide if they want to obtain the complete report.

Body. This is the principal section of the report. It begins with an introduction to activities, including problem identification, background material, and the plan for attacking and solving the problem. If tests were conducted, research completed, and surveys undertaken, the results are recounted and their significance underscored. In essence, the body of the report is the description of the individual or team effort on a project.

Conclusions and recommendations. This section tells why the study was done and explains the purpose of the report. Herein you explain what you now believe to be true as a result of the work discussed in the body and what you recommend be done about it. You must lay the groundwork earlier in the report, and at this point you must sell your idea. If you have done the job carefully and fully, you may make a sale; but don't be discouraged if you don't. There will be other days and other projects.

Appendixes. Appendixes can be used to avoid interrupting your description so that it can flow more smoothly. Those who don't want to know everything about your study can read it without digression.

What is in the appendix completes the story by showing all that was done. But it should not contain information that is essential to one's understanding of the report.

It should be emphasized that all reports do not follow a specific format. For example, lengthy reports should have a summary section placed near the beginning. This one- or two-page section should include a brief statement of the problem, the proposed solution, the anticipated costs, and the benefits. The summary is for the use of higher-level management who in general do not have the time to read your entire report.

In many instances your instructor or supervisor will have specific requirements for a report. Each report is designed to accomplish a specific goal.

Student reports oftentimes must follow an instructor's directions regarding form and topics that must be included. If you are asked to write a report in an introductory design course, refer to Fig. 15.1 to make sure that all of the steps in the design process have been successfully completed. You might also study the bias of your instructor and make sure that you have done especially the steps that he or she considers most important. This may sound as though pleasing the instructor and getting a good grade is all that is important. But perhaps in this respect the academic situation is something like that in industry or private practice: Your report must take into account the audience—its biases and its expectations, whether professor in the classroom or supervisor in the business world.

15.10.3 The Oral Presentation

The objective of the oral presentation is the same as that of the written report—to furnish information and convince the listener. However, the methods and techniques are quite different. The written report is designed to be glanced at, read, and then studied. The oral presentation is a one-shot deal that must be done quickly. So it must be simple. There is no time to go into detail, to show complicated graphs and tables of data or many of the things that are given in a written report. What can you do to make a good presentation?

First, you must be prepared. No audience listens to people who have not bothered to prepare themselves. So you should rehearse with a timer, a mirror, and a tape recorder.

Stand in such a way that you don't detract from what you are saying or showing.

Look at your audience and maintain eye contact. You will be receiving cues from those who are listening, so be prepared to react to these cues.

Project your voice by consciously speaking to the back row. The audience quickly loses interest if it has to struggle to hear.

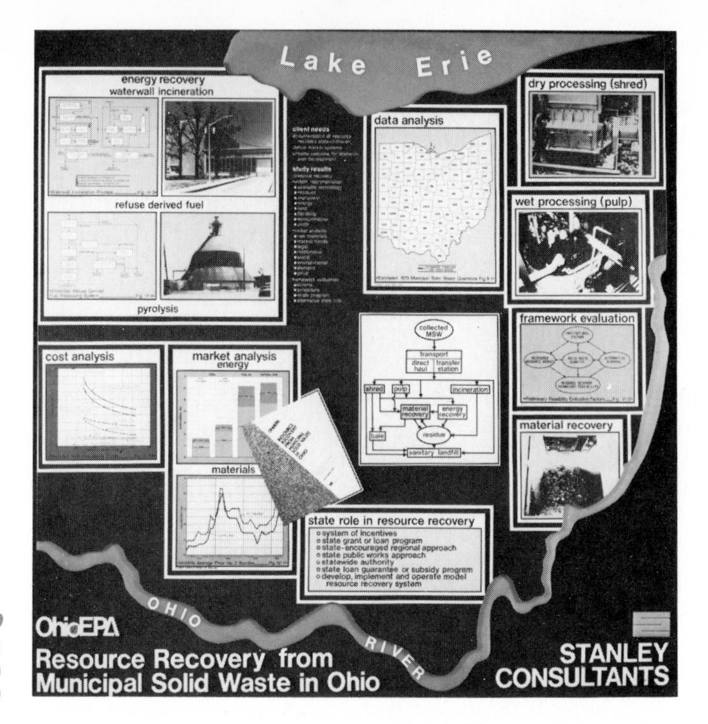

Figure 15.29
Visual aids used by a consulting engineer during an oral presentation to a client. (*Stanley Consultants.*)

Speak clearly. We all have some problems with our voices—they are either too high, too low, or too accented and certain words or sounds are hard for us; but always be concerned for the listener.

Preparation obviously includes being thoroughly familiar with the material. It should also include determining the nature, size, and technical competence of the audience. You must know how much time will be allotted to your presentation and what else, if anything, is to be presented before or after your speech. It is essential that you know what the room is like, because the physical conditions of the room—its size, lighting, acoustics, and seating arrangements—may very well control your use of slides, transparencies, records, and microphones.

The quality of your graphic displays can often influence the opinion of your audience. Again, consider to whom you are speaking carefully as you choose which and how many graphics to use. Be certain that they can be read and understood or don't use them at all. Don't clutter your displays with so many details that the message is obscured. Don't try to make a single visual aid accomplish too many tasks: It is good to change the center of emphasis. By all means, test your visual aids before the meeting and never apologize for their quality. (If they aren't good, don't use them.) Figure 15.29 shows a number of visual aids that were used by an engineering consultant in presenting views on resource recovery. Included are photographs, flow diagrams, and several types of graphs and tables.

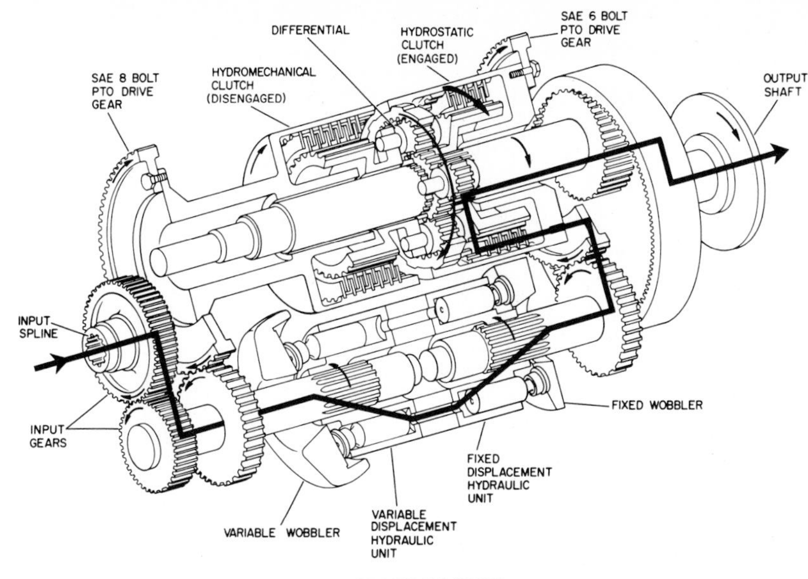

Figure 15.30
A cutaway sectional view of a transmission. Visual aids are a necessity when discussing a complicated device such as this with potential users. (*Sundstrand Hydro-Transmission Division.*)

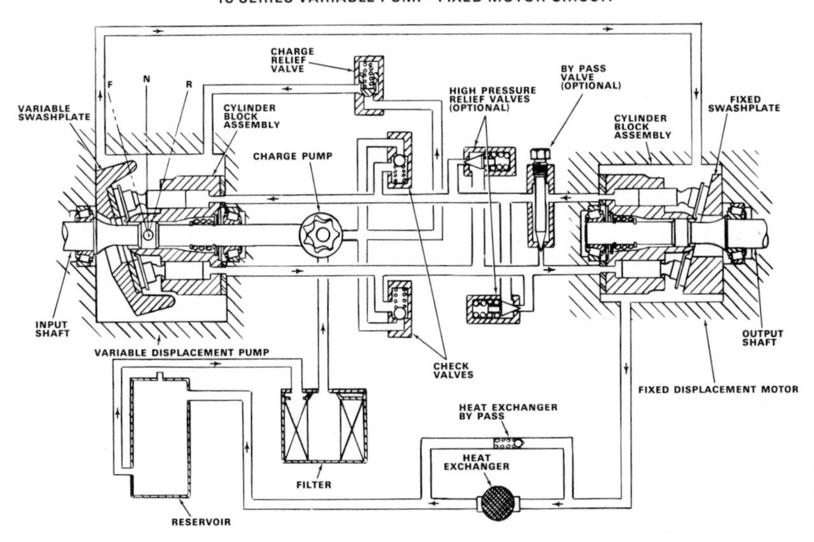

Figure 15.31
A schematic used to augment a presentation. (*Sundstrand Hydro-Transmission Division.*)

Figure 15.30 is a good, clear cutaway drawing that is excellent for use as a visual aid during an oral presentation. Figure 15.31 is a schematic drawing that can be used to explain a process. The quality of your visual aids can influence many people for you or against you before they hear all you have to say.

Have a good finish. Save something important for the last and make sure everyone knows when the end has come. By all means, don't end with "Well, I guess that's about all I have to say." You have much more to say, you just don't have the time to say it.

15.10.4 The Chapter Example—Step 9

The written and oral reports prepared and presented by the students were significant parts of their design experience. Both reports were regulated somewhat by their professor in much the same way that reports are in industry. They were told who would be reading the report and who would judge the oral presentation. They were given copies of the written report grading sheet and the oral presentation judging card. They correctly accepted these constraints as real (not imagined), and they were given high ratings by their evaluators.

Suggested Projects

Problems 1 and 2 involve synthesis similar to that illustrated in Example Prob. 15.3.

1. For the beam configuration of Fig. 15.15 determine the dimensions b, h for least mass under the following conditions:

   The deflection cannot exceed 5 cm.

   The height h cannot be greater than b.

   $E = 2.0 \times 10^{11}$ Pa

   $L = 6.0$ m

   $P = 1.0 \times 10^{5}$ N

   Produce a brief report containing a plot similar to Fig. 15.24 and a discussion of the design space and how the solution was found.

2. A company transfers packages from point to point across the country. The limit on package size is that the girth plus the longest dimension (measured on the package) cannot exceed 60 in. Consider two kinds of packages, a rectangular-prism shape with square ends and a cylindrical shape where the cylinder height is greater than the diameter (girth = circumference). Determine for each shape the largest package volume that can be shipped and the package dimensions at the maximum volume.

   Suggested procedure:
   a. Write the constraint equation (60-in limit) and the payoff function (volume). Eliminate one of the unknowns in the payoff function by substituting the constraint equation.
   b. Plot the payoff function against the remaining variable.
   c. Determine the optimum values.
   d. Prepare a report of your findings.

3. The following list of potential projects can be approached in the manner used by the student design team featured in this chapter.

   Headlights that follow the wheels' direction

   A protective "garage" that can be stored in the car's trunk

   A means of preventing body rust

   A device to prevent theft of helmets left on motorcycles

A conversion kit for winter operation of motorcycles

An improved rack for carrying packages or books on a motorcycle or bicycle

A child's seat for motorcycle or bicycle

A tray for eating, writing, and playing games in the back seat of a car

A system for improving traction on ice without studs or chains

An inexpensive built-in jack for raising a car

An auto-engine warmer

A better way of informing motorists of speed limits, road conditions, hazards, etc.

Theft- and vibration-proof hub caps

A better way to check engine oil level

A device to permit easier draining of the oil pan by weekend mechanics

A heated steering wheel for cold weather

A less-expensive replacement for auto air-cleaner elements

An overdrive system for a trail bike

A sun shield for an automobile

A well-engineered, efficient automobile instrument panel

An SOS sign for cars stalled on freeways

A remote car-starting system for warmup

A car-door positioner for windy days

A bicycle trailer

Automatic rate-sensitive windshield wipers

A corn detasseler

An improved wall outlet

A beverage holder for a card table

A car wash for pickups

A better rural mailbox

A home safe

An improved automobile traffic pattern on campus

An alert for drowsy or sleeping drivers

An improved automobile headlight

An improved bicycle for recreation or racing

Improved bicycle brakes

A transit system for campus

A pleasure boat with retractable trailer wheels

Improved pedestrian crossings at busy intersections

A transportation system within large airports

An improved baggage-handling system at airports

Improved parking facilities in and around campus

A simple but effective device to assist in cleaning clogged drains

A device to attach to a paint can for pouring

An improved soap dispenser

A better method of locking weights to a barbell shaft

A shoestring fastener to replace the knot

An automatic moisture-sensitive lawn waterer

A better harness for seeing-eye dogs

A better jar opener

A system or device to improve efficiency of limited closet space

A shoe transporter and storer

A pen and pencil holder for college students

An acceptable rack for mounting electric fans in dormitory windows

A device to pit fruit without damage

A riot-quelling device to subdue participants without injury

An automatic device for selectively admitting and releasing an auxiliary door for pets

A device to permit a person to open a door when loaded with packages

A more efficient toothpaste tube

A fingernail catcher for fingernail clippers

A more effective alarm clock for reluctant students

An alarm clock with a display to show it has been set to go off

A device to help a mother monitor small children's presence and activity in and around the house

A chair that can rotate, swivel, rock, or stay stationary

A simple pocket alarm that is difficult to shut off, for discouraging muggers

An improved storage system for luggage, books, etc., in dormitories

A lampshade designed to permit one to study while his or her roommate is asleep

A device that would permit blind people to vote in an otherwise conventional voting booth

A one-cup coffee maker

A solar greenhouse

A quick-connect garden-hose coupling

A device for recycling household water

A silent wakeup alarm

Home aids for the blind (or deaf)

A safer, more efficient, and quieter air mover for room use

A lock that can be opened by secret method without a key

A can crusher

A rain-sensitive house window

A better grass catcher for a riding lawnmower

A winch for hunters of large game

Gauges for water, transmission fluid, etc., in autos

A built-in auto refrigerator

A better camp cooler

A dormitory-room cooler

A device for raising and lowering TV racks in classroom

An impact hammer adapter for electric drills

An improved method of detecting and controlling level position of the bucket on a bucket loader

Shields to prevent corn spillage where the drag line dumps into sheller elevator (angle varies)

An automatic tractor-trailer-hitch aligning device

A jack designed expressly for motorcycle use (special problems involved)

A motorbike using available (junk) materials

Improved road signs for speed limits, curves, deer crossings, etc.

More effective windshield wipers

A windshield deicer

Shock-absorbing bumpers for minor accidents

A home fire-alarm device

A means of evacuating buildings in case of fire

Automatic light switches for rooms

A carbon monoxide detector

An indicator to report the need for an oil change

A collector for dust (smoke) particles from stacks

A means of disposing of or recycling soft-drink containers

A way to stop dust storms, resultant soil loss, and air entrainment

An attractive system for handling trash on campus

A self-decaying disposable container

A device for dealing with oil slicks

A means of preventing heat loss from greenhouses

A way of creating energy from waste

A bookshelf with horizontally and vertically adjustable shelves and dividers

A device that would make the working surface of graphics desks adjustable in height and slope, retaining the existing top and pedestal

An egg container (light, strong, compact) for camping and canoeing

Ramps or other facilities for handicapped students

A multifunctional (suitcase/chair/bookshelf, etc.) packing device for students

A self-sharpening pencil for drafting

An adapter to provide tilt and elevation control on existing graphics tables

A compact and inexpensive camp stove for backwoods hiking

A road trailer operable from inside the car

A hood lock for cars to prevent vandalism

A system to prevent car thefts

A keyless lock

APPENDIXES

Selected Topics from Algebra

A.1 Introduction

This appendix includes material on exponents and logarithms, simultaneous equations, and the solution of equations by approximation methods. The material can be used for reference or review. The reader should consult an algebra textbook for more detailed explanations or additional topics for study.

A.2 Exponents and Radicals

The basic laws of exponents are stated below along with an illustrative example.

Law	Example
$a^m a^n = a^{m+n}$	$x^5 x^{-2} = x^3$
$\frac{a^m}{a^n} = a^{m-n} \quad a \neq 0$	$\frac{x^5}{x^3} = x^2$
$(a^m)^n = a^{mn}$	$(x^{-2})^3 = x^{-6}$
$(ab)^m = a^m b^m$	$(xy)^2 = x^2y^2$
$\left(\frac{a}{b}\right)^m = \frac{a^m}{b^m} \quad b \neq 0$	$\left(\frac{x}{y}\right)^2 = \frac{x^2}{y^2}$
$a^{-m} = \frac{1}{a^m} \quad a \neq 0$	$x^{-3} = \frac{1}{x^3}$
$a^0 = 1 \quad a \neq 0$	$2(3x^2)^0 = 2(1) = 2$
$a^1 = a$	$(3x^2)^1 = 3x^2$

These laws are valid for positive and negative integer exponents and for a zero exponent, and can be shown to be valid for rational exponents. Some examples of fractional exponents are illustrated here. Note the use of radical ($\sqrt{\ }$) notation as an alternative to fractional exponents.

Law	Example
$a^{m/n} = \sqrt[n]{a^m}$	$x^{2/3} = \sqrt[3]{x^2}$
$\frac{\sqrt[n]{a}}{\sqrt[n]{b}} = \sqrt[n]{\frac{a}{b}} \quad b \neq 0$	$\frac{\sqrt[3]{16}}{\sqrt[3]{2}} = \sqrt[3]{8} = 2$
$a^{1/2} = \sqrt[2]{a^1} = \sqrt{a} \quad a \geq 0$	$\sqrt{25} = 5 \quad$ (not $\pm$ 5)

A.3

Exponential and Power Functions

Functions involving exponents occur in two forms—power and exponential. The power function contains the base as the variable and the exponent is a rational number. An exponential function has a fixed base and variable exponent.

The simplest exponential function is of the form

$$y = b^x \qquad b \geq 0$$

where b is a constant. Note that this function involves a power but is fundamentally different from the power function $y = x^b$.

The inverse of a function is an important concept for the development of logarithmic functions from exponential functions. Consider a function $y = f(x)$. If this function could be solved for x, the result would be expressed as $x = g(y)$. For example, the power function $y = x^2$ has as its inverse $x = \pm \sqrt{y}$. Note that in $y = x^2$, y is a single-valued function of x, whereas the inverse is a double-valued function. For $y = x^2$, x can take on any real value, whereas the inverse $x = \pm \sqrt{y}$ restricts y to only positive values or zero. This result is important in the study and application of logarithmic functions.

A.4

The Logarithmic Function

The definition of a logarithm may be stated as follows:

A number L is said to be the logarithm of a positive real number N to the base b (where b is real, positive, and different from 1), *if L is the exponent to which b must be raised to obtain N.*

Symbolically, the logarithm function is expressed as

$$L = \log_b N$$

for which the inverse is

$$N = b^L$$

For instance,

$$\log_2 8 = 3 \qquad \text{since } 8 = 2^3$$

$$\log_{10} 0.01 = -2 \qquad \text{since } 0.01 = 10^{-2}$$

$$\log_5 5 = 1 \qquad \text{since } 5 = 5^1$$

$$\log_b 1 = 0 \qquad \text{since } 1 = b^0$$

Several properties of logarithms and exponential functions can be identified when plotted on a graph.

Example problem A.1 Plot graphs of $y = \log_2 x$ and $x = 2^y$ that are inverse functions.

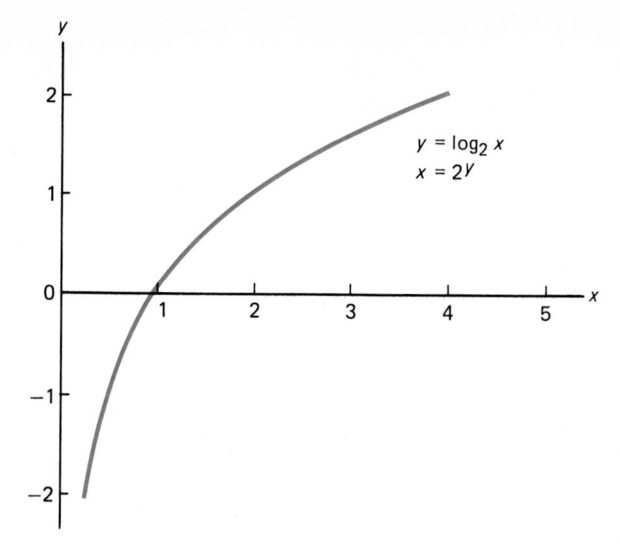

Figure A.1
The logarithmic function.

Solution Since $y = \log_2 x$ and $x = 2^y$ are equivalent by definition, they will graph into the same line. Choosing values of y and computing x from $x = 2^y$ yields Fig. A.1.

Some properties of logarithms that can be generalized from Fig. A.1 are

1. $\log_b x$ is not defined for negative or zero values of x.
2. $\log_b 1 = 0$.
3. If $x > 1$, then $\log_b x > 0$.
4. If $0 < x < 1$, then $\log_b x < 0$.

Other properties of logarithms that can be proved as a direct consequence of the laws of exponents are, with P and Q being real and positive numbers,

1. $\log_b PQ = \log_b P + \log_b Q$.
2. $\log_b \frac{P}{Q} = \log_b P - \log_b Q$.
3. $\log_b (P)^m = m \log_b P$.
4. $\log_b \sqrt[n]{P} = \frac{1}{n} \log_b P$.

The base b, as stated in the definition of a logarithm, can be any real number greater than 0 but not equal to 1, since 1 to any power remains 1. When using logarithmic notation, the base is always indicated, with the exception of base 10, in which case the base is frequently omitted. In the expression $y = \log x$, the base is understood to be 10. A somewhat different notation is used for the natural (Naperian) logarithms discussed in the Sec. A.5.

Sometimes it is desirable to change the base of logarithms. The procedure is shown by the following example.

Example problem A.2 Given that $y = \log_a N$, find $\log_b N$.

Solution

$$y = \log_a N$$

$$N = a^y \qquad \text{(inverse function)}$$

$$\log_b N = y \log_b a \qquad \text{(taking logs to base } b\text{)}$$

$$\log_b N = (\log_a N)(\log_b a) \qquad \text{(substitution for } y\text{)}$$

$$= \frac{\log_a N}{\log_a b} \qquad \left(\text{since } \log_b a = \frac{1}{\log_a b}\right)$$

A.5 Natural Logarithms and e

In advanced mathematics, the base e is usually chosen for logarithms to achieve simpler expressions. Logarithms to the base e are called natural, or Naperian, logarithms. The constant e is defined in the calculus as

$$e = \lim_{n \to 0} (1 + n)^{1/n} = 2.7182818284 \cdots$$

For purposes of calculating e to a desired accuracy, an infinite series is used.

$$e = \sum_{n=0}^{\infty} \frac{1}{n!}$$

The required accuracy is obtained by summing sufficient terms. For example,

$$\sum_{n=0}^{6} \frac{1}{n!} = 1 + 1 + \frac{1}{2} + \frac{1}{6} + \frac{1}{24} + \frac{1}{120} + \frac{1}{720}$$

$$= 2.718\ 055$$

which is accurate to four significant figures.

Natural logarithms are denoted by the symbol ln, and all the properties defined previously for logarithms apply to natural logarithms. The inverse of $y = \ln x$ is $x = e^y$. The following examples illustrate applications of natural logarithms.

Example problem A.3

$\ln 1 = 0 \qquad \text{since } e^0 = 1$

$\ln e = 1 \qquad \text{since } e^1 = e$

Example problem A.4 Solve for x:

$$2^x = 3^{x-1}$$

Specify answer to four significant figures.

Taking natural logarithms of both sides of the equation and using a calculator for evaluation of numerical quantities,

$$x \ln 2 = (x - 1)\ln 3$$

$$\frac{x}{x-1} = \frac{\ln 3}{\ln 2} = 1.585\,0$$

$$x = 2.709 \qquad \text{(four significant figures)}$$

This problem could have been solved by choosing any base for taking logarithms. However, in general, base e or 10 should be chosen so that a scientific calculator can be used for numerical work.

A.6 Simultaneous Equations

Several techniques exist for finding the common solution to a set of n algebraic equations in n unknowns. A formal method for solution of a system of linear equations is known as Cramer's rule, which requires a knowledge of determinants.

A second-order determinant is defined and evaluated as

$$\begin{vmatrix} a_1 b_1 \\ a_2 b_2 \end{vmatrix} = a_1 b_2 - a_2 b_1$$

A third-order determinant is defined and evaluated as

$$\begin{vmatrix} a_1 b_1 c_1 \\ a_2 b_2 c_2 \\ a_3 b_3 c_3 \end{vmatrix} = a_1 \begin{vmatrix} b_2 c_2 \\ b_3 c_3 \end{vmatrix} - a_2 \begin{vmatrix} b_1 c_1 \\ b_3 c_3 \end{vmatrix} + a_3 \begin{vmatrix} b_1 c_1 \\ b_2 c_2 \end{vmatrix}$$

where the second-order determinants are evaluated as indicated above. The procedure may be extended to higher-order determinants.

Cramer's rule for a system of n equations in n unknowns can be stated as follows:

1. Arrange the equations to be solved so that the unknowns x, y, z, and so forth appear in the same order in each equation; if any unknown is missing from an equation, it is to be considered as having a coefficient of zero in that equation.
2. Place all terms that do not involve the unknowns in the right member of each equation.
3. Designate by D the determinant of the coefficients of the unknowns in the same order as they appear in the equations. Designate by D_i the determinant obtained by replacing the elements of the ith column of D by the terms in the right member of the equations.

4. Then, if $D \neq 0$, the values of the unknowns x, y, z, and so forth, are given by

$$x = \frac{D_1}{D} \qquad y = \frac{D_2}{D} \qquad z = \frac{D_3}{D} \qquad \cdots$$

Example problem A.5 Solve the following system of equations that have already been written in proper form for application of Cramer's rule.

$$3x + y - z = 2$$

$$x - 2y + z = 0$$

$$4x - y + z = 3$$

Solution

$$x = \frac{\begin{vmatrix}2 & 1 & -1\\ 0 & -2 & 1\\ 3 & -1 & 1\end{vmatrix}}{\begin{vmatrix}3 & 1 & -1\\ 1 & -2 & 1\\ 4 & -1 & 1\end{vmatrix}} = \frac{2\begin{vmatrix}-2 & 1\\ -1 & 1\end{vmatrix} - 0\begin{vmatrix}1 & -1\\ -1 & 1\end{vmatrix} + 3\begin{vmatrix}1 & -1\\ -2 & 1\end{vmatrix}}{3\begin{vmatrix}-2 & 1\\ -1 & 1\end{vmatrix} - 1\begin{vmatrix}1 & -1\\ -1 & 1\end{vmatrix} + 4\begin{vmatrix}1 & -1\\ -2 & 1\end{vmatrix}}$$

$$= \frac{2(-2 + 1) - 0(1 - 1) + 3(1 - 2)}{3(-2 + 1) - 1(1 - 1) + 4(1 - 2)}$$

$$= \frac{-5}{-7}$$

$$= \frac{5}{7}$$

The reader may verify the solutions $y = 6/7$ and $z = 1$.

There are several other methods of solution for systems of equations that are illustrated by the following examples.

Example problem A.6 Solve the system of equations:

$$9x^2 - 16y^2 = 144$$

$$x - 2y = 4$$

Solution The common solution represents the intersection of a hyperbola and straight line. The method used is substitution. Solving the linear equation for x yields

$$x = 2y + 4$$

Substitution into the second-order equation gives

$$9(2y + 4)^2 - 16y^2 = 144$$

which reduces to

$$20y^2 + 144y = 0$$

Factoring gives

$$4y(5y + 36) = 0$$

which yields

$$y = 0, \frac{-36}{5}$$

Substitution into the linear equation $x = 2y + 4$ gives the corresponding values of x:

$$x = 4, -\tfrac{52}{5}$$

The solutions are thus the coordinates of intersection of the line and the hyperbola:

$$(4,0), (-\tfrac{52}{5}, -\tfrac{36}{5})$$

which can be verified by graphical construction.

Example problem A.7 Solve the system of equations:

(a) $3x + y = 7$

(b) $x + z = 4$

(c) $y - z = -1$

Solution Systems of equations similar to these arise frequently in engineering applications. Obviously, they can be solved by Cramer's rule. However, a more rapid solution can be obtained directly by elimination.

From Eq. (c),

$$y = z - 1$$

From Eq. (a),

$$y = 7 - 3x$$

From Eq. (b),

$$x = 4 - z$$

Successive substitution yields

$$z - 1 = 7 - 3x$$

$$z - 1 = 7 - 3(4 - z)$$

$$-2z = -4$$

$$z = +2$$

Continued substitution gives

$$y = 1$$

$$x = -2$$

Every system of equations should first be carefully investigated before a method of solution is chosen so that the most direct method, requiring the minimum amount of time, is used.

A.7 Approximate Solutions

Many equations developed in engineering applications do not lend themselves to direct solution by standard methods. These equations must be solved by approximation methods to the accuracy dictated by the problem conditions. Experience is helpful in choosing the numerical technique for solution.

Example problem A.8 Find to three significant figures the solution to the equation

$$2 - x = \ln x$$

Solution One method of solution is graphical. If the equations $y = 2 - x$ and $y = \ln x$ are plotted, the common solution would be the intersection of the two lines. This would not likely give three-significant-figure accuracy, however. A more accurate method requires use of a scientific calculator.

Inspection of the equation reveals that the desired solution must lie between 1 and 2. It is then a matter of setting up a routine that will continue to bracket the solution between two increasingly accurate numbers. Table A.1 shows the intermediate steps and indicates that the solution is $x = 1.56$ to three significant figures.

For greater accuracy and/or more complex problems, a computer or programmable calculator could easily be used to determine a solution by the method just described. The time available and equipment on hand will always influence the numerical technique to be used.

Table A.1 Solution of $2 - x = 1n\ x$

x	1	2	1.5	1.6	1.55	1.56	1.557
$2 - x$	1	0	0.500	0.400	0.450	0.440	0.443
$\ln x$	0	0.693	0.405	0.470	0.438	0.445	0.443

Trigonometry

B.1 Introduction

This material is intended to be a brief review of concepts from plane trigonometry that are commonly used in engineering calculations. The section deals only with plane trigonometry and furnishes no information about spherical trigonometry. The reader is referred to standard texts in trigonometry for more detailed coverage and analysis.

B.2 Trigonometric Function Definitions

The trigonometric functions are defined for an angle contained within a right triangle, as shown in Fig. B.1.

$$\text{sine } \theta = \sin \theta = \frac{\text{opposite side}}{\text{hypotenuse}} = \frac{y}{r}$$

$$\text{cosine } \theta = \cos \theta = \frac{\text{adjacent side}}{\text{hypotenuse}} = \frac{x}{r}$$

$$\text{tangent } \theta = \tan \theta = \frac{\text{opposite side}}{\text{adjacent side}} = \frac{y}{x}$$

$$\text{cotangent } \theta = \cot \theta = \frac{\text{adjacent side}}{\text{opposite side}} = \frac{x}{y} = \frac{1}{\tan \theta}$$

$$\text{secant } \theta = \sec \theta = \frac{\text{hypotenuse}}{\text{adjacent side}} = \frac{r}{x} = \frac{1}{\cos \theta}$$

$$\text{cosecant } \theta = \csc \theta = \frac{\text{hypotenuse}}{\text{opposite side}} = \frac{r}{y} = \frac{1}{\sin \theta}$$

The angle θ is by convention measured positive in the counterclockwise direction from the positive x axis.

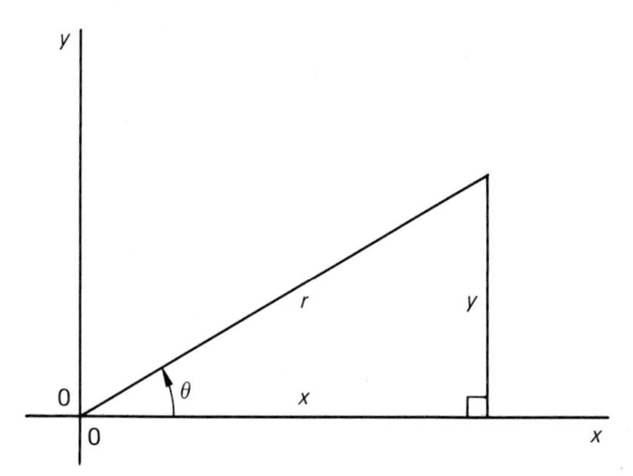

Figure B.1 Coordinate definition.

B.3

Signs of Trigonometric Functions by Quadrant

Table B.1

Quadrant 2	Quadrant 1
$x(-), y(+)$ sin and csc (+) cos and sec (−) tan and cot (−)	$x(+), y(+)$ sin and csc (+) cos and sec (+) tan and cot (+)
Quadrant 3	**Quadrant 4**
$x(-), y(-)$ sin and csc (−) cos and sec (−) tan and cot (+)	$x(+), y(-)$ sin and csc (−) cos and sec (+) tan and cot (−)

B.4

Radians and Degrees

Angles may be measured in either degrees or radians (see Fig. B.2). By definition,

$$1 \text{ degree } (^\circ) = \tfrac{1}{360} \text{ of the central angle of a circle}$$

$$1 \text{ radian (rad)} = \text{angle subtended at center 0 of a circle by an arc equal to the radius}$$

The central angle of a circle is 2π rad or 360°. Therefore,

$$1^\circ = \frac{2\pi}{360^\circ} = \frac{\pi}{180^\circ} = 0.017\,453\,29 \cdots \text{ rad}$$

and

$$1 \text{ rad} = \frac{360^\circ}{2\pi} = \frac{180^\circ}{\pi} = 57.295\,78 \cdots {}^\circ$$

It follows that the conversion of θ in degrees to θ in radians is given by

$$\theta \text{ (rad)} = \theta\,(^\circ)\,\frac{\pi}{180^\circ}$$

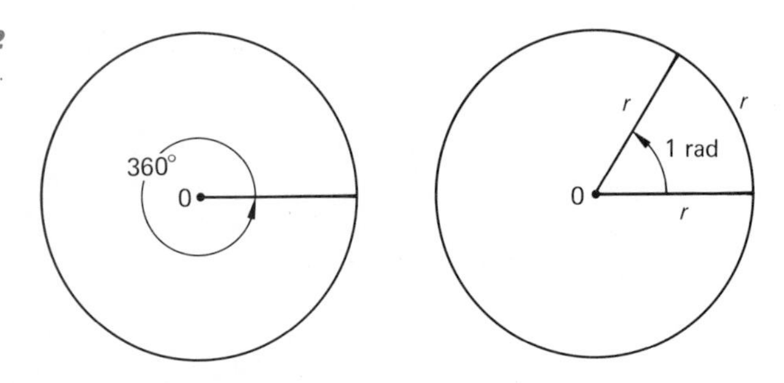

Figure B.2
Definition of degrees and radians.

and in like manner,

$$\theta\,(^\circ) = \theta\,(\text{rad})\,\frac{180^\circ}{\pi}$$

B.5 Plots of Trigonometric Functions

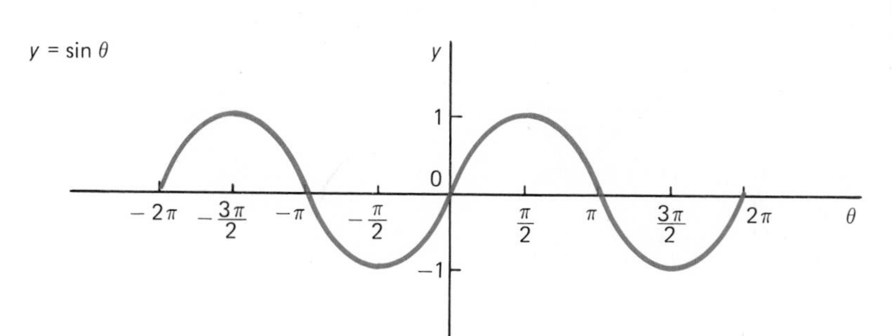

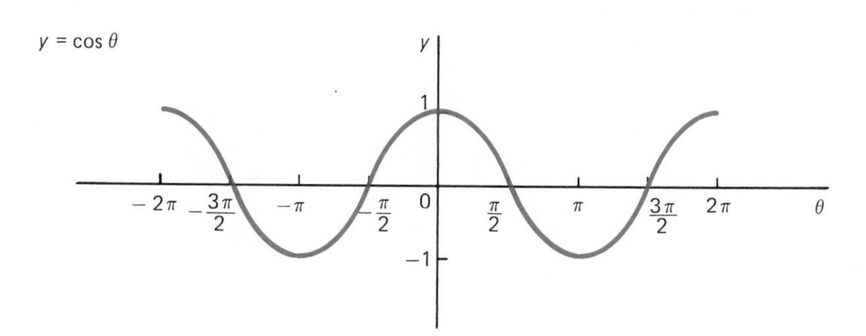

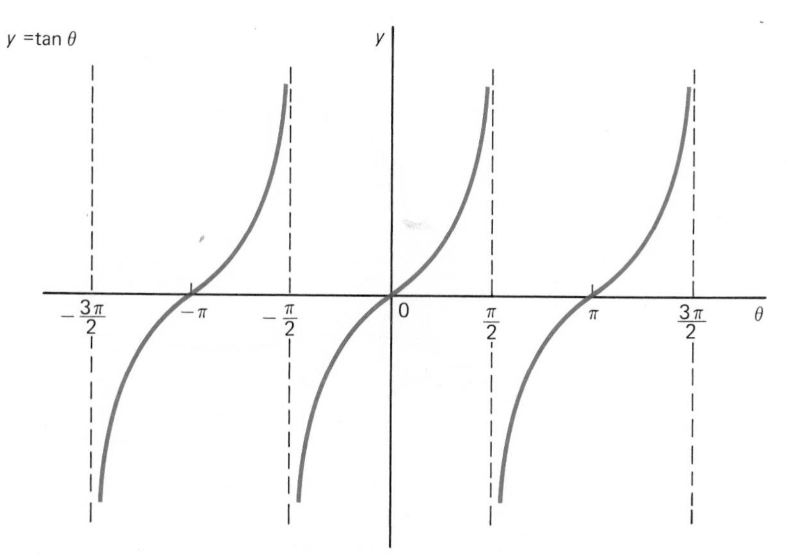

Figure B.3
Plots of trigonometric functions.

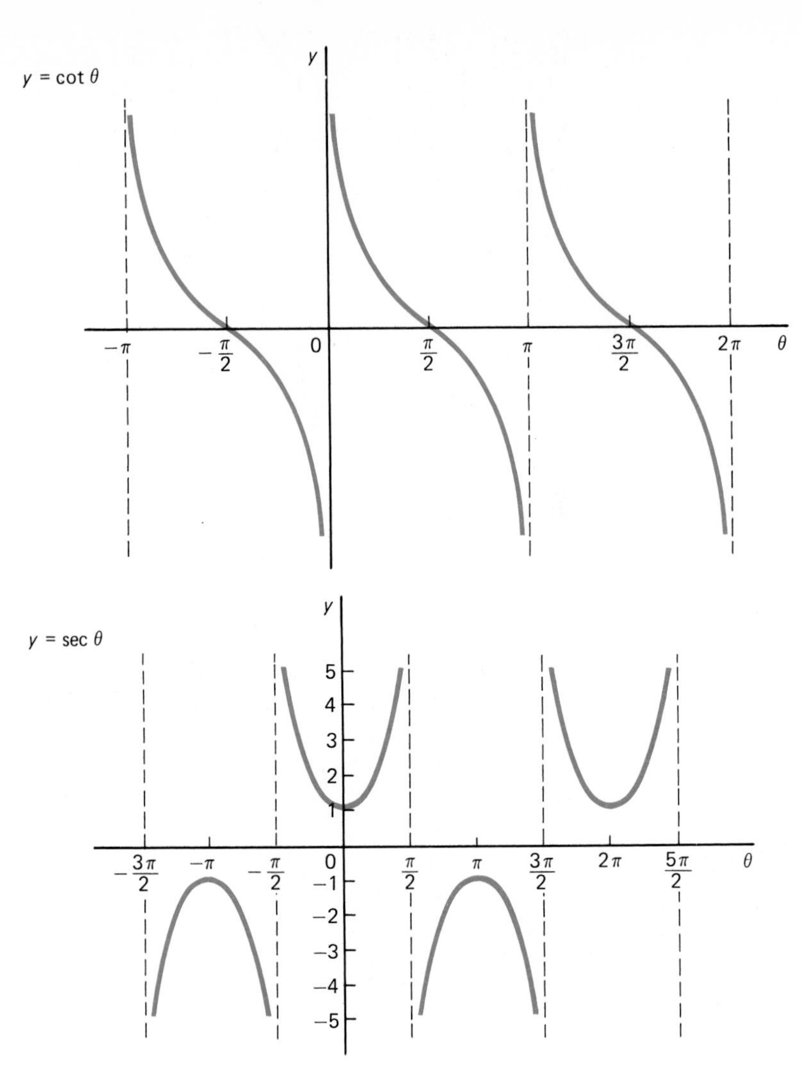

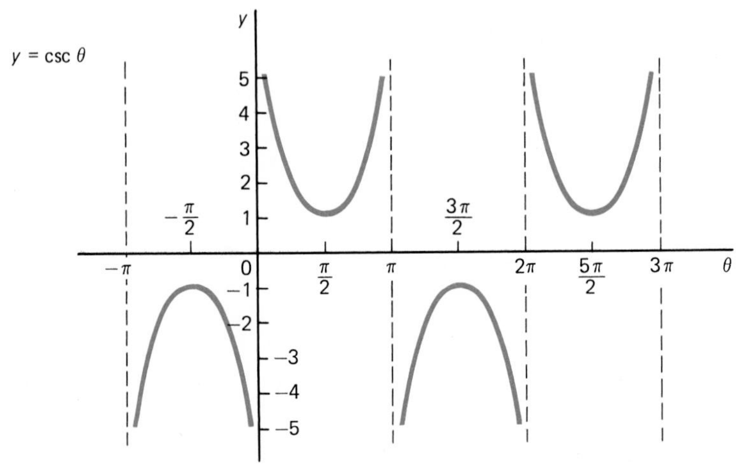

Figure B.3
(continued)

B.6

Standard Values of Often-Used Angles

From three basic triangles, it is possible to compute the values of the trigonometric functions for many standard angles such as 30°, 45°, 60°, 120°, 135°, etc. It is only necessary for us to recall that sin 30° = cos 60° = $\frac{1}{2}$, and tan 45° = 1 to construct the necessary triangles from which values can be taken to obtain the other functions.

The functions for 0°, 90°, 180°, etc., can be found directly from the function definitions and a simple line sketch. See Table B.2.

2
1
30°

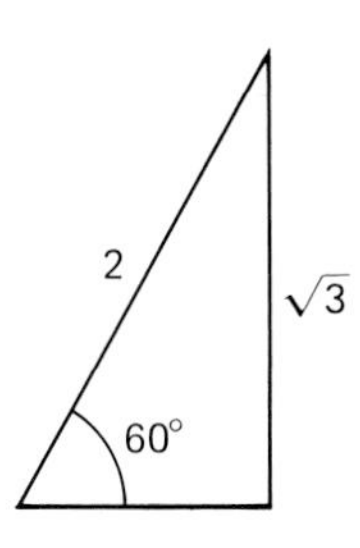

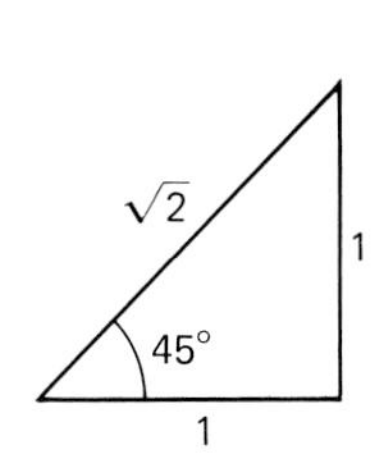

Figure B.4 Common triangles.

Table B.2 Functions of common angles

Function \ Angle	0°	30°	45°	60°	90°
sin	0	$1/2$	$\sqrt{2}/2$	$\sqrt{3}/2$	1
cos	1	$\sqrt{3}/2$	$\sqrt{2}/2$	$1/2$	0
tan	0	$\sqrt{3}/3$	1	$\sqrt{3}$	∞
cot	∞	$\sqrt{3}$	1	$\sqrt{3}/3$	0
sec	1	$2\sqrt{3}/3$	$\sqrt{2}$	2	∞
csc	∞	2	$\sqrt{2}$	$2\sqrt{3}/3$	1

B.7

Inverse Trigonometric Functions

Definition

If $y = \sin\theta$, *then* θ *is an angle whose sine is* y. The symbols ordinarily used to denote an inverse function are

$$\theta = \arcsin y$$

or

$$\theta = \sin^{-1} y$$

Note:

$$\sin^{-1} y \neq \frac{1}{\sin y}$$

This is an exception to the conventional use of exponents.

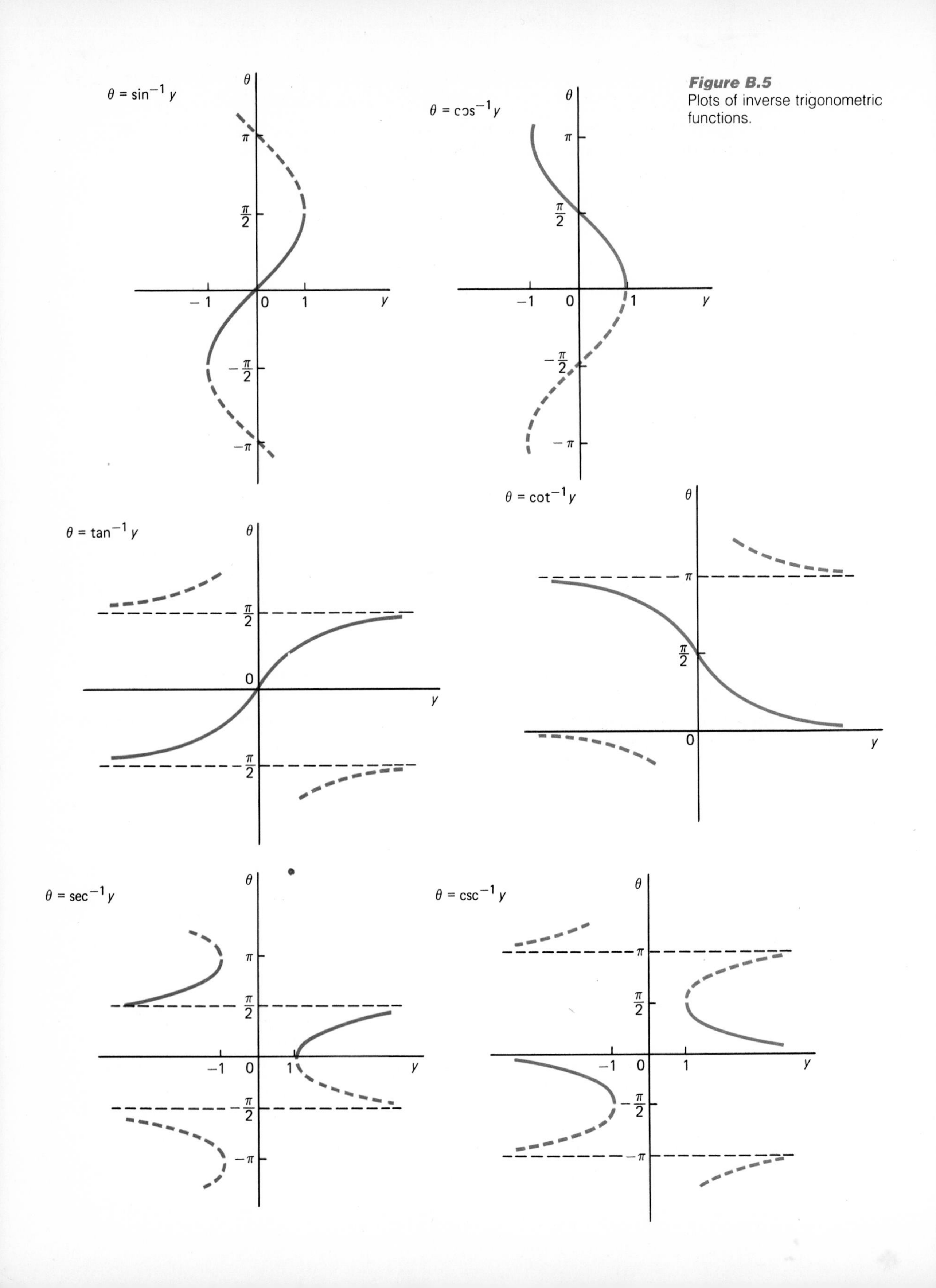

Figure B.5
Plots of inverse trigonometric functions.

Inverse functions $\cos^{-1} y$, $\tan^{-1} y$, $\cot^{-1} y$, $\sec^{-1} y$, and $\csc^{-1} y$ are similarly defined. Each of these is a many-valued function of y. The values are grouped into collections called *branches*. One of these branches is defined to be the principal branch, and the values found there are the principal values.

The principal values are as follows:

$$-\frac{\pi}{2} \leq \sin^{-1} y \leq \frac{\pi}{2}$$

$$0 \leq \cos^{-1} y \leq \pi$$

$$-\frac{\pi}{2} < \tan^{-1} y < \frac{\pi}{2}$$

$$0 < \cot^{-1} y < \pi$$

$$0 \leq \sec^{-1} y \leq \pi \qquad \left(\sec^{-1} y \neq \frac{\pi}{2}\right)$$

$$-\frac{\pi}{2} \leq \csc^{-1} y \leq \frac{\pi}{2} \qquad (\csc^{-1} y \neq 0)$$

B.8 Plots of Inverse Trigonometric Functions

All angles are given in radians. Principal branches are shown as solid lines. See Fig. B.5.

B.9 Polar-Rectangular Coordinate Conversion

See Fig. B.6.

Conversion from polar to rectangular coordinates $(r,\theta) \rightarrow (x,y)$ is given by the following equations:

$$x = r \cos \theta$$

$$y = r \sin \theta$$

Conversion from rectangular to polar coordinates $(x,y) \rightarrow (r,\theta)$ requires the following equations:

$$r = [x^2 + y^2]^{1/2}$$

$$\theta = \tan^{-1}\left(\frac{y}{x}\right)$$

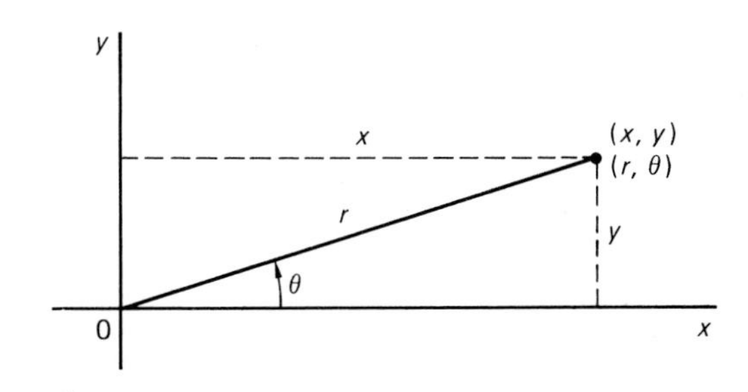

Figure B.6
Rectangular and polar coordinate definitions.

The conversion from polar to rectangular coordinates can also be thought of as the determination of the x and y components of a vector (r, θ). Likewise, conversion from rectangular to polar coordinates is the same as finding the resultant vector (r, θ) from its x and y components.

B.10 Laws of Sines and Cosines

The fundamental definitions of sine, cosine, etc., apply strictly to right triangles. Solutions needed for oblique triangles must then be accomplished by appropriate constructions that reduce the problem to a series of solutions to right triangles.

Two formulas have been derived for oblique triangles that are much more convenient to use than the construction technique. They are the law of sines and the law of cosines, which apply to any plane triangle. See Fig. B.7.

The *law of sines* states

$$\frac{\sin A}{a} = \frac{\sin B}{b} = \frac{\sin C}{c}$$

The *law of cosines* is

$$a^2 = b^2 + c^2 - 2bc \cos A$$

or

$$b^2 = a^2 + c^2 - 2ac \cos B$$

or

$$c^2 = a^2 + b^2 - 2ab \cos C$$

Application of the law of sines is most convenient in the case where two angles and one side are known and a second side is to be found.

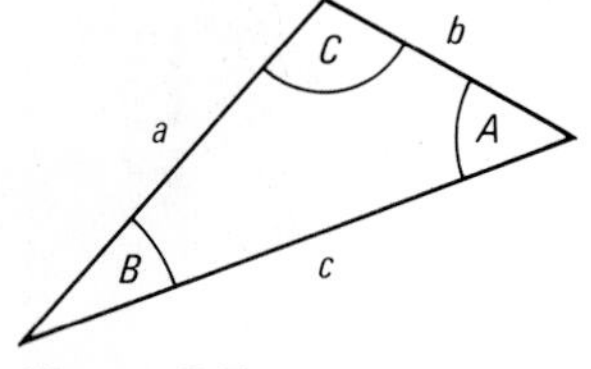

Figure B.7
Angle and side designations.

Example problem B.1 Determine the length of side x for the triangle with base 12.0 m as shown in Fig. B.8.

Solution The sum of the interior angles must be 180°; therefore, $\alpha = 180° - 43° - 59° = 78°$.

Applying the law of sines,

$$\frac{\sin 78°}{12.0 \text{ m}} = \frac{\sin 43°}{x}$$

$$x = \frac{\sin 43°}{\sin 78°} 12.0 \text{ m}$$

$$= 8.37 \text{ m}$$

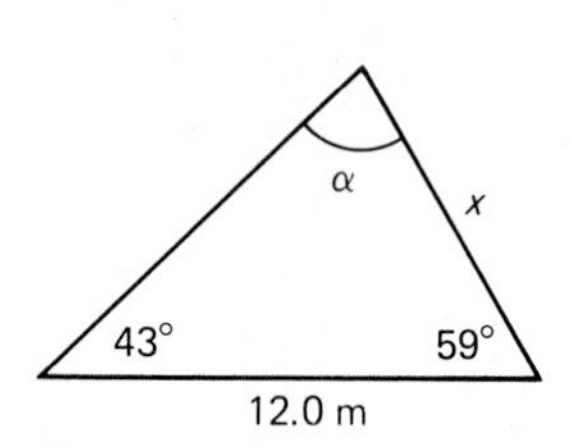

Figure B.8

The law of cosines is most convenient to use when two sides and the included angle are known for a triangle.

Example problem B.2 Calculate the length of side y of the triangle shown in Fig. B.9. Its base is 14.0 m.

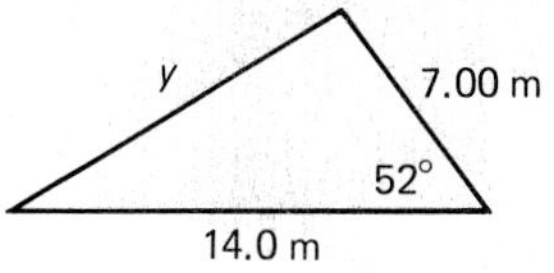

Figure B.9

Solution Substitute into the law of cosines:

$$y^2 = (14.0\text{ m})^2 + (7.00\text{ m})^2 - 2(14.0\text{ m})(7.00\text{ m})(\cos 52°)$$

$$= 124.33\text{ m}^2$$

$$y = 11.2\text{ m}$$

B.11 Area of a Triangle

Formulas for the area of a triangle (see Fig. B.7) in terms of two sides and their included angle are

$$\text{Area} = \frac{1}{2} ab \sin C$$

$$\text{Area} = \frac{1}{2} ac \sin B$$

$$\text{Area} = \frac{1}{2} bc \sin A$$

Formulas written in terms of one side and three angles are

$$\text{Area} = \frac{1}{2} a^2 \frac{\sin B \sin C}{\sin A}$$

$$\text{Area} = \frac{1}{2} b^2 \frac{\sin A \sin C}{\sin B}$$

$$\text{Area} = \frac{1}{2} c^2 \frac{\sin A \sin B}{\sin C}$$

The formula for the area in terms of the sides is

$$\text{Area} = [s(s - a)(s - b)(s - c)]^{1/2}$$

where $s = \frac{1}{2}(a + b + c)$.

Example problem B.3 Determine the areas of the two triangles defined in Sec. B.10.

Solution For the 12.0 m base triangle,

$$\text{Area} = \frac{1}{2}(12.0\text{ m})^2 \frac{\sin 59° \sin 43°}{\sin 78°}$$

$$= 43.0\text{ m}^2$$

For the 14.0 m base triangle,

$$\text{Area} = \frac{1}{2}(7.00\text{ m})(14.0\text{ m})\sin 52°$$

$$= 38.6\text{ m}^2$$

B.12

Series Representation of Trigonometric Functions

Infinite-series representations exist for each of the trigonometric functions. Those for sine, cosine, and tangent are

$$\sin\theta = \theta - \frac{\theta^3}{3!} + \frac{\theta^5}{5!} + \cdots + (-1)^{n-1}\frac{\theta^{2n-1}}{(2n-1)!} + \cdots$$

$$\theta \text{ in radians} \qquad (-\infty < \theta < +\infty)$$

$$\cos\theta = 1 - \frac{\theta^2}{2!} + \frac{\theta^4}{4!} + \cdots + (-1)^{n-1}\frac{\theta^{2n-2}}{(2n-2)!} + \cdots$$

$$\theta \text{ in radians} \qquad (-\infty < \theta < +\infty)$$

$$\tan\theta = \theta + \frac{\theta^3}{3} + \frac{2\theta^5}{15} + \cdots + \frac{2^{2n}(2^{2n}-1)\,B_n\theta^{2n-1}}{(2n)!} + \cdots$$

$$\theta \text{ in radians} \qquad \left(-\frac{\pi}{2} < \theta < \frac{\pi}{2}\right)$$

where B_n are the Bernoulli numbers

$$B_1 = \frac{1}{6}$$

$$B_2 = \frac{1}{30}$$

$$B_3 = \frac{1}{42}$$

$$B_4 = \frac{1}{30}$$

$$B_5 = \frac{5}{66}$$

$$B_n = \frac{(2n)!}{2^{2n-1}(\pi)^{2n}}\left(1 + \frac{1}{2^{2n}} + \frac{1}{3^{2n}} + \cdots\right)$$

and where the factorial symbol (!) is defined as

$$n! = n(n-1)(n-2)\cdots(3)(2)(1)$$

Trigonometric functions to any accuracy can be calculated from the series if enough terms are used.

For small angles, on the order of 5° or less, it may be sufficient to use only the first term of each series.

$$\sin\theta \cong \theta$$

$$\cos\theta \cong 1$$

$$\tan\theta \cong \theta$$

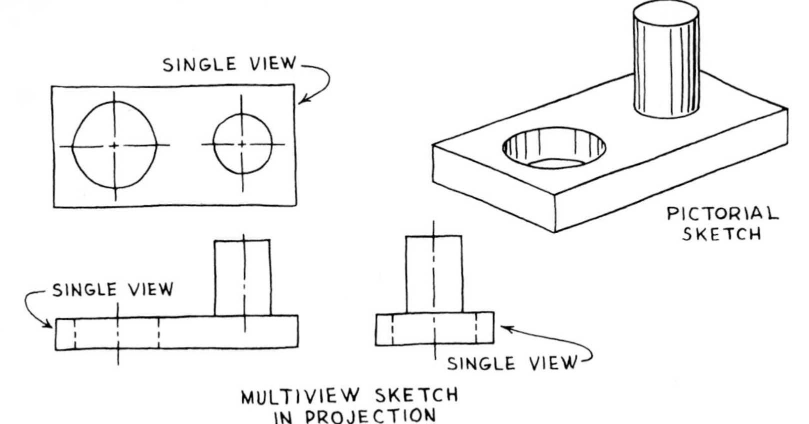

Figure C.2
Freehand drawings.

C.3.2 Types of Freehand Drawing

Three types of freehand drawings will be considered, as illustrated in Fig. C.2.

1. Single view: Objects with primarily two dimensions: that is, maps, charts, diagrams, graphs, etc. Included in this definition are single orthographic views or one view of multiview drawings.
2. Pictorial: Objects with three dimensions illustrated: that is, length, width, and height. It is an attempt to show in a single drawing what the eye would see.
3. Multiview: Separate, single orthographic views oriented in adjacent related positions to describe an object completely. Multiview drawings are discussed in Sec. C.4.

C.3.3 Construction of Single Views

The construction of single views primarily requires the ability to draw parallel lines and circles.

Circles should be sketched without construction lines only if the sketch is very rough or if the circles are small. See Fig. C.3.

A helpful construction technique is illustrated in Fig. C.4. The center of the circle is at point 0, so an arc can be drawn from A through E to B as follows. Construct a diagonal from A to B and then divide the line DC again. This will locate point E, which is very close to the exact location for the circular arc.

Figure C.3
Construction of a circle.

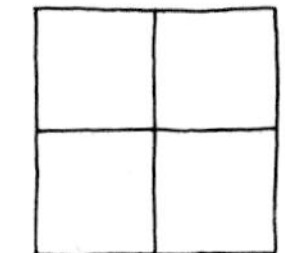
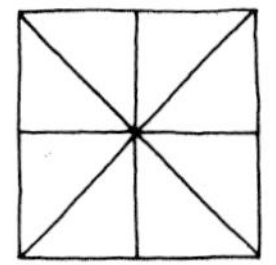
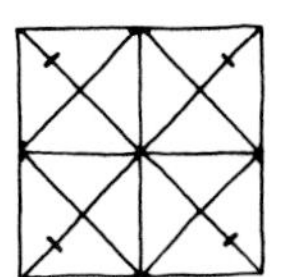
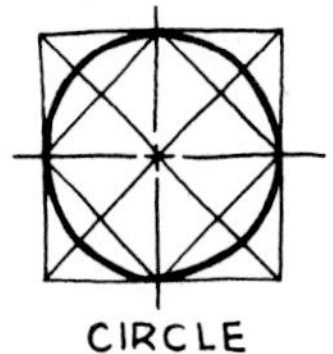

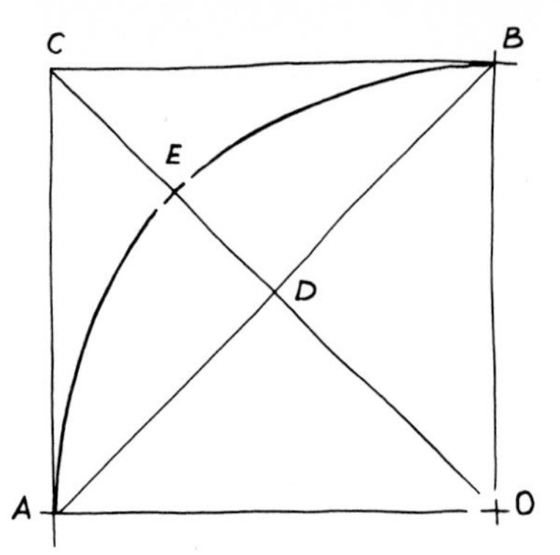

Figure C.4
Circle construction technique.

C.3.4
Construction of Pictorials

The correct selection of the axes is critical when constructing a pictorial. If all faces require equal emphasis, an isometric selection would be appropriate, as seen in Fig. C.5*a*, whereas oblique would better illustrate major features in one or parallel planes (Fig. C.5*b*).

A key element in the construction of pictorial sketches is the ellipse. Circles in the major plane of an oblique drawing are circles, but in the receding planes of most pictorials they are elliptical. The construction of an ellipse is illustrated in Fig. C.6.

Note the construction of step 4 in Fig. C.6. This procedure is identical to the construction technique developed in Fig. C.4. Although it is not precisely correct mathematically, it is helpful when sketching an ellipse.

Figure C.5
Axes selections.

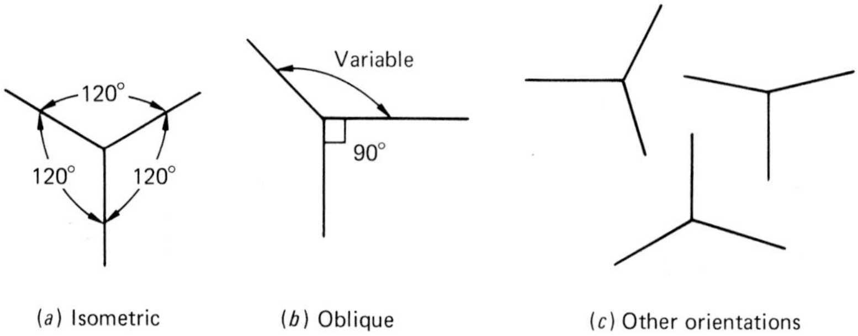

120° 120° 120° Variable 90°

(*a*) Isometric (*b*) Oblique (*c*) Other orientations

Figure C.6
Ellipse construction.

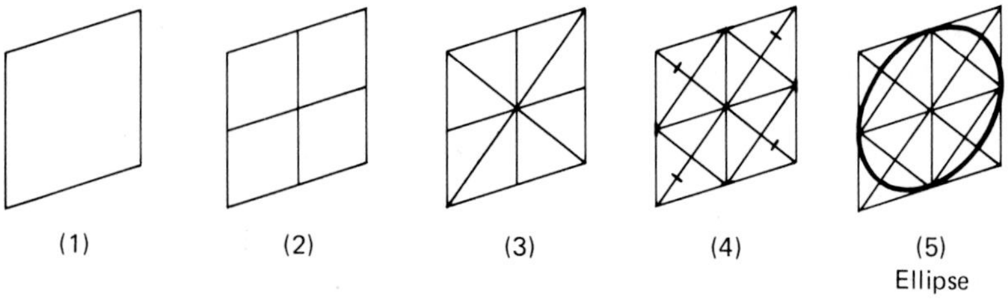

(1) (2) (3) (4) (5) Ellipse

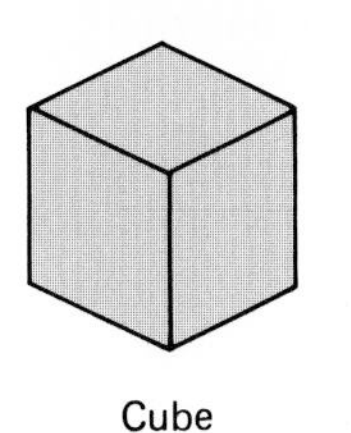

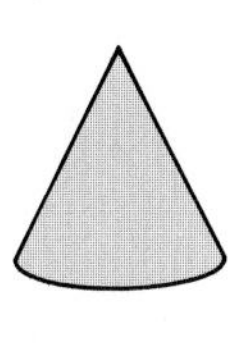

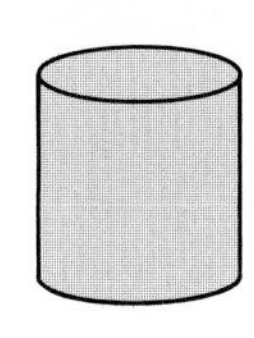

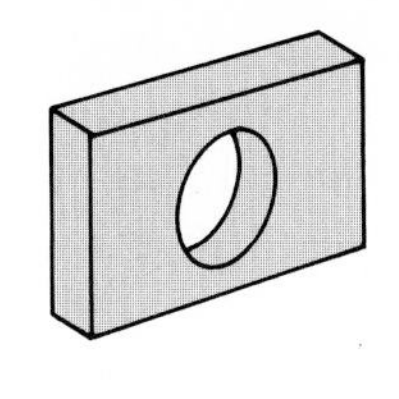

Figure C.7
Basic shapes.

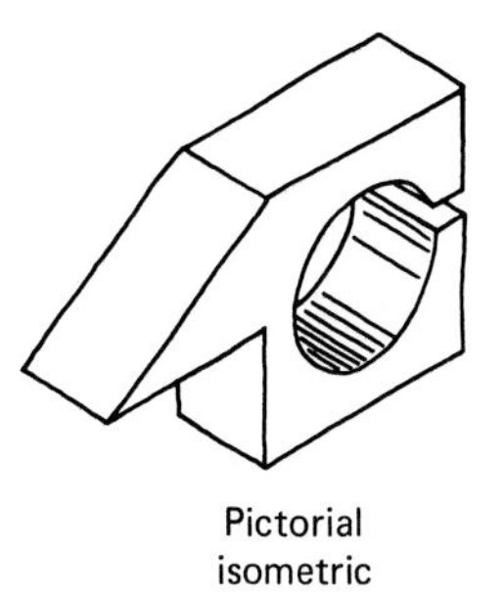

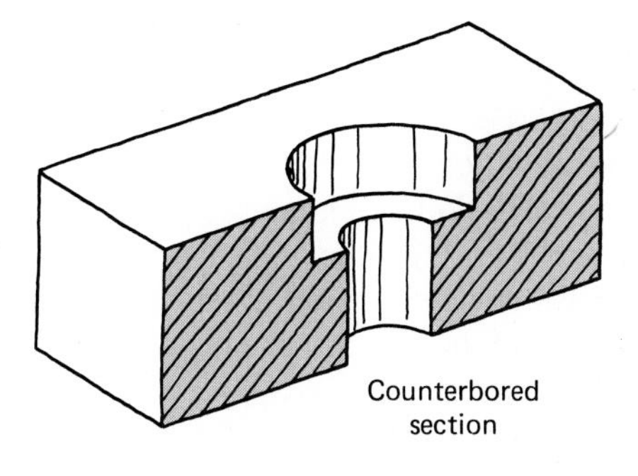

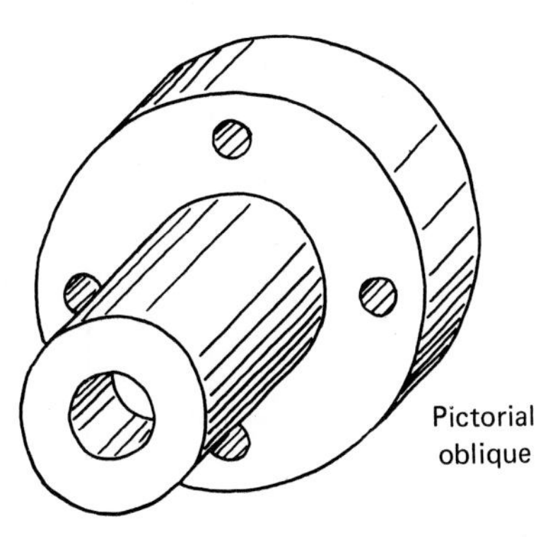

Figure C.8
Pictorials.

The construction of many pictorials is a simple adaptation of the four basic shapes illustrated in Fig. C.7.

Practical application of the foregoing construction principles and six steps of sketching are demonstrated in the examples of Fig. C.8.

C.4 Multiview Drawings

The precise definition and delineation of objects can best be represented by a series of carefully selected single views. Although pictorials are an excellent method of conveying a visual image of the object, they do not provide the detail needed for manufacturing.

Multiview drawings define an object by placing the correct number of properly constructed single views in correct orthographic alignment. Figure C.9 illustrates a simple object showing four of its six principal views, two of which are not essential for complete description of the object but are shown only to clarify various correct orthographic positions.

The number of views necessary to describe any object obviously varies with the complexity of the object, but constructing views that are not needed is a waste of time and money.

Objects with more detail may require more careful study because all contours must be represented in the correct construction of multiviews. Figure C.10*a* illustrates the use of centerlines; and Fig. C.10*b* shows proper use of both hidden lines and centerlines. Figure C.11 illustrates object lines, hidden lines, and centerlines on a single object.

The correct or necessary number of views that adequately describe an object is illustrated in Fig. C.12. The top and front views, in this case, do not completely describe a single object, but the five different right profile views, each taken independently with the given front view, should adequately describe five different objects.

An optimum number of properly constructed views is the way to communicate objects graphically without confusion and misunderstanding.

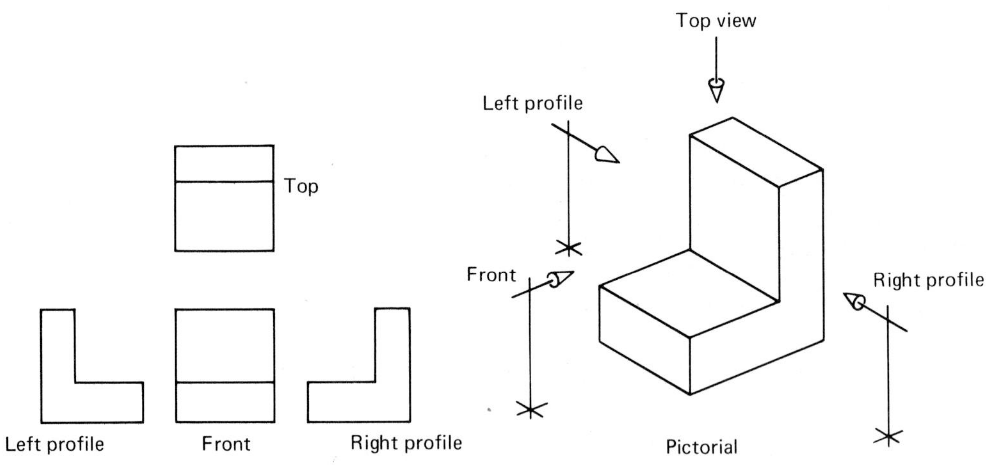

Figure C.9
Multiview drawings.

Figure C.10
Centerlines and hidden lines.

(*a*) Cylinder

(*b*) Negative cylinder

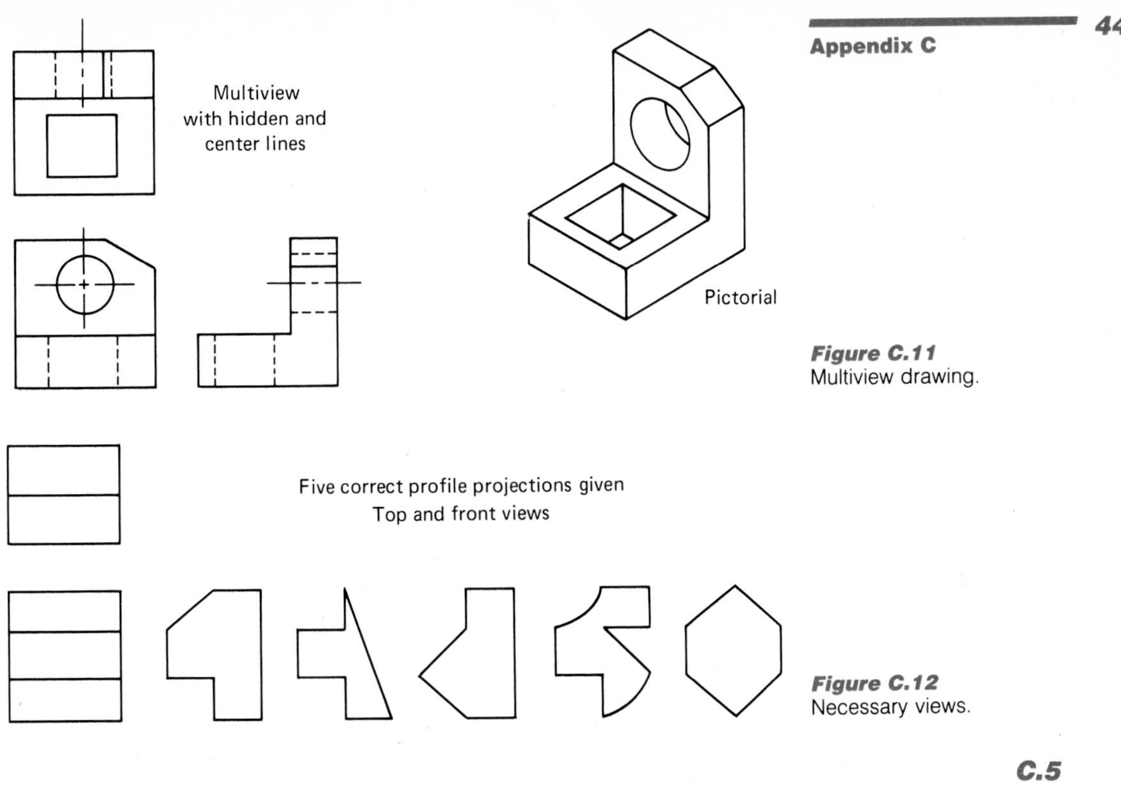

Figure C.11
Multiview drawing.

Figure C.12
Necessary views.

C.5 Scales

Graduations on metric scales are identified as a ratio: 1:100, 1:20, etc. This ratio signifies the drawing reduction from actual size; for example, 1:100 indicates that 1 unit on the drawing represents 100 units on the object.

A metric scale labeled with a ratio of 1:100 signifies that the distance from 0 to the number 1 is to represent a meter, as illustrated in Fig. C.13.

If you change the ratio from 1:100 to 1:10 but use the same scale,

Table C.1

Ratio	Distance from 0 to 1.0 is equal to
1:100	1 m = 10 dm = 100 cm = 1 000 mm
1:10	0.1 m = 1 dm = 10 cm = 100 mm
1:1	0.01 m = 0.1 dm = 1 cm = 10 mm
1:0.1	0.001 m = 0.01 dm = 0.1 cm = 1 mm

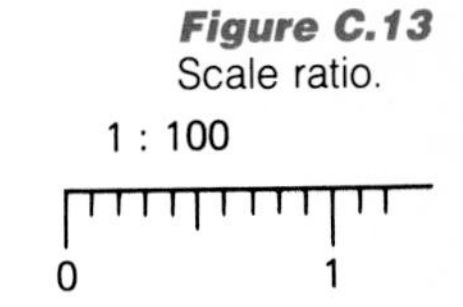

Figure C.13
Scale ratio.

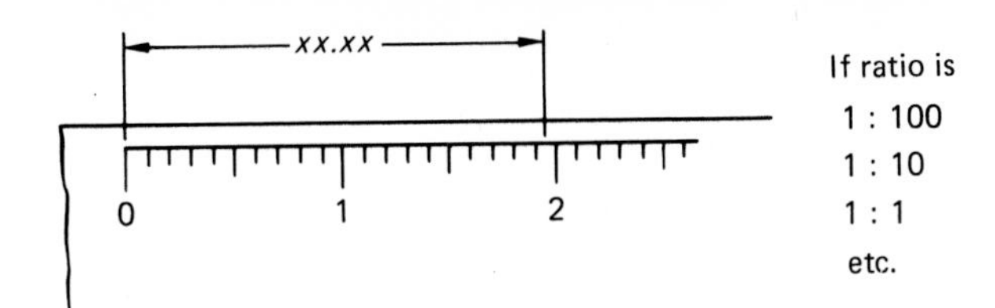

If ratio is	Scale reading is
1 : 100	1.95 m, 19.5 dm
1 : 10	0.195 m, 1.95 dm
1 : 1	1.95 cm, 19.5 mm
etc.	

Figure C.14
Relation between ratio and scale reading.

the distance from the 0 to the 1 now represents 0.1 m. Table C.1 demonstrates this functional characteristic of metric scales.

Table C.1 can be constructed for any metric scale or ratio. Figure C.14 illustrates how the scale varies with different ratios.

C.6 Dimensions

In order that objects might be perceived in terms of precise physical measurement, dimensions must be correctly indicated. The size of each geometric shape, together with necessary location dimensions, has to be clearly understood.

Many standards that provide considerable uniformity to dimensioning practice have been recognized and adopted. One of the most widely used is that issued by the American National Standards Institute.

C.6.1 Definition of Terms

Dimensions are the numbers expressed in consistent units and used to indicate the physical lengths.

Dimension lines and arrowheads indicate the extent of the measurement. Arrowheads should be closed and consistently uniform in size and shape. Dimension lines are placed about 10 mm from the object and 6 mm from each other, as illustrated in Fig. C.15.

Extension lines are used to locate the extension of the surface, and leaders are used to dimension circles or to direct notes to a specific place.

Geometric shapes

A prism, cylinder, cone, and right pyramid are correctly dimensioned in Fig. C.16.

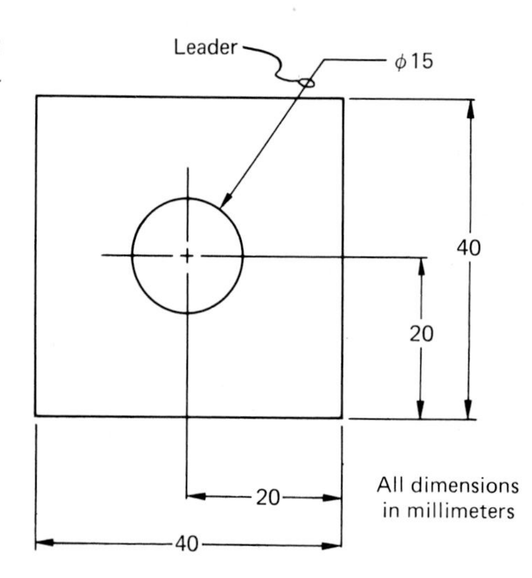

Figure C.15 Proper spacing.

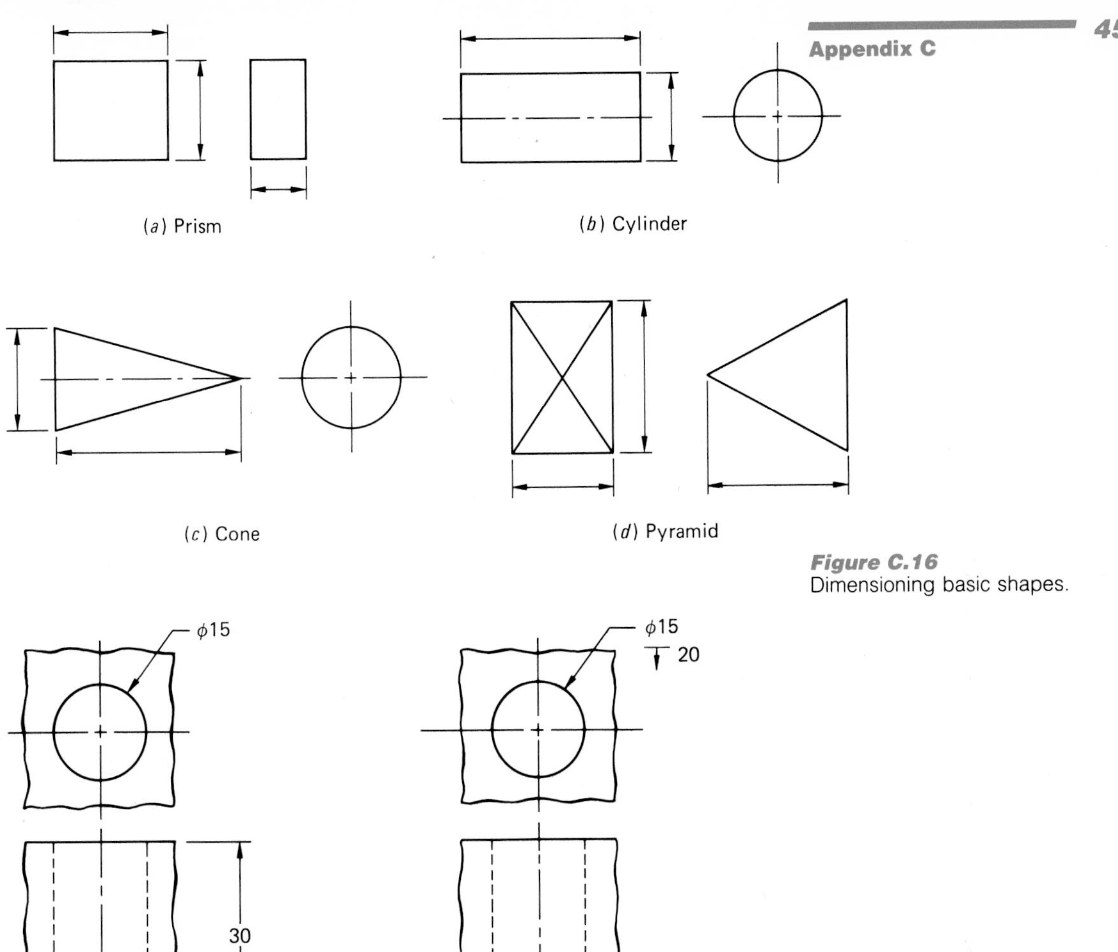

Figure C.16
Dimensioning basic shapes.

Figure C.17
Negative cylinders.

Holes, radii, and arcs

The illustrations in Fig. C.17 demonstrate correct dimensioning procedures for *through* holes and *blind* holes.

Circular arcs are dimensioned as shown in Fig. C.18. For castings with a large number of fillets and rounds, it is customary to indicate all radii by use of a general note: for example, "all radii 8 mm unless otherwise specified."

Counterbored and countersunk holes (see Fig. C.19)

One of the most important guides to follow when dimensioning is *always* to *place the dimension in the view that shows the most characteristic feature*. Select a scale for the drawing, and use a minimum but sufficient number of views to completely describe and accurately dimension the object. Generally, three overall dimensions are included on all parts.

Figure C.20 illustrates three correctly dimensioned views of an object.

C.6.2 Recommended Dimensioning Practices

1. Never duplicate dimensions.
2. Don't crowd dimensions.
3. Place dimensions *between* views, not *on* views.
4. Avoid crossing dimension lines.
5. Break dimension lines at numerals.
6. Don't dimension to hidden lines.
7. Arrange numerals and notes to be read from bottom of sheet.
8. Use standard height for letters and numerals.
9. Provide dimensions so that calculations are not necessary.
10. Measure lengths in metric units.
11. Don't allow extension lines to touch object.
12. Use centerlines as extension lines when appropriate.

C.7 Sections

At times, visualization of an object that has considerable interior complexity can be enhanced by taking an appropriate section view. An imaginary plane, as illustrated in Fig. C.21, is cut through the

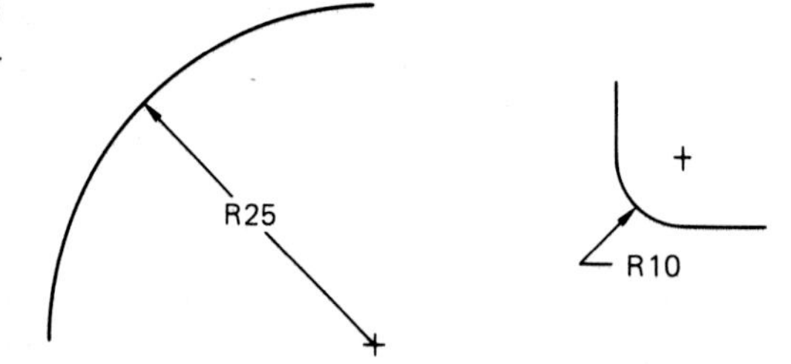

Figure C.18
Radii.

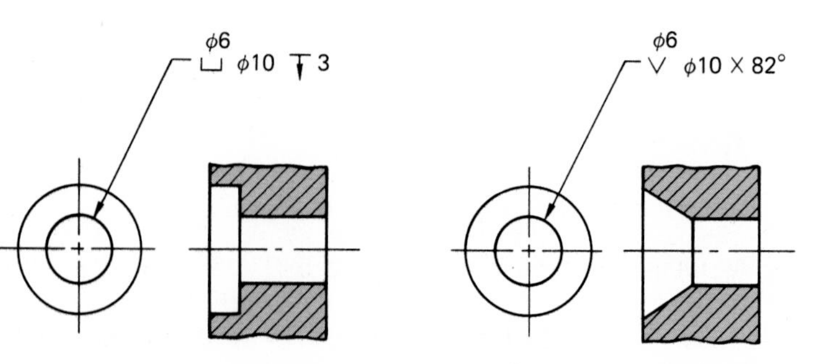

Figure C.19
Counterbore and countersink.

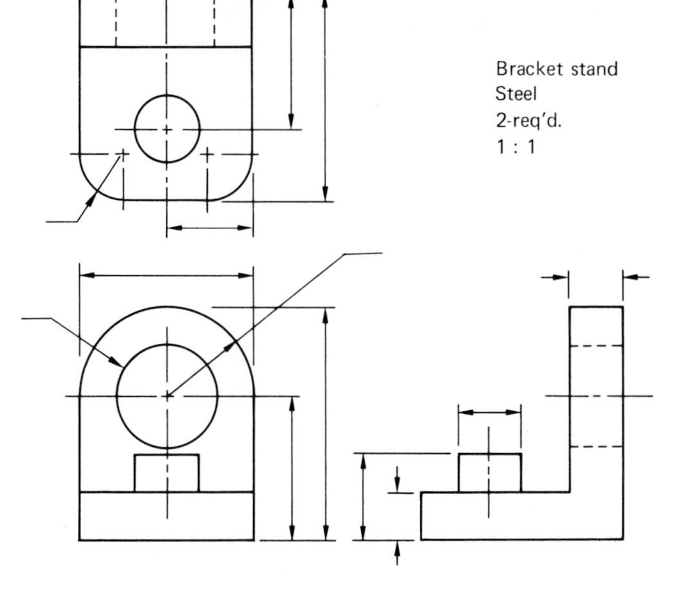

Figure C.20
Detail drawing with complete title.

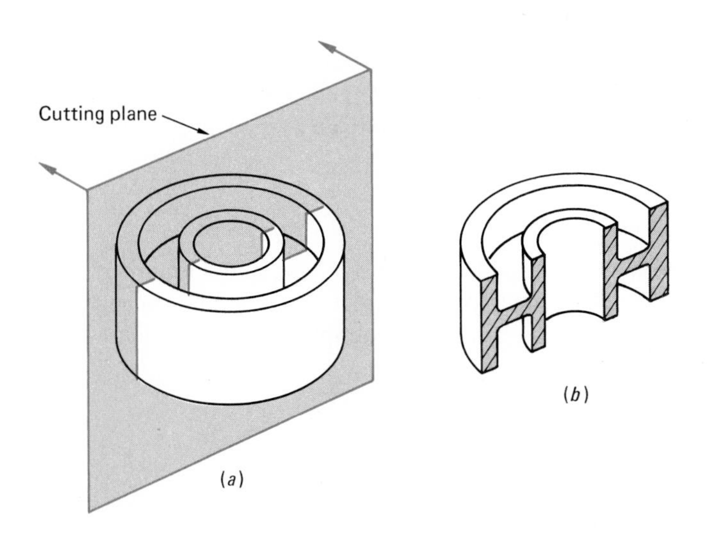

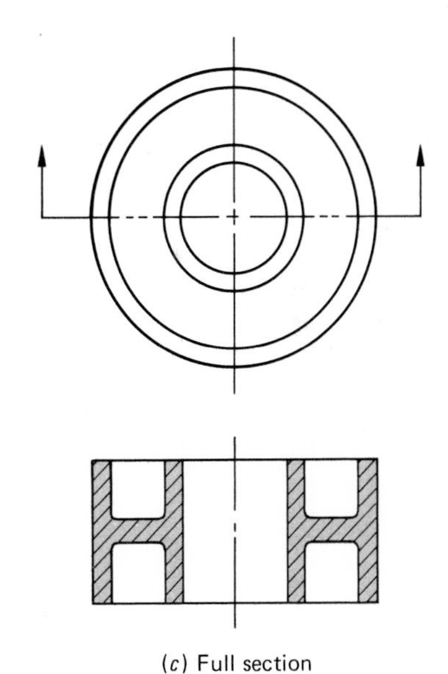

Figure C.21
Cutting plane.

object and the near portion is taken away, thereby exposing the interior (Fig. C.21*b*).

Solids that are cut by the imaginary plane are cross-hatched according to preestablished standards. Three of these are illustrated in Fig. C.22.

Six different types of section views are defined and illustrated here.

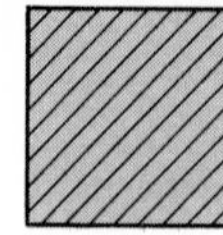

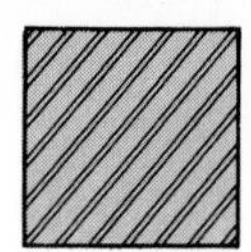

Wood

Figure C.22 Section lining.

Full section

A full section results when the cutting plane passes completely through the object, as illustrated in Fig. C.21*c*. Hidden lines behind the cutting plane are normally omitted unless essential.

Half sections

Half sections can be used on symmetrical objects when it is desirable to show an internal detail as well as a view of the exterior. Note that separation of the inside and outside is by a centerline, as seen in Fig. C.23.

Offset section

An offset plane goes completely through the object but is staggered rather than straight. A convention normally practiced allows the lines of demarcation (change in direction of cutting plane) to be omitted in the sectioned view. The offset section is illustrated in Fig. C.24.

Revolved section

Revolved sections are used to show the shape and contour of ribs, spokes, etc. The plane is passed through the object and the cut area is revolved 90° and cross-hatched. Examples of this section can be seen in Fig. C.25*a*, *b*, and *c*.

Figure C.23 Half-section.

Half-section

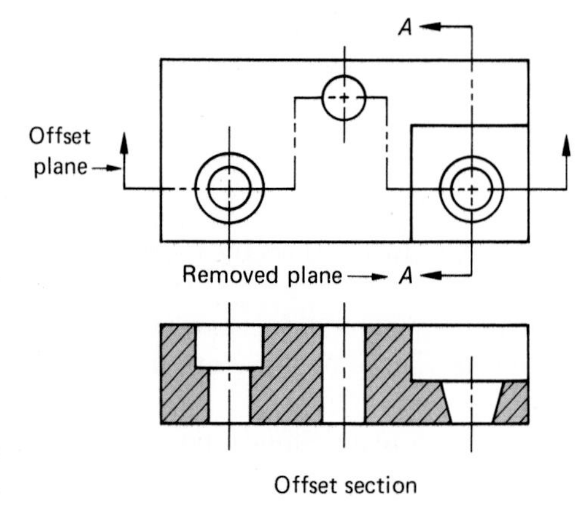

Figure C.24 Offset section.

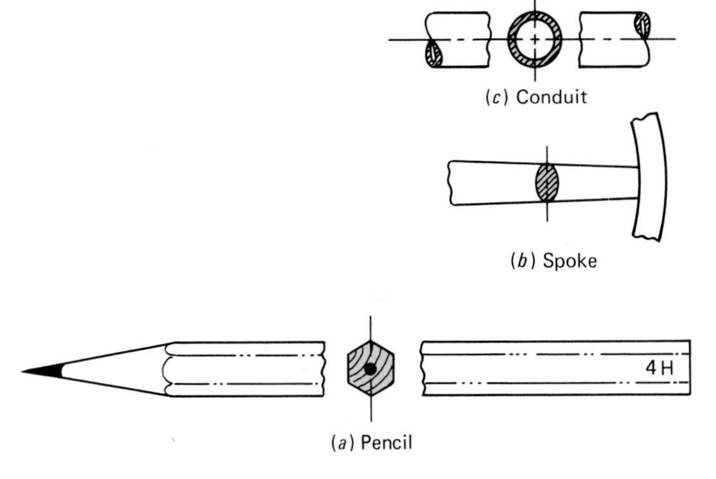

Figure C.25
Revolved section.

Removed section

There are two significant advantages to a removed section. First, it can be moved to a separate location. A second advantage is that the scale can be changed, e.g., drawn at a larger scale. The cutting plane for the removed section illustrated in Fig. C.26 is found in Fig. C.24.

Figure C.26
Removed section.

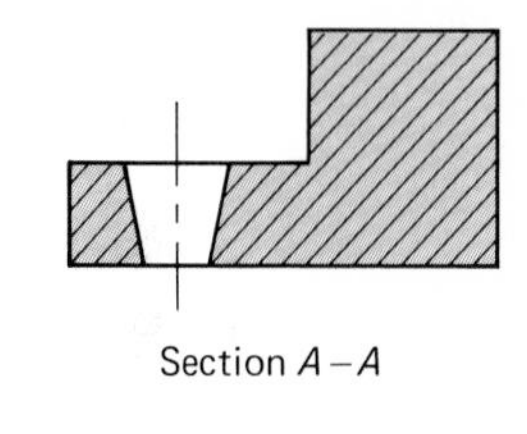

Partial section

A partial section is used when a portion of an orthographic view or pictorial is broken away to expose the interior (see Fig. C.27). The cross-hatching indicates the type of material from which the object is constructed.

C.8 Design Drawings

Although each of the previous sections can be used at different times in the design process, they must be tied together to describe a total system more completely. It is the purpose of this section to outline some of the graphics necessary for a design to be delineated.

C.8.1 Layout Drawings

A layout is a very accurate, scaled instrument drawing used to determine operational characteristics, such as clearances, and the relation of one part to another.

Figure C.29 is a pictorial of an assembly, whereas Fig. C.28 is a layout to determine critical clearances at points 1 and 2. Because of the drawing time involved, the layout drawing shows a minimum of information. Centerlines and key features are used to verify operation.

Figure C.27
Partial section.

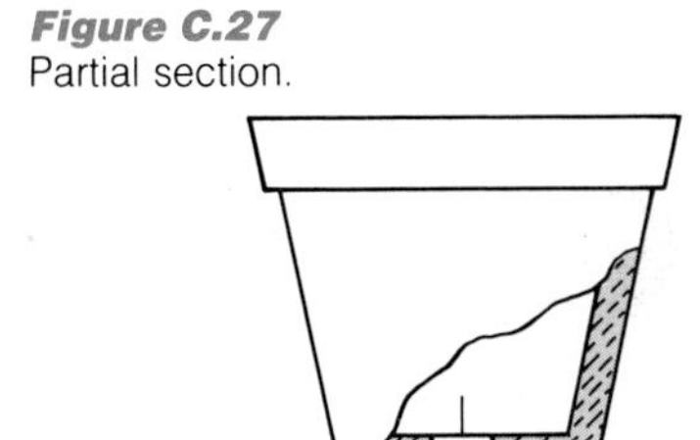

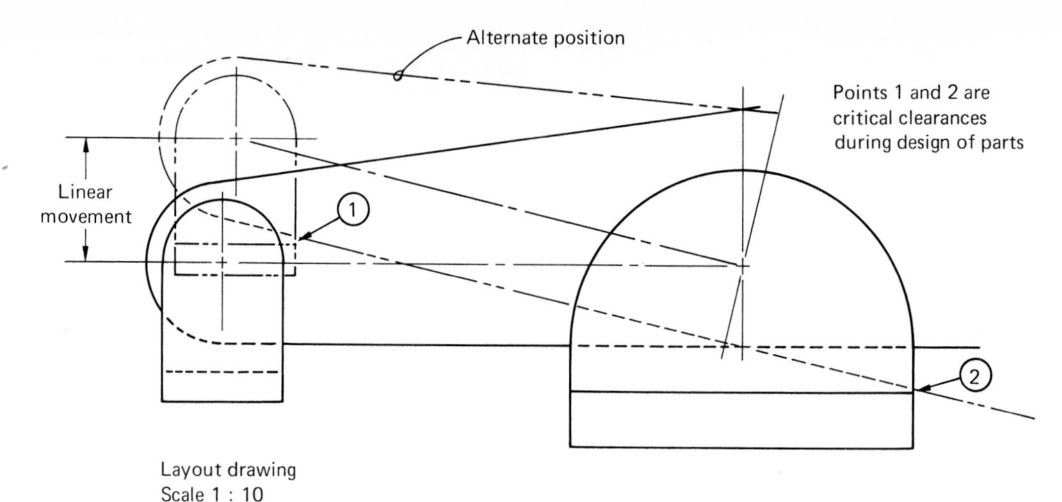

Figure C.28
Layout drawing.

C.8.2 Assembly Drawings

Assembly drawings illustrate, either in pictorial or orthographic form, the individual parts assembled. Figures C.29 and C.30 are both examples of assembly drawings. This type of drawing has several important functions. It demonstrates how the entire collection of individual items fit together. Critical dimensions, centerline dimensions, and overall dimensions are normally included to specify clearances, etc. The drawing provides an opportunity to investigate the relationship of individual parts as they move or rotate to alternate positions.

Figure C.29
Pictorial assembly.

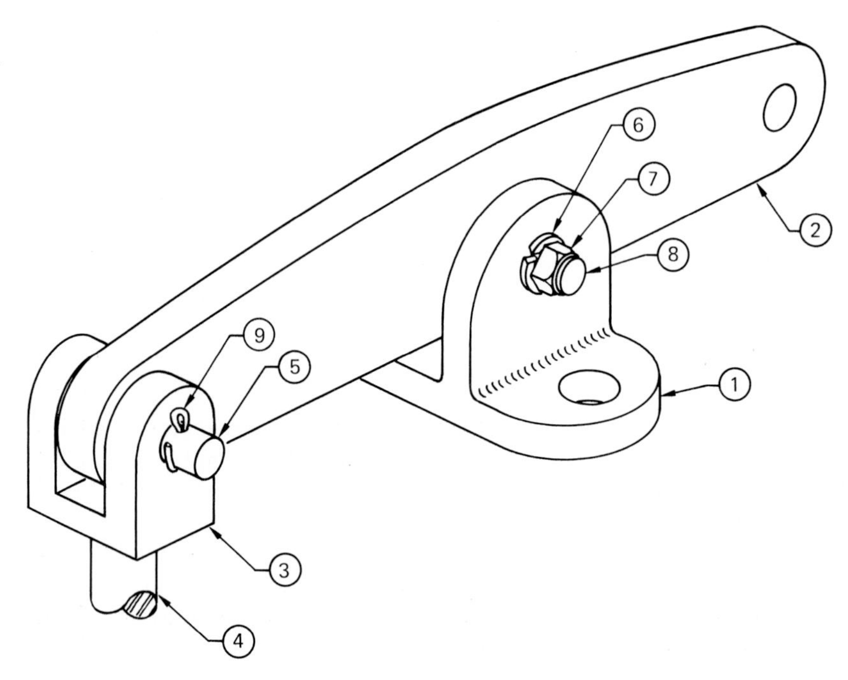

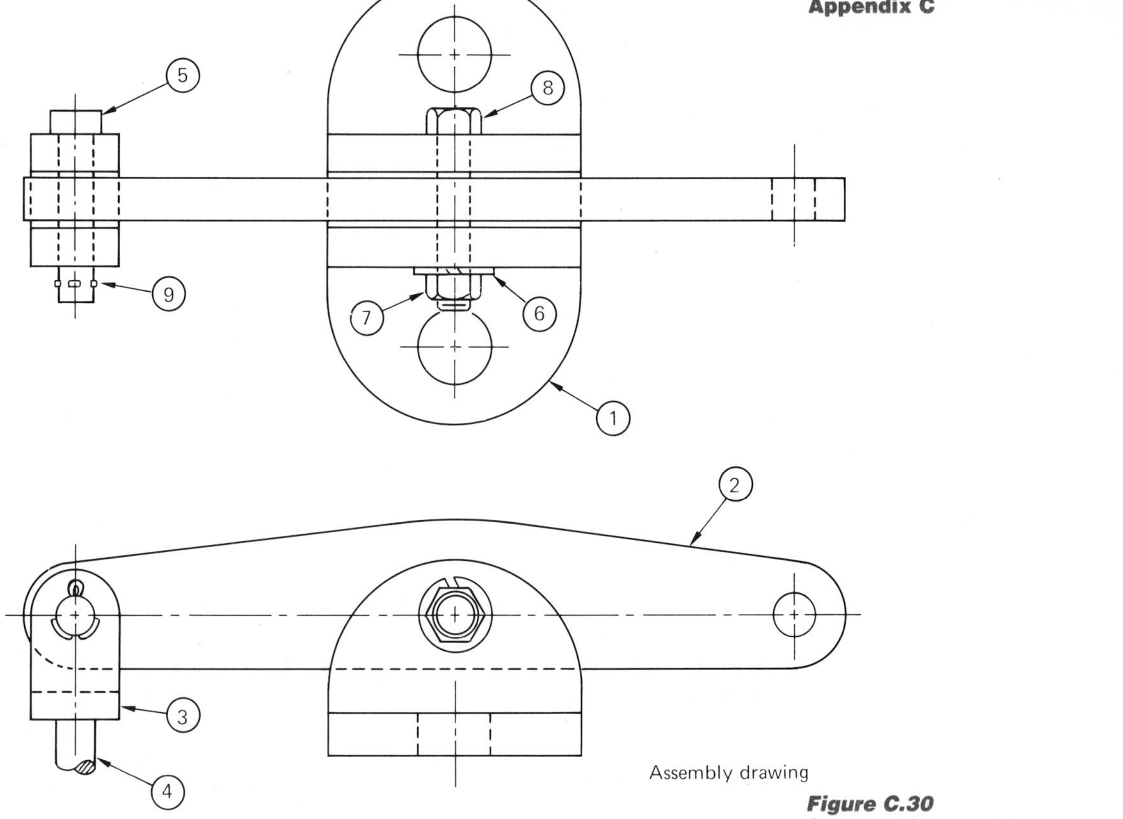

Figure C.30
Orthographic assembly.

Balloons (numbers within circles) and leaders pointing to the individual parts establish an identification system for the assembly of parts. Detail drawings (Sec. C.8.3) are keyed to the assembly by use of these identification numbers. The bill of material (Sec. C.8.4) also uses the same identification system.

C.8.3
Detail Drawings

A detail drawing, as illustrated in Fig. C.31, is a drawing of a *single* part completely specified. By working with the assembly and layout drawings the detail can be completed so that the individual part can be made. Standard parts, i.e., nuts, bolts, etc., are not customarily detailed.

A detail drawing will consist of sufficient views, completely dimensioned, with appropriate section views, and a title that includes the identification number, name, number required in assembly, material, scale, and necessary notes.

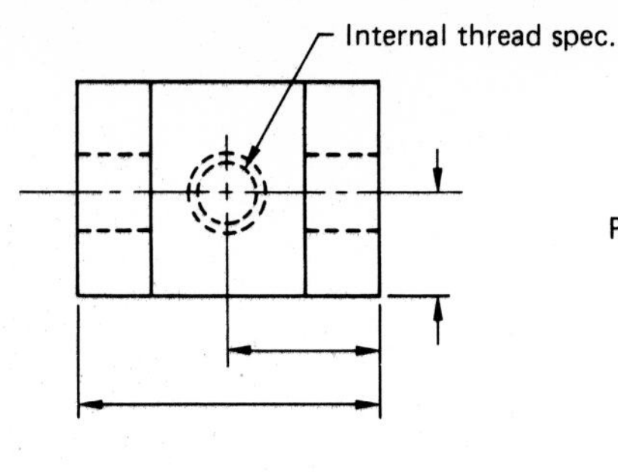

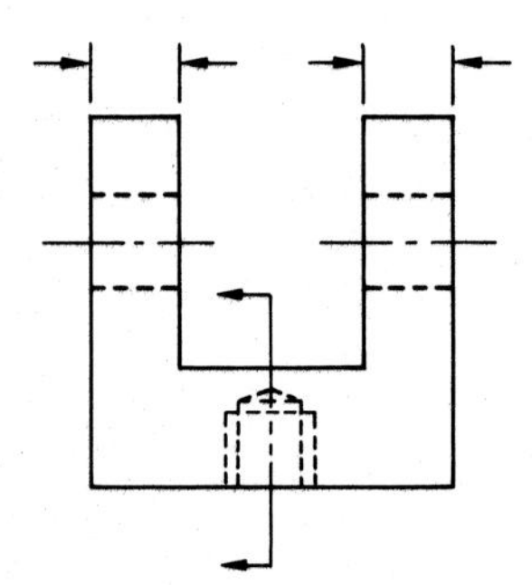

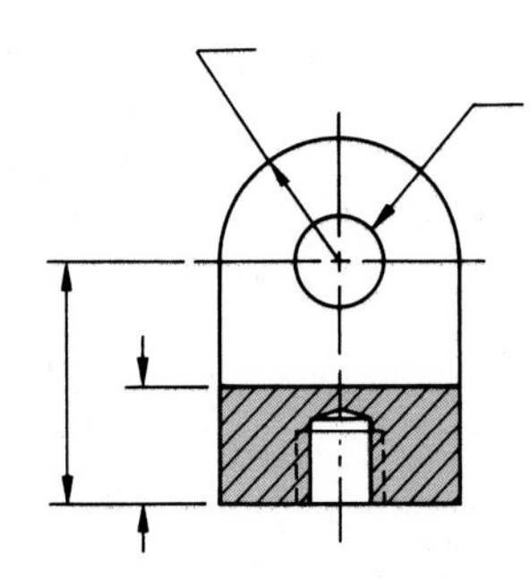

Figure C.31
Detail drawing.

C.8.4
Bill of Material

Every item in the assembly should appear in the bill of materials (see Table C.2). Materials, sizes, notes, etc., are added as the assembly is formalized.

C.9
Presentation Drawings

It is often necessary for the engineer to present data to people who are not familiar with technical graphs. A few of the many methods available to do this are included below.

Pie diagrams

Pie diagrams are most popular when representing items that total 100 percent. All lettering and percentages should be placed on the

Table C.2

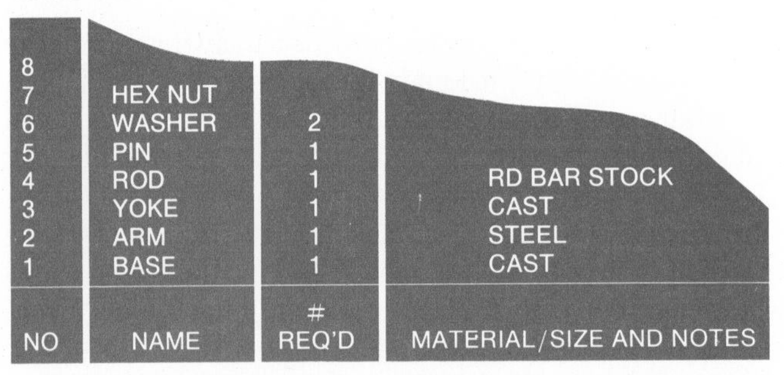

NO	NAME	# REQ'D	MATERIAL/SIZE AND NOTES
8			
7	HEX NUT		
6	WASHER	2	
5	PIN	1	
4	ROD	1	RD BAR STOCK
3	YOKE	1	CAST
2	ARM	1	STEEL
1	BASE	1	CAST

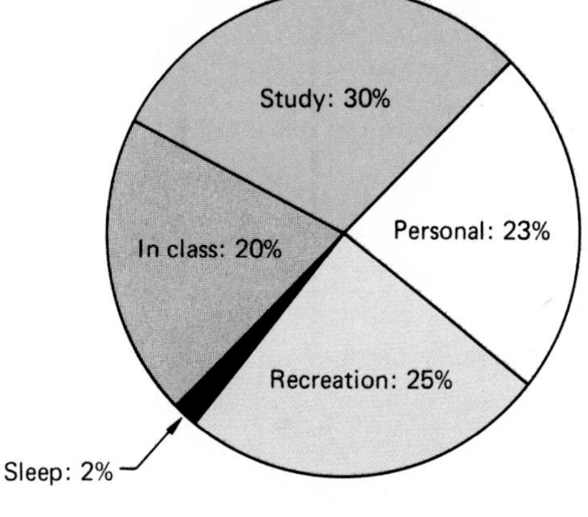

Figure C.32
Pie diagram. (Data collected at Iowa State University.)

sector or immediately adjacent. The circle should not contain more than five or six categories and each should be cross-hatched or marked differently. A title with pertinent information concerning source, etc., should always be included. See Fig. C.32.

Column charts

One of the most common nontechnical methods of representing information is the column chart, illustrated in Fig. C.33.

The quantity to be graphed is illustrated by bars, each of whose length is proportional to the value represented.

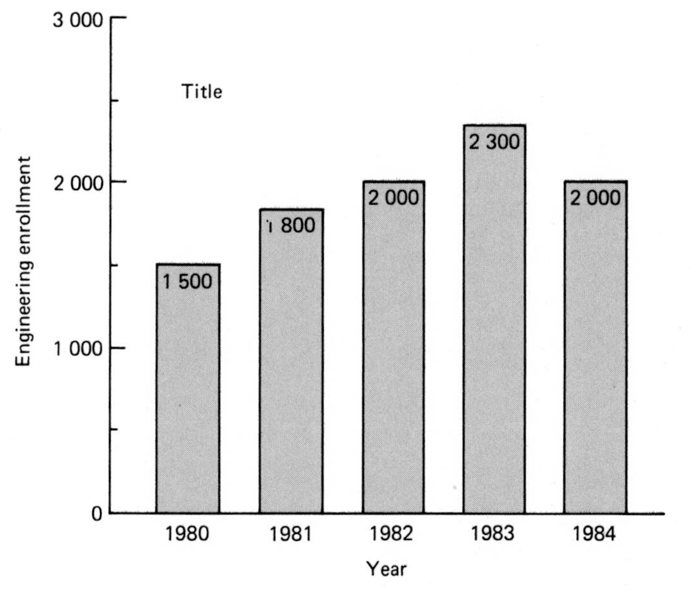

Figure C.33
Column chart.

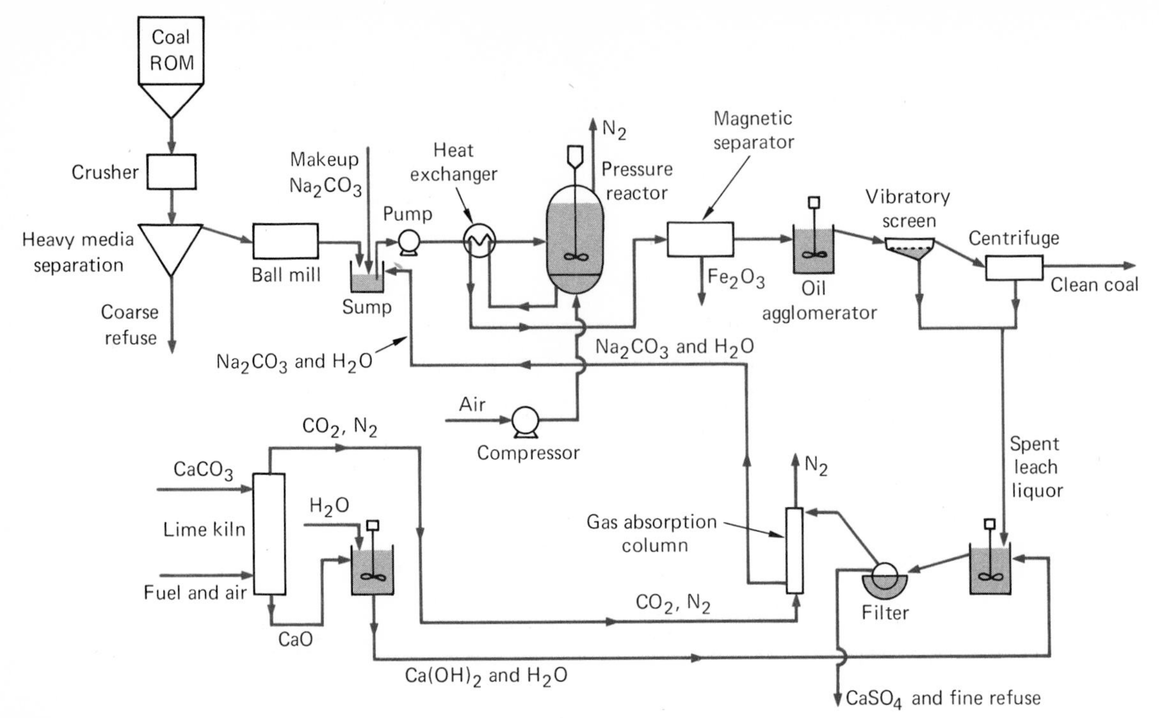

Figure C.34
Block diagram.

Block diagrams

A block diagram, as represented in Fig. C.34, is an excellent technique to show in a simple fashion the overall process. Different symbols denote certain processes with size and balance considerations important for ease of understanding.

Schematic diagrams

Combination block and schematic diagrams, as illustrated in Fig. C.35, and the schematic diagram in Fig. C.36 convey the operational characteristics of electrical as well as other complex systems. Symbols, layout, and labels provide a graphical expression of the functional relationship or interconnection of component parts.

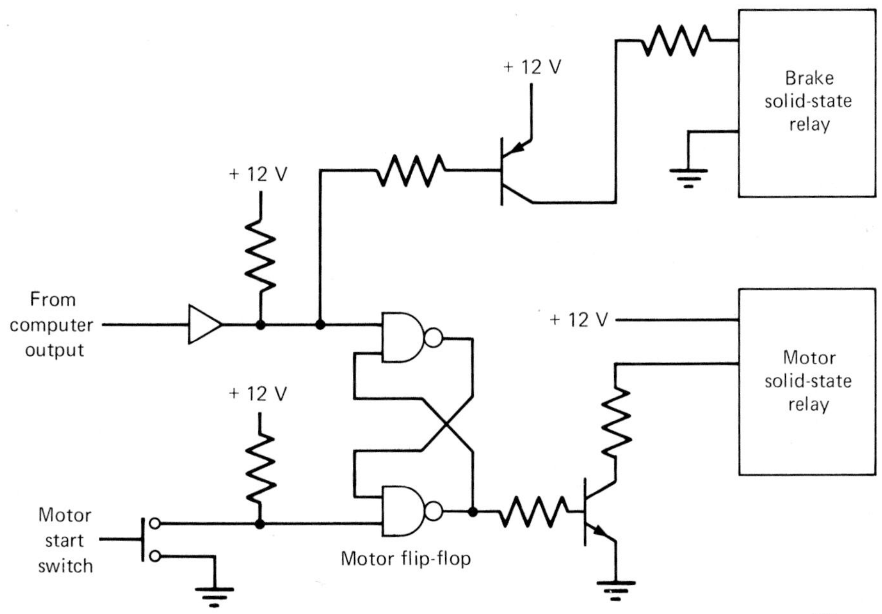

Figure C.35
Block and schematic diagram.

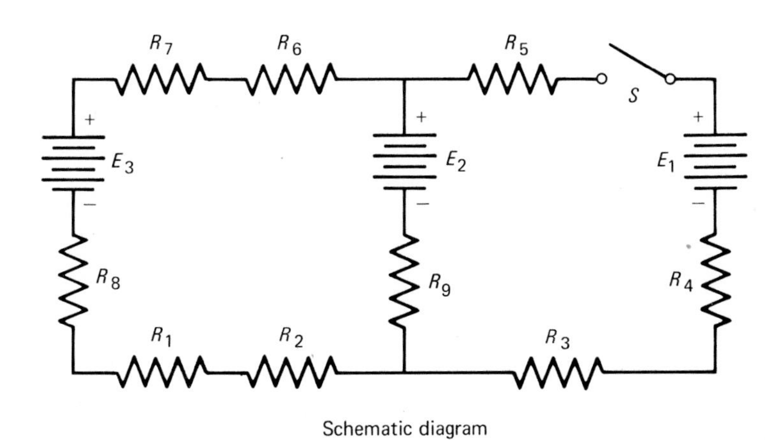

Figure C.36
Schematic diagram.

General

Approximate Specific Gravities and Densities

Gases (0°C and 1 atm)

Material	Specific gravity	Average density lbm/ft³	Average density kg/m³
Air		0.080 18	1.284
Ammonia		0.048 13	0.771 0
Carbon dioxide		0.123 4	1.977
Carbon monoxide		0.078 06	1.251
Ethane		0.084 69	1.357
Helium		0.011 14	0.178 4
Hydrogen		0.005 611	0.089 88
Methane		0.044 80	0.717 6
Nitrogen		0.078 07	1.251
Oxygen		0.089 21	1.429
Sulfur dioxide		0.182 7	2.927

Liquids (20°C)

Material	Specific gravity	Average density lbm/ft³	Average density kg/m³
Alcohol, ethyl	0.79	49	790
Alcohol, methyl	0.80	50	800
Benzene	0.88	55	880
Gasoline	0.67	42	670
Heptane	0.68	42	680
Hexane	0.66	41	660
Octane	0.71	44	710
Oil	0.88	55	880
Toluene	0.87	54	870
Water	1.00	62.4	1 000

Metals (20°C)

Material	Specific gravity	Average density lbm/ft³	Average density kg/m³
Aluminum	2.55–2.80	165	2 640
Brass, cast	8.4–8.7	535	8 570
Bronze	7.4–8.7	510	8 170
Copper, cast	8.9	555	8 900
Gold, cast	19.3	1 210	19 300
Iron, cast	7.04–7.12	440	7 050
Iron, wrought	7.6–7.9	485	7 770
Iron ore	5.2	325	5 210
Lead	11.3	705	11 300
Manganese	7.4	462	7 400
Mercury	13.6	849	13 600
Nickel	8.9	556	8 900
Silver	10.4–10.6	655	10 500
Steel, cold drawn	7.83	489	7 830
Steel, machine	7.80	487	7 800
Steel, tool	7.70	481	7 700
Tin, cast	7.30	456	7 300
Titanium	4.5	281	4 500
Uranium	18.7	1 170	18 700
Zinc, cast	6.9–7.2	440	7 050

Nonmetallic Solids (20°C)

Material	Specific gravity	Average density lbm/ft³	kg/m³
Brick, common	1.80	112	1 800
Cedar	0.35	22	350
Clay, damp	1.8–2.6	137	2 200
Coal, bituminous	1.2–1.5	84	1 350
Concrete	2.30	144	2 300
Douglas fir	0.50	31	500
Earth, loose	1.2	75	1 200
Glass, common	2.5–2.8	165	2 650
Gravel, loose	1.4–1.7	97	1 550
Gypsum	2.31	144	2 310
Limestone	2.0–2.9	153	2 450
Mahogany	0.54	34	540
Marble	2.6–2.9	172	2 750
Oak	0.64–0.87	47	750
Paper	0.7–1.2	58	925
Rubber	0.92–0.96	59	940
Salt	0.8–1.2	62	1 000
Sand, loose	1.4–1.7	97	1 550
Sugar	1.61	101	1 610
Sulfur	2.1	131	2 100

Unit Prefixes

Multiple and submultiple	Prefix	Symbol
1 000 000 000 000 = 10^{12}	tera	T
1 000 000 000 = 10^{9}	giga	G
1 000 000 = 10^{6}	mega	M
1 000 = 10^{3}	kilo	k
100 = 10^{2}	hecto	h
10 = 10	deka	da
0.1 = 10^{-1}	deci	d
0.01 = 10^{-2}	centi	c
0.001 = 10^{-3}	milli	m
0.000 001 = 10^{-6}	micro	μ
0.000 000 001 = 10^{-9}	nano	n
0.000 000 000 001 = 10^{-12}	pico	p
0.000 000 000 000 001 = 10^{-15}	femto	f
0.000 000 000 000 000 001 = 10^{-18}	atto	a

Chemical Elements

Element	Symbol	Atomic No.	Atomic Weight
Actinium	Ac	89	
Aluminum	Al	13	26.981 5
Americium	Am	95	
Antimony	Sb	51	121.750
Argon	Ar	18	39.948
Arsenic	As	33	74.921 6
Astatine	At	85	
Barium	Ba	56	137.34
Berkelium	Bk	97	
Beryllium	Be	4	9.012 2
Bismuth	Bi	83	208.980
Boron	B	5	10.811
Bromine	Br	35	79.904
Cadmium	Cd	48	112.40
Calcium	Ca	20	40.08
Californium	Cf	98	

Element	Symbol	Atomic No.	Atomic Weight
Carbon	C	6	12.011 15
Cerium	Ce	58	140.12
Cesium	Cs	55	132.905
Chlorine	Cl	17	35.453
Chromium	Cr	24	51.996
Cobalt	Co	27	58.933 2
Columbium (see Niobium)			
Copper	Cu	29	63.546
Curium	Cm	96	
Dysprosium	Dy	66	162.50
Einsteinium	Es	99	
Erbium	Er	68	167.26
Europium	Eu	63	151.96
Fermium	Fm	100	
Fluorine	F	9	18.998 4
Francium	Fr	87	
Gadolinium	Gd	64	157.25
Gallium	Ga	31	69.72
Germanium	Ge	32	72.59
Gold	Au	79	196.967
Hafnium	Hf	72	178.49
Helium	He	2	4.002 6
Holmium	Ho	67	164.930
Hydrogen	H	1	1.007 97
Indium	In	49	114.82
Iodine	I	53	126.904 4
Iridium	Ir	77	192.2
Iron	Fe	26	55.847
Krypton	Kr	36	83.80
Lanthanum	La	57	138.91
Lead	Pb	82	207.19
Lithium	Li	3	6.939
Lutetium	Lu	71	174.97
Magnesium	Mg	12	24.312
Manganese	Mn	25	54.938 0
Mendelevium	Md	101	
Mercury	Hg	80	200.59
Molybdenum	Mo	42	95.94
Neodymium	Nd	60	144.24
Neon	Ne	10	20.183
Neptunium	Np	93	
Nickel	Ni	28	58.71
Niobium	Nb	41	92.906
Nitrogen	N	7	14.006 7
Nobelium	No	102	
Osmium	Os	76	109.2
Oxygen	O	8	15.999 4
Palladium	Pd	46	106.4
Phosphorus	P	15	30.973 8
Platinum	Pt	78	195.09
Plutonium	Pu	94	
Polonium	Po	84	
Potassium	K	19	39.102
Praseodymium	Pr	59	140.907
Promethium	Pm	61	
Protactinium	Pa	91	
Radium	Ra	88	
Radon	Rn	86	
Rhenium	Re	75	186.2

Element	Symbol	Atomic	
		No.	Weight
Rhodium	Rh	45	102.905
Rubidium	Rb	37	85.47
Ruthenium	Ru	44	101.07
Samarium	Sm	62	150.35
Scandium	Sc	21	44.956
Selenium	Se	34	78.96
Silicon	Si	14	28.086
Silver	Ag	47	107.868
Sodium	Na	11	22.989 8
Strontium	Sr	38	87.62
Sulphur	S	16	32.064
Tantalum	Ta	73	180.948
Technetium	Tc	43	
Tellurium	Te	52	127.60
Terbium	Tb	65	158.924
Thallium	Tl	81	204.37
Thorium	Th	90	232.038
Thulium	Tm	69	168.934
Tin	Sn	50	118.69
Titanium	Ti	22	47.90
Tungsten	W	74	183.85
Uranium	U	92	238.03
Vanadium	V	23	50.942
Xenon	Xe	54	131.30
Ytterbium	Yb	70	173.04
Yttrium	Y	39	88.905
Zinc	Zn	30	65.37
Zirconium	Zr	40	91.22

Greek Alphabet

Alpha	A	α
Beta	B	β
Gamma	Γ	γ
Delta	Δ	δ
Epsilon	E	ϵ
Zeta	Z	ζ
Eta	H	η
Theta	Θ	θ
Iota	I	ι
Kappa	K	κ
Lambda	Λ	λ
Mu	M	μ
Nu	N	ν
Xi	Ξ	ξ
Omicron	O	o
Pi	Π	π
Rho	P	ρ
Sigma	Σ	σ
Tau	T	τ
Upsilon	Υ	υ
Phi	Φ	ϕ
Chi	X	χ
Psi	Ψ	ψ
Omega	Ω	ω

Physical Constants

Avogadro's number = $6.022\,57 \times 10^{23}$/mol

Density of dry air at 0°C, 1 atm = 1.293 kg/m^3

Density of water at 3.98°C = $9.999\,973 \times 10^2$ kg/m^3

Equatorial radius of the earth = 6 378.39 km = 3 963.34 mi

Gravitational acceleration (standard) at sea level = 9.806 65 m/s^2 = 32.174 ft/s^2

Gravitational constant = 6.672×10^{-11} N·m^2/kg^2

Heat of fusion of water, 0°C = $3.337\,5 \times 10^5$ J/kg = 143.48 Btu/lbm

Heat of vaporization of water, 100°C = $2.259\,1 \times 10^6$ J/kg = 971.19 Btu/lbm

Mass of hydrogen atom = $1.673\,39 \times 10^{-27}$ kg

Mean density of the earth = 5.522×10^3 kg/m^3 = 344.7 lbm/ft^3

Molar gas constant = 8.314 4 J/(mol·K)

Planck's constant = $6.625\,54 \times 10^{-34}$ J/Hz

Polar radius of the earth = 6 356.91 km = 3 949.99 mi

Velocity of light in a vacuum = $2.997\,9 \times 10^8$ m/s

Velocity of sound in dry air at 0°C = 331.36 m/s = 1 087.1 ft/s

Areas Under the Standard Normal Curve from 0 to z

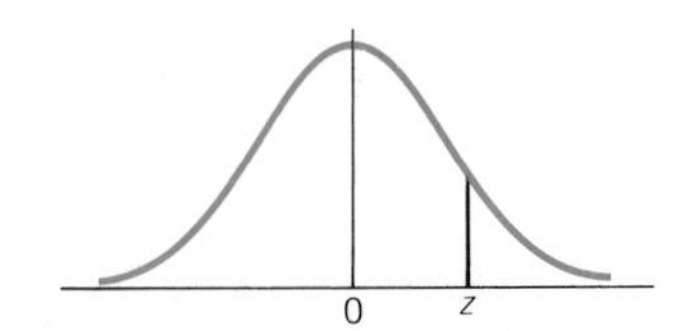

z	.00	.01	.02	.03	.04	.05	.06	.07	.08	.09
0.0	.000 0	.004 0	.008 0	.012 0	.016 0	.019 9	.023 9	.027 9	.031 9	.035 9
0.1	.039 8	.043 3	.047 8	.051 7	.055 7	.059 6	.063 6	.067 5	.071 4	.075 4
0.2	.079 3	.083 2	.087 1	.091 0	.094 8	.098 7	.102 6	.106 4	.110 3	.114 1
0.3	.117 9	.121 7	.125 5	.129 3	.133 1	.136 8	.140 6	.144 3	.148 0	.151 7
0.4	.155 4	.159 1	.162 8	.166 4	.170 0	.173 6	.177 2	.180 8	.184 1	.187 9
0.5	.191 5	.195 0	.198 5	.201 9	.205 4	.208 8	.212 3	.215 7	.219 0	.222 4
0.6	.225 8	.229 1	.232 4	.235 7	.238 9	.242 2	.245 4	.248 6	.251 8	.254 9
0.7	.258 0	.261 2	.264 2	.267 3	.270 4	.273 4	.276 4	.279 4	.282 3	.285 2
0.8	.288 1	.291 0	.293 9	.296 7	.299 6	.302 3	.305 1	.307 8	.310 6	.313 3
0.9	.315 9	.318 6	.321 2	.323 8	.326 4	.328 9	.331 5	.334 0	.336 5	.338 9
1.0	.341 3	.343 8	.346 1	.348 5	.350 8	.353 1	.355 4	.357 7	.359 9	.362 1
1.1	.364 3	.366 5	.368 6	.370 8	.372 9	.374 9	.377 0	.379 0	.381 0	.383 0
1.2	.384 9	.386 9	.388 8	.390 7	.392 5	.394 4	.396 2	.398 0	.399 7	.401 5
1.3	.403 2	.404 9	.406 6	.408 2	.409 9	.411 5	.413 1	.414 7	.416 2	.417 7
1.4	.419 2	.420 7	.422 2	.423 6	.425 1	.426 5	.427 9	.429 2	.430 6	.431 9
1.5	.433 2	.434 5	.435 7	.437 0	.438 2	.439 4	.440 6	.441 8	.442 9	.444 1
1.6	.445 2	.446 3	.447 4	.448 4	.449 5	.450 5	.451 5	.452 5	.453 5	.454 5
1.7	.455 5	.456 4	.457 3	.458 2	.459 1	.459 9	.460 8	.461 6	.462 5	.463 3
1.8	.464 1	.464 9	.465 6	.466 4	.467 1	.467 8	.468 6	.469 3	.469 9	.470 6
1.9	.471 3	.471 9	.472 6	.473 2	.473 8	.474 4	.475 0	.475 6	.476 1	.476 7

z	.00	.01	.02	.03	.04	.05	.06	.07	.08	.09
2.0	.477 2	.477 8	.478 3	.478 8	.479 3	.479 8	.480 3	.480 8	.481 2	.481 7
2.1	.482 1	.482 6	.483 0	.483 4	.483 8	.484 2	.484 6	.485 0	.485 4	.485 7
2.2	.486 1	.486 4	.486 8	.487 1	.487 5	.487 8	.488 1	.488 4	.488 7	.489 0
2.3	.489 3	.489 6	.489 8	.490 1	.490 4	.490 6	.490 9	.491 1	.491 3	.491 6
2.4	.491 8	.492 0	.492 2	.492 5	.492 7	.492 9	.493 1	.493 2	.493 4	.493 6
2.5	.493 8	.494 0	.494 1	.494 3	.494 5	.494 6	.494 8	.494 9	.495 1	.495 2
2.6	.495 3	.495 5	.495 6	.495 7	.495 9	.496 0	.496 1	.496 2	.496 3	.496 4
2.7	.496 5	.496 6	.496 7	.496 8	.496 9	.497 0	.497 1	.497 2	.497 3	.497 4
2.8	.497 4	.497 5	.497 6	.497 7	.497 7	.497 8	.497 9	.497 9	.498 0	.498 1
2.9	.498 1	.498 2	.498 2	.498 3	.498 4	.498 4	.498 5	.498 5	.498 6	.498 6
3.0	.498 7	.498 7	.498 7	.498 8	.498 8	.498 9	.498 9	.498 9	.499 0	.499 0
3.1	.499 0	.499 1	.499 1	.499 1	.499 2	.499 2	.499 2	.499 2	.499 3	.499 3
3.2	.499 3	.499 3	.499 4	.499 4	.499 4	.499 4	.499 4	.499 5	.499 5	.499 5
3.3	.499 5	.499 5	.499 5	.499 6	.499 6	.499 6	.499 6	.499 6	.499 6	.499 7
3.4	.499 7	.499 7	.499 7	.499 7	.499 7	.499 7	.499 7	.499 7	.499 7	.499 8
3.5	.499 8	.499 8	.499 8	.499 8	.499 8	.499 8	.499 8	.499 8	.499 8	.499 8
3.6	.499 8	.499 8	.499 9	.499 9	.499 9	.499 9	.499 9	.499 9	.499 9	.499 9
3.7	.499 9	.499 9	.499 9	.499 9	.499 9	.499 9	.499 9	.499 9	.499 9	.499 9
3.8	.499 9	.499 9	.499 9	.499 9	.499 9	.499 9	.499 9	.499 9	.499 9	.499 9
3.9	.500 0	.500 0	.500 0	.500 0	.500 0	.500 0	.500 0	.500 0	.500 0	.500 0

CODE OF ETHICS
For Engineers*

Preamble

Engineering is an important and learned profession. The members of the profession recognize that their work has a direct and vital impact on the quality of life for all people. Accordingly, the services provided by engineers require honesty, impartiality, fairness and equity, and must be dedicated to the protection of the public health, safety and welfare. In the practice of their profession, engineers must perform under a standard of professional behavior which requires adherence to the highest principles of ethical conduct on behalf of the public, clients, employers and the profession.

I. Fundamental Canons

Engineers, in the fulfillment of their professional duties, shall:

1. Hold paramount the safety, health and welfare of the public in the performance of their professional duties.
2. Perform services only in areas of their competence.
3. Issue public statements only in an objective and truthful manner.
4. Act in professional matters for each employer or client as faithful agents or trustees.
5. Avoid improper solicitation of professional employment.

II. Rules of Practice

1. Engineers shall hold paramount the safety, health and welfare of the public in the performance of their professional duties.

a. Engineers shall at all times recognize that their primary obligation is to protect the safety, health, property and welfare of the public. If their professional judgment is overruled under circumstances where the safety, health, property or welfare of the public are endangered, they shall notify their employer or client and such other authority as may be appropriate.

*Code of the National Society of Professional Engineers

b. Engineers shall approve only those engineering documents which are safe for public health, property and welfare in conformity with accepted standards.

c. Engineers shall not reveal facts, data or information obtained in a professional capacity without the prior consent of the client or employer except as authorized or required by law or this Code.

d. Engineers shall not permit the use of their name or firm name nor associate in business ventures with any person or firm which they have reason to believe is engaging in fraudulent or dishonest business or professional practices.

e. Engineers having knowledge of any alleged violation of this Code shall co-operate with the proper authorities in furnishing such information or assistance as may be required.

2. Engineers shall perform services only in the areas of their competence.

a. Engineers shall undertake assignments only when qualified by education or experience in the specific technical fields involved.

b. Engineers shall not affix their signatures to any plans or documents dealing with subject matter in which they lack competence, nor to any plan or document not prepared under their direction and control.

c. Engineers may accept assignments and assume responsibility for coordination of an entire project and sign and seal the engineering documents for the entire project, provided that each technical segment is signed and sealed only by the qualified engineers who prepared the segment.

3. Engineers shall issue public statements only in an objective and truthful manner.

a. Engineers shall be objective and truthful in professional reports, statements or testimony. They shall include all relevant and pertinent information in such reports, statements or testimony.

b. Engineers may express publicly a professional opinion on technical subjects only when that opinion is founded upon adequate knowledge of the facts and competence in the subject matter.

c. Engineers shall issue no statements, criticisms or arguments on technical matters which are inspired or paid for by interested parties, unless they have prefaced their comments by explicitly identifying the interested parties on whose behalf they are speaking, and by revealing the existence of any interest the engineers may have in the matters.

4. Engineers shall act in professional matters for each employer or client as faithful agents or trustees.

a. Engineers shall disclose all known or potential conflicts of interest to their employers or clients by promptly informing them of any business association, interest, or other circumstances which could influence or appear to influence their judgment or the quality of their services.

b. Engineers shall not accept compensation, financial or otherwise, from more than one party for services on the same project, or for services pertaining to the same project, unless the circumstances are fully disclosed to, and agreed to, by all interested parties.

c. Engineers shall not solicit or accept financial or other valuable consideration, directly or indirectly, from contractors, their agents, or other parties in connection with work for employers or clients for which they are responsible.

d. Engineers in public service as members, advisors or employees of a governmental body or department shall not participate in decisions with respect to professional services solicited or provided by them or their organizations in private or public engineering practice.

e. Engineers shall not solicit or accept a professional contract from a governmental body on which a principal or officer of their organization serves as a member.

5. Engineers shall avoid improper solicitation of professional employment.

a. Engineers shall not falsify or permit misrepresentation of their, or their associates, academic or professional qualifications. They shall not misrepresent or exaggerate their degree of responsibility in or for the subject matter of prior assignments. Brochures or other presentations incident to the solicitation of employment shall not misrepresent pertinent facts concerning employers, employees, associates, joint venturers or past accomplishments with the intent and purpose of enhancing their qualifications and their work.

b. Engineers shall not offer, give, solicit or receive, either directly or indirectly, any political contribution in an amount intended to influence the award of a contract by public authority, or which may be reasonably construed by the public of having the effect or intent to influence the award of a contract. They shall not offer any gift, or other valuable consideration in order to secure work. They shall not pay a commission, percentage or brokerage fee in order to secure work except to a bona fide employee or bona fide established commercial or marketing agencies retained by them.

III. Professional Obligations

1. Engineers shall be guided in all their professional relations by the highest standards of integrity.

a. Engineers shall admit and accept their own errors when proven wrong and refrain from distorting or altering the facts in an attempt to justify their decisions.

b. Engineers shall advise their clients or employers when they believe a project will not be successful.

c. Engineers shall not accept outside employment to the detriment of their regular work or interest. Before accepting any outside employment they will notify their employers.

d. Engineers shall not attempt to attract an engineer from another employer by false or misleading pretenses.

e. Engineers shall not actively participate in strikes, picket lines, or other collective coercive action.

f. Engineers shall avoid any act tending to promote their own interest at the expense of the dignity and integrity of the profession.

2. Engineers shall at all times strive to serve the public interest.

a. Engineers shall seek opportunities to be of constructive service in civic affairs and work for the advancement of the safety, health and well-being of their community.

b. Engineers shall not complete, sign, or seal plans and/or specifications that are not of a design safe to the public health and welfare and in conformity with accepted engineering standards. If the client or employer insists on such unprofessional conduct, they shall notify the proper authorities and withdraw from further service on the project.

c. Engineers shall endeavor to extend public knowledge and appreciation of engineering and its achievements and to protect the engineering profession from misrepresentation and misunderstanding.

3. Engineers shall avoid all conduct or practice which is likely to discredit the profession or deceive the public.

a. Engineers shall avoid the use of statements containing a material misrepresentation of fact or omitting a material fact necessary to keep statements from being misleading; statements intended or likely to create an unjustified expectation; statements containing prediction of future success; statements containing an opinion as to the quality of the Engineers' services; or statements intended or likely to attract clients by the use of showmanship, puffery, or self-laudation, including the use of slogans, jingles, or sensational language or format.

b. Consistent with the foregoing; Engineers may advertise for recruitment of personnel.

c. Consistent with the foregoing; Engineers may prepare articles for the lay or technical press, but such articles shall not imply credit to the author for work performed by others.

4. Engineers shall not disclose confidential information concerning the business affairs or technical processes of any present or former client or employer without his consent.

a. Engineers in the employ of others shall not without the consent of all interested parties enter promotional efforts or negotiations for work or make arrangements for other employment as a principal or to practice in connection with a specific project for which the Engineer has gained particular and specialized knowledge.

b. Engineers shall not, without the consent of all interested parties, participate in or represent an adversary interest in connection with a specific project or proceeding in which the Engineer has gained particular specialized knowledge on behalf of a former client or employer.

5. Engineers shall not be influenced in their professional duties by conflicting interests.

a. Engineers shall not accept financial or other considerations, including free engineering designs, from material or equipment suppliers for specifying their product.

b. Engineers shall not accept commissions or allowances, directly or indirectly, from contractors or other parties dealing with clients or employers of the Engineer in connection with work for which the Engineer is responsible.

6. Engineers shall uphold the principle of appropriate and adequate compensation for those engaged in engineering work.

a. Engineers shall not accept remuneration from either an employee or employment agency for giving employment.

b. Engineers, when employing other engineers, shall offer a salary according to professional qualifications and the recognized standards in the particular geographical area.

7. Engineers shall not compete unfairly with other engineers by attempting to obtain employment or advancement or professional engagements by taking advantage of a salaried position, by criticizing other engineers, or by other improper or questionable methods.

a. Engineers shall not request, propose, or accept a professional commission on a contingent basis under circumstances in which their professional judgment may be compromised.

b. Engineers in salaried positions shall accept part-time engineering work only at salaries not less than that recognized as standard in the area.

c. Engineers shall not use equipment, supplies, laboratory, or office facilities of an employer to carry on outside private practice without consent.

8. Engineers shall not attempt to injure, maliciously or falsely, directly or indirectly, the professional reputation, prospects, practice or employment of other engineers, nor indiscriminately criticize other engineers' work. Engineers who believe others are guilty of unethical or illegal practice shall present such information to the proper authority for action.

a. Engineers in private practice shall not review the work of another engineer for the same client, except with the knowledge of such engineer, or unless the connection of such engineer with the work has been terminated.

b. Engineers in governmental, industrial or educational employ are entitled to review and evaluate the work of other engineers when so required by their employment duties.

c. Engineers in sales or industrial employ are entitled to make engineering comparisons of represented products with products of other suppliers.

9. Engineers shall accept personal responsibility for all professional activities.

a. Engineers shall conform with state registration laws in the practice of engineering.

b. Engineers shall not use association with a nonengineer, a corporation, or partnership, as a "cloak" for unethical acts, but must accept personal responsibility for all professional acts.

10. Engineers shall give credit for engineering work to those to whom credit is due, and will recognize the proprietary interests of others.

a. Engineers shall, whenever possible, name the person or persons who may be individually responsible for designs, inventions, writings, or other accomplishments.

b. Engineers using designs supplied by a client recognize that the designs remain the property of the client and may not be duplicated by the Engineer for others without express permission.

c. Engineers, before undertaking work for others in connection with which the Engineer may make improvements, plans, designs, inventions, or other records which may justify copyrights or patents, should enter into a positive agreement regarding ownership.

d. Engineers' designs, data, records, and notes referring exclusively to an employer's work are the employer's property.

11. Engineers shall cooperate in extending the effectiveness of the profession by interchanging information and experience with other engineers and students, and will endeavor to provide opportunity for the professional development and advancement of engineers under their supervision.

a. Engineers shall encourage engineering employees' efforts to improve their education.

b. Engineers shall encourage engineering employees to attend and present papers at professional and technical society meetings.

c. Engineers shall urge engineering employees to become registered at the earliest possible date.

d. Engineers shall assign a professional engineer duties of a nature to utilize full training and experience, insofar as possible, and delegate lesser functions to subprofessionals or to technicians.

e. Engineers shall provide a prospective engineering employee with complete information on working conditions and proposed status of employment, and after employment will keep employees informed of any changes.

As Revised, March 1985

"By order of the United States District Court for the District of Columbia, former Section 11(c) of the NSPE Code of Ethics prohibiting competitive bidding, and all policy statements, opinions, rulings or other guidelines interpreting its scope, have been rescinded as unlawfully interfering with the legal right of engineers, protected under the antitrust laws, to provide price information to prospective clients; accordingly, nothing contained in the NSPE Code of Ethics, policy statements, opinions, rulings or other guidelines prohibits the submission of price quotations or competitive bids for engineering services at any time or in any amount."

Statement by NSPE Executive Committee

In order to correct misunderstandings which have been indicated in some instances since the issuance of the Supreme Court decision and the entry of the Final Judgment, it is noted that in its decision of April 25, 1978, the Supreme Court of the United States declared: "The Sherman Act does not require competitive bidding."

It is further noted that as made clear in the Supreme Court decision:

1. Engineers and firms may individually refuse to bid for engineering services.
2. Clients are not required to seek bids for engineering services.
3. Federal, state, and local laws governing procedures to procure engineering services are not affected, and remain in full force and effect.
4. State societies and local chapters are free to actively and aggressively seek legislation for professional selection and negotiation procedures by public agencies.
5. State registration board rules of professional conduct, including rules prohibiting competitive bidding for engineering services, are not affected and remain in full force and effect. State registration boards with authority to adopt rules of professional conduct may adopt rules governing procedures to obtain engineering services.
6. As noted by the Supreme Court, "nothing in the judgment prevents NSPE and its members from attempting to influence governmental action. . . ."

Note: In regard to the question of application of the Code to corporations vis-a-vis real persons, business form or type should not negate nor influence conformance of individuals to the Code. The Code deals with professional services, which services must be performed by real persons. Real persons in turn establish and implement policies within business structures. The Code is clearly written to apply to the Engineer and it is incumbent on a member of NSPE to endeavor to live up to its provisions. This applies to all pertinent sections of the Code.

Unit Conversions

To convert from	To	Multiply by
acres	ft^2	4.356×10^4
acres	ha	$4.046\,9 \times 10^{-1}$
acres	m^2	$4.046\,9 \times 10^3$
amperes	C/s	1
ampere hours	C	3.6×10^3
angstroms	cm	1×10^{-8}
angstroms	in	$3.937\,0 \times 10^{-9}$
atmospheres	bars	1.013 3
atmospheres	in of Hg	$2.992\,1 \times 10^1$
atmospheres	lbf/in^2	$1.469\,6 \times 10^1$
atmospheres	mm of Hg	7.6×10^2
atmospheres	Pa	$1.013\,3 \times 10^5$
barrels (petroleum, US)	gal (US liquid)	4.2×10^1
bars	atm	$9.869\,2 \times 10^{-1}$
bars	in of Hg	$2.953\,0 \times 10^1$
bars	lbf/in^2	$1.450\,4 \times 10^1$
bars	Pa	1×10^5
Btu	ft·lbf	$7.776\,5 \times 10^2$
Btu	hp·h	$3.927\,5 \times 10^{-4}$
Btu	J	$1.055\,1 \times 10^3$
Btu	kWh	$2.928\,8 \times 10^{-4}$
Btu per hour	ft·lbf/s	$2.160\,1 \times 10^{-1}$
Btu per hour	hp	$3.927\,5 \times 10^{-4}$
Btu per hour	W	$2.928\,8 \times 10^{-1}$
Btu per minute	ft·lbf/min	$7.776\,5 \times 10^2$
Btu per minute	hp	$2.356\,5 \times 10^{-2}$
Btu per minute	kW	$1.757\,3 \times 10^{-2}$
bushels (US)	ft^3	1.244 5
bushels (US)	L	$3.523\,9 \times 10^1$
bushels (US)	m^3	$3.523\,9 \times 10^{-2}$
candelas	lm/sr	1
candelas per square foot	lamberts	$3.381\,6 \times 10^{-3}$
centimeters	Å	1×10^8
centimeters	ft	$3.280\,8 \times 10^{-2}$
centimeters	in	$3.937\,0 \times 10^{-1}$
centipoises	g/(cm·s)	1×10^{-2}
circular mils	cm^2	$5.067\,1 \times 10^{-6}$
circular mils	in^2	$7.854\,0 \times 10^{-7}$
coulombs	A·s	1
cubic centimeters	in^3	$6.102\,4 \times 10^{-2}$
cubic centimeters	ft^3	$3.531\,5 \times 10^{-5}$
cubic centimeters	gal (US liquid)	$2.641\,7 \times 10^{-4}$
cubic centimeters	L	1×10^{-3}
cubic centimeters	oz (US fluid)	$3.381\,4 \times 10^{-2}$
cubic centimeters per gram	ft^3/lbm	$1.601\,8 \times 10^{-2}$
cubic centimeters per second	ft^3/min	$2.118\,9 \times 10^{-3}$

To convert from	To	Multiply by
cubic centimeters per second	gal (US liquid)/min	$1.585\,0 \times 10^{-2}$
cubic feet	acre·ft	$2.295\,7 \times 10^{-5}$
cubic feet	bushels (US)	$8.035\,6 \times 10^{-1}$
cubic feet	gal (US liquid)	7.480 5
cubic feet	in^3	1.728×10^{3}
cubic feet	L	$2.831\,7 \times 10^{1}$
cubic feet	m^3	$2.831\,7 \times 10^{-2}$
cubic feet per minute	gal (US liquid)/min	7.480 5
cubic feet per minute	L/s	$4.719\,5 \times 10^{-1}$
cubic feet per pound-mass	cm^3/g	$6.242\,8 \times 10^{1}$
cubic feet per second	gal (US liquid)/min	$4.488\,3 \times 10^{2}$
cubic feet per second	L/s	$2.831\,7 \times 10^{1}$
cubic inches	bushels (US)	$4.650\,3 \times 10^{-4}$
cubic inches	cm^3	$1.638\,7 \times 10^{1}$
cubic inches	gal (US liquid)	$4.329\,0 \times 10^{-3}$
cubic inches	L	$1.638\,7 \times 10^{-2}$
cubic inches	m^3	$1.638\,7 \times 10^{-5}$
cubic inches	oz (US fluid)	$5.541\,1 \times 10^{-1}$
cubic meters	acre·ft	$8.107\,1 \times 10^{-4}$
cubic meters	bushels (US)	$2.837\,8 \times 10^{1}$
cubic meters	ft^3	$3.531\,5 \times 10^{1}$
cubic meters	gal (US liquid)	$2.641\,7 \times 10^{2}$
cubic meters	L	1×10^{3}
cubic yards	bushels (US)	$2.169\,6 \times 10^{1}$
cubic yards	gal (US liquid)	$2.019\,7 \times 10^{2}$
cubic yards	L	$7.645\,5 \times 10^{2}$
cubic yards	m^3	$7.645\,5 \times 10^{-1}$
dynes	N	1×10^{-5}
dynes per square centimeter	atm	$9.869\,2 \times 10^{-7}$
dynes per square centimeter	bars	1×10^{-6}
dynes per square centimeter	lbf/in^2	$1.450\,4 \times 10^{-5}$
dyne centimeters	ft·lbf	$7.375\,6 \times 10^{-8}$
dyne centimeters	N·m	1×10^{-7}
ergs	dyne·cm	1
fathoms	ft	6
feet	cm	3.048×10^{1}
feet	in	1.2×10^{1}
feet	km	3.048×10^{-4}
feet	m	3.048×10^{-1}
feet	mi	$1.893\,9 \times 10^{-4}$
feet	rods	$6.060\,6 \times 10^{-2}$
feet per second	km/h	1.097 3
feet per second	m/min	$1.828\,8 \times 10^{1}$
feet per second	mi/h	$6.818\,2 \times 10^{-1}$
feet per second squared	m/s^2	3.048×10^{-1}
foot-candles	lm/ft^2	1
foot-candles	lux	$1.076\,4 \times 10^{1}$
foot pounds-force	Btu	$1.285\,9 \times 10^{-3}$
foot pounds-force	dyne·cm	$1.355\,8 \times 10^{7}$
foot pounds-force	hp·h	$5.050\,5 \times 10^{-7}$
foot pounds-force	J	1.355 8
foot pounds-force	kWh	$3.766\,2 \times 10^{-7}$
foot pounds-force	N·m	1.355 8
foot pounds-force per hour	Btu/min	$2.143\,2 \times 10^{-5}$
foot pounds-force per hour	ergs/min	$2.259\,7 \times 10^{5}$
foot pounds-force per hour	hp	$5.050\,5 \times 10^{-7}$
foot pounds-force per hour	kW	$3.766\,2 \times 10^{-7}$
furlongs	ft	6.6×10^{2}
furlongs	m	$2.011\,7 \times 10^{2}$
gallons (US liquid)	ft^3	$1.336\,8 \times 10^{-1}$
gallons (US liquid)	in^3	2.31×10^{2}

To convert from	To	Multiply by
gallons (US liquid)	L	3.785 4
gallons (US liquid)	m^3	$3.785\,4 \times 10^{-3}$
gallons (US liquid)	oz (US fluid)	1.28×10^{2}
gallons (US liquid)	pt (US liquid)	8
gallons (US liquid)	qt (US liquid)	4
grams	lbm	$2.204\,6 \times 10^{-3}$
grams per centimeter second	poises	1
grams per cubic centimeter	lbm/ft^3	$6.242\,8 \times 10^{1}$
hectares	acres	2.471 1
hectares	ares	1×10^{2}
hectares	ft^2	$1.076\,4 \times 10^{5}$
hectares	m^2	1×10^{4}
horsepower	Btu/h	$2.546\,1 \times 10^{3}$
horsepower	ft·lbf/s	5.5×10^{2}
horsepower	kW	$7.457\,0 \times 10^{-1}$
horsepower	W	$7.457\,0 \times 10^{2}$
horsepower hours	Btu	$2.546\,1 \times 10^{3}$
horsepower hours	ft·lbf	1.98×10^{6}
horsepower hours	J	$2.684\,5 \times 10^{6}$
horsepower hours	kWh	$7.457\,0 \times 10^{-1}$
hours	min	6×10^{1}
hours	s	3.6×10^{3}
inches	Å	2.54×10^{8}
inches	cm	2.54
inches	ft	$8.333\,3 \times 10^{-2}$
inches	mils	1×10^{3}
inches	yd	$2.777\,8 \times 10^{-2}$
joules	Btu	$9.484\,5 \times 10^{-4}$
joules	ft·lbf	$7.375\,6 \times 10^{-1}$
joules	hp·h	$3.725\,1 \times 10^{-7}$
joules	kWh	$2.777\,8 \times 10^{-7}$
joules	W·s	1
joules per second	Btu/min	$5.690\,7 \times 10^{-2}$
joules per second	ergs/s	1×10^{7}
joules per second	ft·lbf/s	$7.375\,6 \times 10^{-1}$
joules per second	hp	$1.341\,0 \times 10^{-3}$
joules per second	W	1
kilograms	lbm	2.204 6
kilograms	slugs	$6.852\,2 \times 10^{-2}$
kilograms	t	1×10^{-3}
kilometers	ft	$3.280\,8 \times 10^{3}$
kilometers	mi	$6.213\,7 \times 10^{-1}$
kilometers	nmi (nautical mile)	$5.399\,6 \times 10^{-1}$
kilometers per hour	ft/min	$5.468\,1 \times 10^{1}$
kilometers per hour	ft/s	$9.113\,4 \times 10^{-1}$
kilometers per hour	knots	$5.399\,6 \times 10^{-1}$
kilometers per hour	m/s	$2.777\,8 \times 10^{-1}$
kilometers per hour	mi/h	$6.213\,7 \times 10^{-1}$
kilowatts	Btu/h	$3.414\,4 \times 10^{3}$
kilowatts	ergs/s	1×10^{10}
kilowatts	ft·lbf/s	$7.375\,6 \times 10^{2}$
kilowatts	hp	1.341 0
kilowatts	J/s	1×10^{3}
kilowatt hours	Btu	$3.414\,4 \times 10^{3}$
kilowatt hours	ft·lbf	$2.655\,2 \times 10^{6}$
kilowatt hours	hp·h	1.341 0
kilowatt hours	J	3.6×10^{6}
knots	ft/s	1.687 8
knots	mi/h	1.150 8
liters	bushels (US)	$2.837\,8 \times 10^{-2}$
liters	ft^3	$3.531\,5 \times 10^{-2}$

To convert from	To	Multiply by
liters	gal (US liquid)	2.6417×10^{-1}
liters	in^3	6.1024×10^{1}
liters per second	ft^3/min	2.118 9
liters per second	gal (US liquid)/min	1.5850×10^{1}
lumens	candle power	7.9577×10^{-2}
lumens per square foot	foot-candles	1
lumens per square meter	foot-candles	9.2903×10^{-2}
lux	lm/m^2	1
meters	Å	1×10^{10}
meters	ft	3.280 8
meters	in	3.9370×10^{1}
meters	mi	6.2137×10^{-4}
meters per minute	cm/s	1.666 7
meters per minute	ft/s	5.4681×10^{-2}
meters per minute	km/h	6×10^{-2}
meters per minute	knots	3.2397×10^{-2}
meters per minute	mi/h	3.7282×10^{-2}
microns	Å	1×10^{4}
microns	ft	3.2808×10^{-6}
microns	m	1×10^{-6}
miles	ft	5.28×10^{3}
miles	furlongs	8
miles	km	1.609 3
miles	nmi (nautical mile)	8.6898×10^{-1}
miles per hour	cm/s	4.4704×10^{1}
miles per hour	ft/min	8.8×10^{1}
miles per hour	ft/s	1.466 7
miles per hour	km/h	1.609 3
miles per hour	knots	8.6898×10^{-1}
miles per hour	m/min	2.6822×10^{1}
nautical miles	mi	1.150 8
newtons	dynes	1×10^{5}
newtons	lbf	2.2481×10^{-1}
newton meters	dyne·cm	1×10^{7}
newton meters	ft·lbf	7.3756×10^{-1}
ounces (US fluid)	cm^3	2.9574×10^{1}
ounces (US fluid)	gal (US liquid)	7.8125×10^{-3}
ounces (US fluid)	in^3	1.804 7
ounces (US fluid)	L	2.9574×10^{-2}
pascals	atm	9.8692×10^{-6}
pascals	lbf/ft^2	2.0885×10^{-2}
pascals	lbf/in^2	1.4504×10^{-4}
poises	g/(cm·s)	1
pounds-force	N	4.448 2
pounds-mass	g	4.5359×10^{2}
pounds-mass	kg	4.5359×10^{-1}
pounds-mass	slugs	3.1081×10^{-2}
pounds-mass	t	4.5359×10^{-4}
pounds-mass	tons (short)	5×10^{-4}
pounds-force per square foot	atm	4.7254×10^{-3}
pounds-force per square foot	Pa	4.7880×10^{1}
pounds-force per square inch	atm	6.8046×10^{-2}
pounds-force per square inch	bars	6.8948×10^{-2}
pounds-force per square inch	in of Hg	2.036 0
pounds-force per square inch	mm of Hg	5.1715×10^{1}
pounds-force per square inch	Pa	6.8948×10^{3}
pounds-mass per cubic foot	g/cm^3	1.6018×10^{-2}
pounds-mass per cubic foot	kg/m^3	1.6018×10^{1}
radians	°	5.7296×10^{1}
radians	r (revolutions)	1.5915×10^{-1}
radians per second	r/min	9.549 3

To convert from	To	Multiply by
slugs	kg	$1.459\,4 \times 10^{1}$
slugs	lbm	$3.217\,4 \times 10^{1}$
square centimeters	ft^2	$1.076\,4 \times 10^{-3}$
square centimeters	in^2	$1.550\,0 \times 10^{-1}$
square feet	acre	$2.295\,7 \times 10^{-5}$
square feet	cm^2	$9.290\,3 \times 10^{2}$
square feet	ha	$9.290\,3 \times 10^{-6}$
square feet	m^2	$9.290\,3 \times 10^{-2}$
square meters	ft^2	$1.076\,4 \times 10^{1}$
square meters	in^2	$1.550\,0 \times 10^{3}$
square miles	acres	6.4×10^{2}
square miles	ft^2	$2.787\,8 \times 10^{7}$
square miles	ha	$2.590\,0 \times 10^{2}$
square miles	km^2	2.590 0
square millimeters	ft^2	$1.076\,4 \times 10^{-5}$
square millimeters	in^2	$1.550\,0 \times 10^{-3}$
stokes	cm^2/s	1
stokes	in^2/s	$1.550\,0 \times 10^{-1}$
tons (long)	lbm	2.24×10^{3}
tons (long)	t	1.016 0
tons (long)	tons (short)	1.12
watts	Btu/h	3.414 4
watts	ergs/s	1×10^{7}
watts	ft·lbf/min	$4.425\,4 \times 10^{1}$
watts	hp	$1.341\,0 \times 10^{-3}$
watts	J/s	1
watt hours	Btu	3.414 4
watt hours	ft·lbf	$2.655\,2 \times 10^{3}$
watt hours	hp·h	$1.341\,0 \times 10^{-3}$

Answers to Selected Problems

2.2 96 m

2.6 42.2 acres; 2.65×10^3 ft

2.12 82.8°; 37.9 ft

2.18 (*a*) 36.3%; (*b*) 9.97%; (*c*) 0.64%

2.22 (*a*) 905 ft^3; (*b*) 2.03×10^4 gal; (*c*) 84.6 tons; (*d*) 102 days

2.25 376 gal; $v = 94.0h$ (v in gal, h in ft)

3.3 (*b*) emf $= 0.053t + 1.0$

3.7 (*b*) $V = 2.95C^{1.11}$

3.13 (*b*) $P = 23.4e^{0.248(\text{month})}$

3.22 6 poles

4.1 (*a*) 3; (*d*) 4; (*g*) 4

4.3 (*a*) 10.8; (*d*) 4.1×10^6; (*g*) -3.679×10^{-4}

5.3 (*a*) 763.3 cm; (*b*) 1.8×10^{-1} ft^3; (*c*) 68.2°F; (*d*) 1.42×10^7 cm^3; (*e*) 3.51 kW

5.7 (*a*) 1.45×10^3 N; (*b*) 1.45×10^3 N; (*c*) 239 N

5.12 1.4×10^3 gal; 3.5×10^3 kg

5.18 2.51×10^3 J/kg

5.23 (*a*) 1.38×10^3 cm^2; (*b*) 3.01 kg; (*c*) ball will float (water displaced = 4.82 kg)

6.2 (*a*)

Width	Frequency
54–57.9	3
58–61.9	4
62–65.9	9
66–69.9	14
70–73.9	5

(c) Median = 66.9 dm, Mode = 68.2 dm, Mean = 65.8 dm; (d) $s = 4.35$ dm

6.8 (a) 16 percent; (b) 3.5; (c) 14; (d) 73.3

6.12 (b) $V = 2.57t - 14.4$; (d) $r = 0.997\,0$, 99.4 percent of variation in velocity can be accounted for by time differences

6.16 (b) $R = 12.1A^{-0.963}$; (d) $r = -0.9997$

6.22 (b) $V = 197e^{-0.122t}$; (d) $r = -0.9988$

9.1 5.395 kN 8.122°

9.7 4.00 kN 78.7°

9.13 $\Sigma M_A = 25.2$ kNm ↻; $\Sigma M_B = 2.35$ kNm ↻

9.21 $\overline{A} = 1.65$ kN ↓; $\overline{B} = 7.11$ kN ↑

9.26 186 N; 34.2°; ⊥ distance from 0 to equilibrium force = 3.94 m

9.31 (a) 2.15 kN; (b) 2.37 kN 27.1°; (c) 1.50×10^2 MPa; (d) 6.67

10.2 327°C; 620°F

10.8 (a) 18.015 3; (b) 78.114 72; (c) 138.213; (d) 171.35

10.16 2.5

10.22 1.8×10^2 m^3

10.29 0.7754 kg

10.35 $C_{12}H_{25}OH$ = 16.2 t, H_2SO_4 = 8.5 t, NaOH = 3.5 t

10.40 Na = 32.39%, H = 0.71%, P = 21.82%, O = 45.08%

10.46 H_2SO_4

11.2 16.1%

11.6 Q = 2 550 kg/h, ether = 33.6%, oil = 4.65%, inert = 61.8%

11.11 Alcohol = 2.39%, water = 92.2%, inert = 5.39%

11.16 A = 291.1 kg, B = 204.0 kg, C = 156.9 kg

11.21 Mass = 250 kg

11.27 A = 874 kg, B = 398 kg, water = 284 kg

12.2 1.08×10^5 C

12.6 2.4×10^2 W; 9.8 J/impact

12.11 (b) 4.51 Ω; (c) 2.65 A; (d) 1.76 A, 21.0 W; (e) 0.771 J/s

12.15 (a) 59.3 V; (b) 2.63 kW; (c) 551 J/s; (d) 2.07 kW

13.1 4.1×10^5 J

13.5 31.0 m/s

13.9 (*a*) 18 600 J; (*b*) 1.82 hp

13.15 -1.06×10^6 J

13.19 45 percent

13.23 0.80 kW

13.28 % capacity = 52 percent, power input = 2319 W, energy used = 3.64 kWh

14.2 (*a*) $11 640.18; (*b*) $11 534.12; (*c*) $ 11 442.78

14.6 (*a*) $31 861.36; (*b*) $ 61 423.37

14.12 18.9 percent

14.18 Using a present-worth comparison, A = $40 443.80, B = $41 822.71; conclusion: A is cheaper

14.23 (*a*) $985.77; (*b*) $37.02

14.26 $17 422.75, 10.79 percent

Selected Bibliography

Allen, Myron S.: *Morphological Creativity,* Prentice-Hall, Englewood Cliffs, N.J., 1962.

Assaf, Karen, and Said A. Assaf: *Handbook of Mathematical Calculations,* Iowa State University Press, Ames, Iowa, 1974.

Bassin, Milton G., Stanley M. Brodsky, and Harold Woloff: *Statics and Strengths of Materials,* McGraw-Hill, New York, 1969.

Beakley, George C., and H. W. Leach: *Engineering: An Introduction to a Creative Profession,* 4th ed., Macmillan, New York, 1983.

Beakley, George C., and Robert E. Lovell: *Computation, Calculators, and Computers,* Macmillan, New York, 1983.

Bullinger, Clarence E.: *Engineering Economy,* McGraw-Hill, New York, 1958.

DeJong, Paul S., James S. Rising, and Maurice W. Almfeldt: *Engineering Graphics,* 6th ed., Kendall/Hunt, Dubuque, Iowa, 1983.

Drago, Russell S.: *Principles of Chemistry with Practical Perspective,* Allyn and Bacon, Boston, 1974.

Eide, Arvid R., Roland D. Jenison, Lane H. Mashaw, Larry L. Northup, and C. Gordon Sanders: *Engineering Graphics Fundamentals,* McGraw-Hill, New York, 1985.

Erickson, William H., and Nelson H. Bryant: *Electrical Engineering, Theory and Practice,* 2d ed., Wiley, New York, 1959.

Fletcher, Leroy S., and Terry E. Shoup: *Introduction to Engineering Including Fortran Programming,* Prentice-Hall, Englewood Cliffs, N.J., 1978.

Freund, John: *Modern Elementary Statistics,* 3d ed., Prentice-Hall, Englewood Cliffs, N.J., 1967.

Gajda, Walter J., Jr., and William E. Biles: *Engineering: Modeling and Computation,* Houghton Mifflin, Boston, 1978.

Gibson, John E.: *Introduction to Engineering Design,* Holt, New York, 1968.

Glorioso, Robert M., and Francis S. Hill, Jr. (eds.): *Introduction to Engineering,* Prentice-Hall, Englewood Cliffs, N.J., 1975.

Gordon, William J. J.: *Synectics,* Harper, New York, 1961.

Grant, Eugene L., W. Grant Ireson, and Richard S. Leavenworth: *Principles of Engineering Economy,* 6th ed., Ronald, New York, 1976.

Harrisberger, Lee: *Engineersmanship,* 2d ed., Brooks/Cole, Monterey, Calif., 1982.

Hill, Percy H.: *The Science of Engineering,* Holt, New York, 1970.

Jensen, Cecil H., and Jay D. Helsel: *Engineering Drawing and Design,* 2d ed., McGraw-Hill, New York, 1979.

Katz, Donald L., Robert O. Goetz, Edward R. Lady, and Dale C. Ray: *Engineering Concepts and Perspectives,* Wiley, New York, 1968.

Krick, Edward V.: *An Introduction to Engineering: Methods, Concepts, and Issues,* Wiley, New York, 1976.

Longo, Frederick R.: *General Chemistry—Interaction of Matter, Energy, and Mass,* McGraw-Hill, New York, 1974.

Luzadder, Warren J.: *Fundamentals of Engineering Drawing for Design, Product Development, and Numerical Control,* 8th ed., Prentice-Hall, Englewood Cliffs, N.J., 1981.

Mayne, Roger, and Stephen Margolis: *Introduction to Engineering,* McGraw-Hill, New York, 1982.

Osborn, Alex F.: *Applied Imagination,* Scribner, New York, 1963.

Peterson, Ottis: *Your Future in Engineering Careers,* R. Rosen, New York, 1975.

Seeley, Fred B., and Newton E. Ensign: *Analytical Mechanics for Engineers,* Wiley, New York, 1957.

Shoup, Terry E., Leroy S. Fletcher, and Edward V. Mochel: *Introduction to Engineering Design,* Prentice-Hall, Englewood Cliffs, N.J., 1981.

Smith, Gerald W.: *Engineering Economy: Analysis of Capital Expenditures,* Iowa State University Press, Ames, Iowa, 1973.

Smith, Ralph J., Blaine R. Butler, and William K. LeBold: *Engineering as a Career,* 4th ed., McGraw-Hill, New York, 1983.

"Standard for Metric Practice," American National Standards Institute E 388-76 268-1976.

Thuesen, H. G., W. J. Fabrycky, and G. L. Thuesen: *Engineering Economy,* Prentice-Hall, Englewood Cliffs, N.J., 1977.

Wasserman, Leonard S.: *Chemistry: Basic Concepts and Contemporary Applications,* Wadsworth, Belmont, Calif., 1974.

Woodson, Thomas T.: *Introduction to Engineering Design,* McGraw-Hill, New York, 1966.

Index

Index

SOLIDS

Rectangular prism

Volume = $(W)(B)(H)$

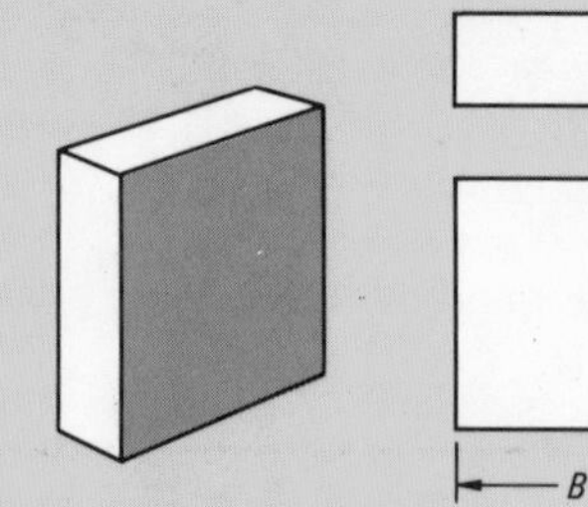

Pyramid

Applicable to base of any shape

Volume = $\frac{1}{3}$(area of base)(H)

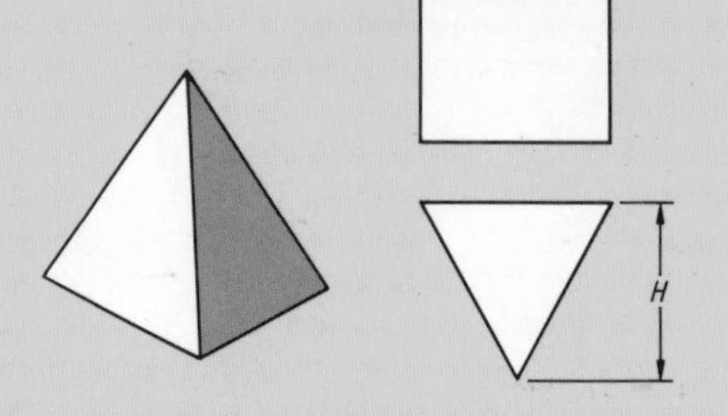

Cone

$$\text{Volume} = \frac{\pi}{12} D^2 H$$

$$= \frac{\pi}{3} R^2 H$$

$$\text{Surface area} = \frac{1}{2}(2\pi R)(S)$$

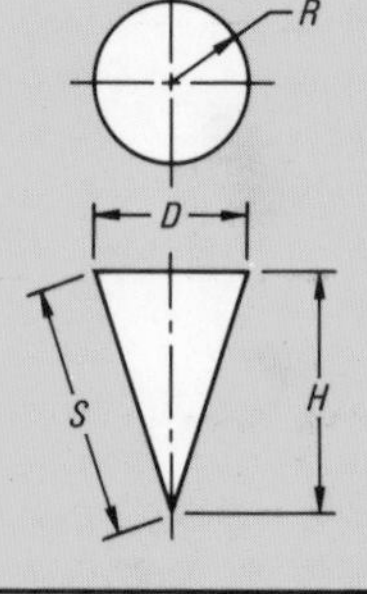

Cylinder

$$\text{Volume} = \pi R^2 H = \frac{\pi D^2}{4} H$$

$$\text{Surface area} = 2\pi R H$$

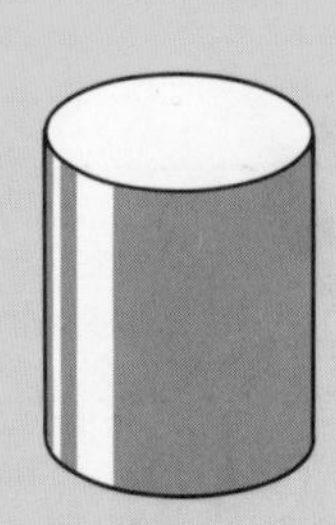

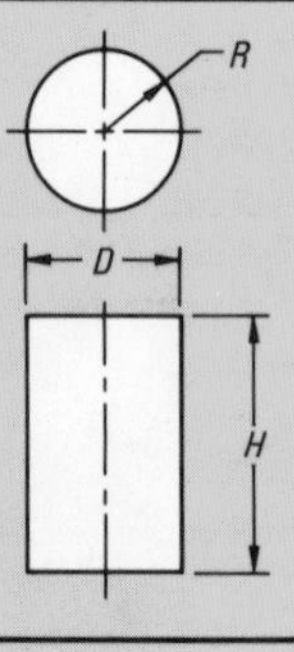

Hollow cylinder

$$\text{Volume} = \frac{\pi H}{4}(D_2^2 - D_1^2)$$

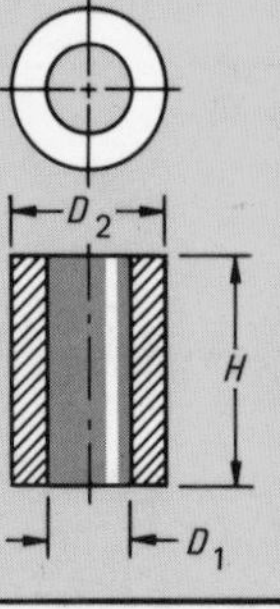